Springer Series in
SOLID-STATE SCIENCES 149

Springer Series in

SOLID-STATE SCIENCES

Series Editors:

M. Cardona P. Fulde K. von Klitzing R. Merlin H.-J. Queisser H. Störmer

The Springer Series in Solid-State Sciences consists of fundamental scientific books prepared by leading researchers in the foield. They strive to communicate, in a systematic and comprehensive way, the basic principles as well as new developments in theoretical and experimental solid-state physics.

136 **Nanoscale Phase Separation and Colossal Magnetoresistance**
The Physics of Manganites and Related Compounds
By E. Dagotto

137 **Quantum Transport in Submicron Devices**
A Theoretical Introduction
By W. Magnus and W. Schoenmaker

138 **Phase Separation in Soft Matter Physics**
Micellar Solutions, Microemulsions, Critical Phenomena
By P.K. Khabibullaev and A.A. Saidov

139 **Optical Response of Nanostructures**
Microscopic Nonlocal Theory
By K. Cho

140 **Fractal Concepts in Condensed Matter Physics**
By T. Nakayama and K. Yakubo

141 **Excitons in Low-Dimensional Semiconductors**
Theory, Numerical Methods, Applications By S. Glutsch

142 **Two-Dimensional Coulomb Liquids and Solids**
By Y. Monarkha and K. Kono

143 **X-Ray Multiple-Wave Diffraction**
Theory and Application
By S.-L. Chang

144 **Physics of Transition Metal Oxides**
By S. Maekawa, T. Tohyama, S.E. Barnes, S. Ishihara, W. Koshibae, and G. Khaliullin

145 **Point-Contact Spectroscopy**
By Yu.G. Naidyuk and I.K. Yanson

146 **Optics of Semiconductors and Their Nanostructures**
Editors: H. Kalt and M. Hetterich

147 **Electron Scattering in Solid Matter**
A Theoretical and Computational Treatise
By J. Zabloudil, R. Hammerling, L. Szunyogh, and P. Weinberger

148 **Physical Acoustics in the Solid State**
By B. Lüthi

149 **Solitary Waves in Complex Dispersive Media**
Theory · Simulation · Applications
By V.Yu. Belashov and S. V. Vladimirov

Volumes 90–135 are listed at the end of the book.

V. Yu. Belashov S.V. Vladimirov

Solitary Waves in Dispersive Complex Media

Theory · Simulation · Applications

With 100 Figures

Springer

Prof. Vasily Yu. Belashov
KSPEU, 51 Krasnosel'skaya St.
Kazan 420066
Russia
E-mail: vbelashov@ns.kgeu.ru

Prof. Sergey V. Vladimirov
School of Physics
University of Sydney
NSW 2006
Australia
E-mail: S.Vladimirov@physics.usyd.edu.au

ISSN 0171-1873
ISBN 3-540-23376-8 Springer Berlin Heidelberg New York

Library of Congress Control Number: 2004114854

Springer is a part of Springer Science+Business Media
springeronline.com

Typesetting by the authors
Cover concept: eStudio Calamar Steinen
Cover production: *design & production* GmbH, Heidelberg
Production: LE-TEX Jelonek, Schmidt & Vöckler GbR, Leipzig

Printed on acid-free paper SPIN: 10867284 57/3141/YL - 5 4 3 2 1 0

Dedication

To our parents:

Yury G. Belashov, a distinguished physicist, and Lyudmila V. Belashova, an outstanding engineer

Vladimir N. Vladimirov and Galina I. Vladimirova, outstanding engineers and teachers

Preface

This book is devoted to one of the most interesting and rapidly developing areas of modern nonlinear physics and mathematics – theoretical, analytical and numerical, study of the structure and dynamics of one-dimensional as well as two- and three-dimensional solitons and nonlinear wave packets described by the Korteweg–de Vries (KdV), Kadomtsev–Petviashvili (KP), nonlinear Schrödinger (NLS) and derivative nonlinear Schrödinger (DNLS) classes of equations. Special attention is paid to generalizations (relevant to various complex physical media) of these equations, accounting for higher-order dispersion corrections, influence of dissipation, instabilities, and stochastic fluctuations of the wave fields.

We present here a coordinated approach to the theory, simulations, and applications of the nonlinear one-, two-, and three-dimensional solitary wave solutions. Overall, the content of the book is a systematic account of results not only already known in the literature, but also those of new original studies related to the theory of models allowing soliton solutions, and analyses of the stability and asymptotics of these solutions. We give significant consideration to numerical methods and results of numerical simulations of the structure and dynamics of solitons and nonlinear wave packets. Together with deep insights into the theory, applications to various branches of modern physics are considered, especially to plasma physics (such as space plasmas including ionospheric and magnetospheric processes), hydrodynamics, and atmosphere dynamics.

Presently, the theory of one-dimensional nonlinear equations of the classes considered by the authors is well developed, and the progress in studies of the structure and evolution of one-dimensional solitons and wave packets is obvious. This progress was especially fast after the discovery of hidden algebraic symmetries of the KdV, NLS, and other (integrable by the inverse scattering transform (IST) method) classes of one-dimensional evolution equations. However, as soon as generalizations of these classes on more complex cases are involved, especially in two and three dimensions, the corresponding systems are often not completely integrable in the generally accepted mathematical sense. Thus an analytic study can provide us, in the best case, with the mere answer on the stability of multidimensional wave solutions and their asymptotics, and, by analyzing the system's behavior in the phase space, can give us qualitative characteristics and classification of the solutions. The detailed study of the structure and dynamics of multidimensional nonlinear waves, not solvable analytically in the general case, thus demands development of proper high-precision and highly effective numerical methods of integration of the related nonlinear systems as well as numerous numerical simulations. This type of research is now mostly represented in uncoordinated journal and conference publications covering particular aspects of the problem, without providing a unified systematic approach.

There is another, in our view, important reason, and that was also one of the major incentives prompting us to write this book. Presently, despite a number of works on mathematical properties of solitons, there is practically no systematic monograph-type literature on the theory of multidimensional KP- and DNLS-class equations, and especially on their applications to various physical situations. Thus important general theoretical results often are not recalled in applied calculations of physical phenomena in various physical systems, leaving these classes of equations as well as their solutions as "exotic objects" for narrow specialists in particular subjects. As an example, we can mention effects related to dissipation, processes leading to developing instabilities and formation of complex turbulent structures, higher-order dispersion effects, influence on the evolution of wave packets of stochastic fluctuations of the wave field, etc. On the other hand, almost all of the above effects are intrinsic properties of nonlinear dispersive complex systems attracting wide attention presently.

We cannot pretend to provide an exhaustive account of all the aspects of the modern theory of "soliton equations" because of its continuing intensive development; however, we hope that the present book does help fill the existing gaps in the literature. Our book is practically the first time for monographic literature to discuss in detail such problems as structure, stability, and dynamics of the nonlinear wave solutions, especially those of KP- and DNLS-classes, consistently taking into account effects, important in the physics of real complex nonlinear media; we study problems of development of relevant classification of allowed solutions in the phase space and by characteristics of their asymptotics; we deeply consider the ideology and realization of new, including so called spectral, approaches to numerical integration of the related multidimensional nonlinear systems, differing from the known methods by their high efficiency and possibility to control evolution of the solutions and soliton interaction in its dynamics.

The spectrum of applications presented in this book is sufficiently wide. Together with well known examples associated with the evolution of hydrodynamic surface waves, ion-acoustic and magnetosonic plasma waves (for the latter we also discuss a number of new results related to the account of relativistic effects as well as the influence of dissipation, stochastic fluctuations of the wave field and higher-order dispersion corrections, which can lead to formation of essentially new types of solutions), we discuss for the first time applications to the analysis of the dynamics of nonlinear solitary internal gravity waves in the Earth's ionosphere, including those generated by fronts of solar terminator and the shadow of a solar eclipse; we investigate the action of Raleigh waves off a seismic source on the plasma dynamics of the ionosphere's F-layer; we study dynamics of multidimensional solitons in media with changing in space and time dispersion characteristics (in particular, we investigate the problem of the structure and deformations of two-dimensional solitons propagating on the surface of a shallow water with changing profile of the bottom). We hope that the considered applications together with analytical and numerical results presented in a consistent and invariant way will help the interested reader apply the corresponding methods to the solution of particular problems in his/her field of research.

We also note that content of this book is a result of generalization of our teaching experience in the development of courses on the theory and numerical modeling of the nonlinear wave dynamics for students at graduate and post-graduate levels. This strongly influenced the structure of the book; we also hope that the methodic selection and presentation of the content will allow to use this publication as a reference book not only for students but also for scientists in various fields of modern physics.

We thus address this publication to researchers working in the theory and numerical simulations of dispersive complex media in such fields as hydrodynamics, plasma physics, and aerodynamics. As a reference book, we expect the monograph to be useful to graduate and post-graduate students majoring in physics and mathematics, as well as to scientists interested in solitons and nonlinear waves in other nonlinear dispersive complex systems.

It is our pleasure to thank many people to whom we owe a great deal. Vasily Belashov is especially thankful to Vladimir Karpman for his stimulation and long term collaboration on many research topics, Vladimir Petviashvili for his kind attention to the research and discussions stimulated this book, as well as Oleg Pokhotelov, Yoshi-Hiko Ichikawa and Masashi Hayakawa for their permanent interest to this work, numerous useful discussions and support. Sergey Vladimirov thanks Vadim Tsytovich, a great mentor, from whom he has drawn much inspiration, as well as Lennart Stenflo and Ming Yu, his valued experts on nonlinear surface and plasma waves, for their influence, support, and close research collaboration over the years. We would also like to thank Yuri Kivshar for his continuing and stimulating interest to the topic.

This book was partially supported by the Australian Research Council and the Russian Foundation for Basic Research (Grant No. 01-02-16116).

Kazan, Sydney, October 2004

V.Yu. Belashov
S.V. Vladimirov

Table of Contents

Introduction

In the recent four decades a new direction related to the investigation of nonlinear phenomena and processes has been actively developing in various areas of physics. The transition from linear to nonlinear description, justified by the necessity to take into account finer details of the observed phenomena, is the natural step in the development of any part of physics.

In the systems described by the wave equations, the nonlinearity, i.e., the dependence of the behavior of the wave packet on its amplitude, related, in particular, to the generation of harmonics with larger wave numbers, can enhance dissipation or trigger instability of the wave packets. If the medium is also dispersive, which leads to the dependence of the group velocity of the waves on the wave numbers, the dispersion, mixing up the phases, can counterbalance the effects of the nonlinearity. As a result, the established level of oscillations in the dispersive nonlinear medium can be rather high. For some branches of oscillations the balance between the nonlinearity and dispersion is established and nonlinear waves and/or nonlinear solitary waves, i.e., solitons [1], appear [2]. *Solitons* are usually defined as wave formations that are localized in space, (locally) stationary, and stable with respect to interactions. They are fundamental wave structures for the nonlinear wave processes in the presence of dispersion and play an important role in the wide spectrum of areas of research related to the wave physics, e.g., in hydrodynamics, plasma physics, condensed matter, and optics [3–11].

The large variety of processes which is necessary to take into account in the theoretical investigation of physical phenomena in dispersive media leads to the high complexity of even relatively simple hydrodynamics models, which in sufficiently complete formulation are described by complex sets of nonlinear partial differential equations. In this context, the decisive influence on the development of the theory of nonlinear waves was exerted by the idea of Korteweg and de Vries [12] that it is possible to significantly simplify the initial equations without compromising the main physics of the phenomena described; this can be done by keeping the nonlinear and dispersive terms of the same order of accuracy. It then often appears that the resulting equations are universal and can describe phenomena in various complex nonlinear dispersive media.

This simplified model equation for surface waves on a shallow water,

$$\partial_t u + \alpha u \partial_x u + \beta \partial_x^3 u = 0, \tag{0.1}$$

where $u = u(t, x)$, was derived for the first times by Korteweg and de Vries in 1895 [12]; however, the term "soliton" was introduced in 1965 by Zabusky and Kruskal [1] who demonstrated that the *Korteweg–de Vries equation* (*KdV equation*) reveals hidden linear properties, allowing a solution in the form of a nonlinear solitary wave propagating without changing its profile. It appeared later that the KdV equation describes a wide class of one-dimensional nonlinear physical systems with "real" dispersion when the wave frequency is given by $\omega \approx c_0 k_x (1 + \delta^2 k_x^2)$ (here, c_0 is the phase velocity of the wave and δ is the "dispersion length"), and, in addition to hydrodynamics, appears in plasma physics, magnetohydrodynamics, theory of lattices, etc. [2,3,13,14], i.e., it is quite universal.

If the considered media are viscous and/or heat conducting, the dispersion is "imaginary", i.e., $\omega \approx c_0 k_x (1 - \mathrm{i}\nu k_x / c_0)$ (here, ν is the viscosity), and the approach of Kortweg and de Vries leads to the *Burgers equation* [15]

$$\partial_t u + \alpha u \partial_x u = \nu \partial_x^2 u. \tag{0.2}$$

Solutions of the Burgers equation describe such formations as shock waves.

In the physics of plasmas, aero- and hydrodynamics, as well as in many other areas of physics, there is obvious interest to the wave dynamics of non-one-dimensional systems with the nonlinearity of the hydrodynamic type where stable stationary structures in the form of non-one-dimensional solitons can exist [7]. Many such systems are described by the class of equations given by

$$\partial_t u + \alpha u \partial_x u + \beta \partial_x^3 u = \mathcal{R}, \tag{0.3}$$

where $u = u(t, x, y, z)$ is the function of the wave field and $\mathcal{R} = \mathcal{R}[u]$ is some linear functional of u. The particular form of the right-hand side of (0.3) is defined by the wave properties of the medium and the sign of its dispersion. For example, sound waves in a plasma with weak dispersion when the wave numbers of the harmonics in the wave packet are small and satisfy the inequalities

$$k\delta \ll 1 \quad \text{and} \quad k_x^2 \gg k_\perp^2, \tag{0.4}$$

while the *dispersion relation* in the linear approximation given by

$$\omega \approx c_0 k_x \left(1 + \frac{k_\perp^2}{2k_x^2} + \delta^2 k_x^2 \right), \tag{0.5}$$

are described by equation of the type (0.3) with $\mathcal{R} = \kappa \nabla_\perp w$ and $\partial_x w = \nabla_\perp u$:

$$\partial_x \left(\partial_t u + c_0 \partial_x u - c_0 \delta^2 \partial_x^3 u + \chi u \partial_x u \right) = \pm \left(c_0/2 \right) \Delta_\perp u. \tag{0.6}$$

Equation (0.6) for $u = u(t, x, y)$ and $\Delta_\perp = \partial_y^2$ was first derived in 1970 by Kadomtsev and Petviashvili [16]; it is called now the *Kadomtsev–Petviashvili equation* (*KP equation*). Later, the KP equation was generalized in the three-dimensional case, with $\Delta_\perp = \partial_y^2 + \partial_z^2$ [7]. For the *ion-acoustic waves* in a plasma, when u stands for the ion velocity, $c_0 = c_s = (T_e/m_i)^{1/2}$ is the ion-acoustic speed (T_e is the electron temperature in the energy units and m_i is the ion mass) and $\delta^2 = r_D^2/2 = T_e/8\pi n_e e^2$ (r_D is the electron Debye length and n_e is the electron number density), we have the positive sign on the right-hand side of (0.6), corresponding to the "negative" dispersion (for a number of other cases the dispersion can be "positive", i.e., the negative sign appears on the right-hand side of (0.6)). This type of dispersion is typical for isotropic media, but can also appear in some anisotropic media. If the characteristic frequencies of the ion-acoustic wave packet in a magnetized plasma significantly exceed the ion cyclotron frequency $\omega_{Bi} = eB_0/m_i c$ (here, B_0 is the external magnetic field and c is the light speed), the anisotropy can be neglected, but in the opposite case $\omega \ll \omega_{Bi}$ this assumption is inadmissible. In the latter case, an additional term proportional to $\omega_{Bi}\mathbf{i} \times \mathbf{v}$ (here, $\mathbf{i}$ is the unit vector in the x-direction, and $\mathbf{v}$ is hydrodynamic velocity) appears on the right-hand side of the corresponding magnetohydrodynamic equations, and the sign of the second term in the dispersion equation (0.5) (and, correspondingly, the sign of the term in the right hand side of (0.6)) changes to the opposite, i.e., negative. We obtain in this case the equation of the class (0.3), but with $\mathcal{R} = \kappa\Delta_\perp\partial_x u$, known as the *Zakharov–Kuznetsov equation* [17].

In a strongly magnetized plasma with $B_0^2 \gg 8\pi nT$ (here, nT stands for the plasma kinetic pressure) in the frequency range $\omega \ll \omega_{Bi}$, the *fast magnetosonic waves* (*FMS waves*) can propagate [3]. For FMS waves, taking into account that $c_0 = v_A = B_0/(4\pi nm)^{1/2}$, where v_A is the Alfvén velocity and nm is the plasma density, the dispersion law is also described by (0.5) and the equation for the dimensionless amplitude of the wave field $h = B_\sim/B_0$ ($B_\sim$ is the magnetic field of the wave) can be also written in the form (0.6). In the reference frame moving along the x-axis with the Alfvén velocity v_A, the wave equation is given by

$$\partial_t h + \frac{3}{2} v_A \sin\theta h \partial_x h - v_A \delta^2 \partial_x^3 h = -\frac{1}{2} v_A \int\limits_{-\infty}^{x} \Delta_\perp h \mathrm{d}x, \tag{0.7}$$

where θ is the angle between the magnetic field $\mathbf{B}_0$ and $\mathbf{k}_x$ and

$$\delta^2 = \frac{c^2}{2\omega_{pi}^2}\left(\cot^2\theta - \frac{m_e}{m_i}\right). \tag{0.8}$$

Here, $\omega_{pi} = (4\pi n_i e^2/m_i)^{1/2}$ is the ion plasma frequency and n_i is the ion number density.

For *gravity-capillary waves* in a shallow water, the equation of the type (0.3) also holds. We have in this case $\mathcal{R} = -(c_0/2)\nabla_\perp w$, $\alpha = 3c_0/2H$, $\beta = -c_0\delta^2$, where $c_0 = (gH)^{1/2}$, H is the depth,

$$\delta^2 = \frac{1}{6}\left(\frac{3\sigma}{\rho g} - H^2\right), \tag{0.9}$$

σ stands for the coefficient of the surface tension, and ρ is the density of the fluid.

Generalizing equations (0.6) and (0.7), we write (0.6) in the reference frame moving along the x-axis with the velocity c_0 and obtain the *KP equation* in the standard form

$$\partial_x\left(\partial_t u + \alpha u \partial_x u + \beta \partial_x^3 u\right) = \kappa \Delta_\perp u. \tag{0.10}$$

Here, the sign of the ratio β/κ defines the character of dispersion. The KP equation (0.10) describes many nonlinear physical systems (where the linear dispersion law is given by (0.5)) and it is thus universal in the same sense as the KdV equation.

The difficulty in the analytical solution of nonlinear problems is in the proper choice of the effective sequence of methods to construct approximate solutions of the nonlinear systems, asymptotic in a small parameter. The perturbation theory is severely impeded in the non-one-dimensional case because nonlinear resonances, instabilities, secular effects, etc., can lead to singularities [18]. The discovery by Gardner, Green, Kruskal, and Miura in 1967 [19] of the *inverse scattering transform* method (*IST method*) for the KdV equation and subsequent developments of this idea (we would single out here the classic paper by Zakharov and Faddeev [20] as well as Refs. [21–24] where non-one-dimensional generalizations of the IST method were considered) has lead to active development of the theory of nonlinear waves [24–28]. In particular, the complete integrability of the KP equation (0.10) for $\Delta_\perp = \partial_y^2$ was proved [29,30], and using the *"dressing method"* the exact two-dimensional soliton solution (first numerically found by Petviashvili [31]) was obtained for $\beta/\kappa > 0$. The latter for $\alpha = -6$, $\beta = -1$, and $\kappa = -3$ is given by

$$u(t, x, y) = 2\partial_x^2 \ln \det B, \tag{0.11}$$

where

$$B_{nm} = \delta_{nm}\left(x - \mathrm{i}\nu_n y - \xi_n - 3\nu_n^2 t\right) + (1 - \delta_{nm})\frac{2}{\nu_n - \nu_m},$$

$$\nu_{K+l} = -\nu_l^*, \quad \xi_{K+l} = \xi_l^*, \quad n, m = 1, \ldots, 2K, \text{ and } l = 1, \ldots, K$$

(ν_n and ξ_n here define the amplitudes, phases, velocity vectors, and other soliton parameters). The invariants were also found [24]:

$$\mathcal{J}_1 = \iint u \mathrm{d}x\mathrm{d}y, \qquad \mathcal{J}_2 \equiv \mathcal{P} = \iint u^2 \mathrm{d}x\mathrm{d}y,$$

$$\mathcal{J}_3 \equiv \mathcal{H} = \iint \left[\frac{1}{2}\beta\left(\partial_x u\right)^2 + \frac{1}{2}\kappa w^2 - u^3\right] \mathrm{d}x\mathrm{d}y, \tag{0.12}$$

where $\partial_x w = \partial_y u$. The first invariant, $\mathcal{J}_1$ is related to the divergence form of the KP equation, the second one, $\mathcal{J}_2$, is due to the translational invariance of the KP equation and is its momentum $\mathcal{P}$, and the third one, $\mathcal{J}_3$, is due to the time invariance of the KP equation and is its Hamiltonian $\mathcal{H}$.

Significant progress was also achieved for a number of other exactly integrable models. In the context of problems considered in this book, the most interest is paid to the non-stationary *derivative nonlinear Schrödinger equation* (*DNLS equation*) given in the one-dimensional case by [32,33]:

$$\mathrm{i}\partial_t h + \mathrm{i}s\partial_x\left(|h|^2 h\right) + \lambda\partial_x^2 h = 0. \tag{0.13}$$

The DNLS equation describes, in the frequency range $\omega \ll \omega_{Bi}$, the evolution of nonlinear finite amplitude *Alfvén waves* [34] propagating in a *magnetized plasma* along the external magnetic field with the ratio of the kinetic plasma pressure to the magnetic pressure $p = 4\pi nT/B_0^2$. Here, the dimensionless function $h(t,x) = (B_y + \mathrm{i}B_z)/2B_0|1-p|^{1/2}$ describes the right circularly-polarized wave when $\lambda = 1$ and $s = \mathrm{sgn}(1-p)$. The change $h \to -sh^*$ with the change of the dispersion sign on the opposite one ($\lambda = -1$) allows description of the left circular-polarized wave. Note that (0.13) is obtained from the full set of the one-fluid magnetohydrodynamics equations (MHD equations) in the dimensionless units $t \to \omega_{Bi}t/2$, $x \to x/r_A$, and $\mathbf{r}_\perp \to \mathbf{r}_\perp\sqrt{2}/r_A$ ($r_A = v_A/\omega_{Bi}$) in the reference frame moving along the x-axis with the Alfvén velocity v_A taking into account $\partial_x = \partial_y = 0$.

The DNLS equation is completely integrable, has infinite number of the conservation laws, and can be solved by the IST method [35]. The evolution of Alfvén waves in the model (0.13) was studied in terms of the sign of the DNLS integral of motion, $\mathcal{H}/2$, where the *Hamiltonian* is given by

$$\mathcal{H} = \int_{-\infty}^{\infty}\left[\frac{1}{2}|h|^4 + s|h|^2\partial_x\varphi\right]\mathrm{d}x, \qquad \varphi = \arg(h).$$

It was established that the wave evolution can lead either to its spreading or to formation of the one-dimensional *Alfvén soliton*, depending on the sign of $\mathcal{H}$. From the physical point of view, this means that there are two types of the nonlinear wave dynamics [33]: the *modulationally-stable case* when $\mathcal{H} > 0$ and an initial pulse is spreading, loosing its structure; and the *modulationally-unstable case* when evolution of the *modulational instability* [36] ends by the formation of the one-dimensional soliton [32]

$$h = \left(\frac{A}{2}\right)^{1/2}\frac{\mathrm{e}^{-Ax} + \mathrm{i}\mathrm{e}^{Ax}}{\cosh^2(2Ax)}\mathrm{e}^{-\mathrm{i}A^2 t}, \tag{0.14}$$

where A is the amplitude of the soliton.

Equation (0.13), similar to the KdV equation, can be generalized in the non-one-dimensional situation. Thus, three-dimensional generalization of the

DNLS equation was obtained [7,37,38] in the form of the set of coupled equations

$$\partial_x \left[\partial_t h^\pm + \partial_x \left(|h|^2 h^\pm\right) \pm \mathrm{i}\partial_x^2 h^\pm\right] = -\nabla_\perp^2 \left(h^+ + h^-\right), \qquad (0.15)$$

where $h^\pm = h_y \pm \mathrm{i}h_z$. In the case of the negative dispersion, solutions in the form of the one-dimensional solitons were numerically found [7,37,38]; and no numerical study of non-one-dimensional solutions was done. The exact two-dimensional solitary solutions of the set (0.15) can in principle be obtained by the IST method, similar to the case of the KP equation; it is necessary to construct the corresponding Lax $\hat{\mathrm{L}}$–$\hat{\mathrm{A}}$ *pair* [24].

Successes in the theoretical investigation of nonlinear physical systems having soliton solutions has stimulated experimental studies in various areas of the physics of nonlinear waves. Here, we would like to mention the laboratory and space experiments on excitation, evolution, and interaction dynamics of the ion-acoustic and Alfvén solitons as well as the experimental studies of the structure of plasma *shock waves* [39–50], the modeling of the solitons' dynamics in the electric circuits [51,52], the experiments on *surface waves* and *internal waves* in rotating vessels and hydro-trays [53,54], etc. This in turn has raised new questions and demonstrated the urgency of the theoretical investigation of such problems as the stability of non-one-dimensional solitons, the dynamics of their interaction, the nonlinear resonances and formation of the bound states, the effects of small corrections in equations of the class (0.3), the influence of dissipation on the structure and evolution of the non-one-dimensional and nonlinear waves and solitons, the *wave self-influence* effects (the *wave collapse* and the *wave self-focusing*), etc.

It was shown, in Refs. [16,55-57] for some particular cases (concrete values of the coefficients) of the equation (0.10) with $\Delta_\perp = \partial_y^2$, and in Ref. [58] for arbitrary α, β, and κ, that for the negative dispersion ($\beta/\kappa < 0$) the one-dimensional solitons are stable whereas for the positive dispersion ($\beta/\kappa > 0$) they are unstable with respect to growth of infinitesimal perturbations. The case of very small k was studied [16] by the *Krylov–Bogolyubov method*; in Ref. [55], the term $\partial^4_{yyxx}u$ was added on the right-hand side of the equation and the problem was solved by the method of the variation of the action with the Lagrangian integrated over x and the test functions in the solitary-like form; in Ref. [56], the *IST method* was used, and in Ref. [57], the *Lyapunov's functional method* was used. In Ref. [58], the numerical investigation was done. It was demonstrated that for $\beta/\kappa < 0$ the perturbation can easily transfer from the soliton into the medium and spread in all directions while for $\beta/\kappa > 0$ it is "locked" in the soliton. In the region of localization of the perturbation, the soliton's velocity differs from the velocity of an unperturbed soliton. This, taking into account the spreading, leads to the growth of the perturbation.

For two-dimensional and three-dimensional solitons of the KP equation, the stability analysis is highly non-trivial. The most illustrative study

[59,60] by the method of investigation of transformational properties of the Hamiltonian (0.12) of the equation (0.10) has demonstrated that the two-dimensional soliton is stable with respect to two-dimensional perturbations. It was also shown [60], using the perturbation method of the linearized (on the background of the solution (0.11) with $n = 1, 2$, $m = 1, 2$, and $l = 1$) equation (0.10), that in the case of translation along the x-axis, the two-dimensional soliton is unstable with respect to the bending of its front (a three-dimensional perturbation). Qualitatively, the reasons for this instability are the same as in the case of the one-dimensional soliton [30,31]. However, the investigation [60] of small three-dimensional perturbations corresponding to the transverse translation demonstrates that the two-dimensional solitons of KP equation (0.10) are stable in the long-wavelength limit . The consequence of the stability of the two-dimensional KP solitons with respect to two-dimensional perturbations is the fact that, as can be seen from expression (0.11) for $n, m = 1, \ldots, N$, $N = 4, 6, \ldots$, and from numerical simulations [61], they elastically collide in their interactions, and the phase shifts after the collisions (well known in the one-dimensional problems [12]) are identically zero [24].

The character of the evolution of three-dimensional solitons in the model (0.10), as was demonstrated for *FMS waves* [59], is determined by the sign of the ratio β/κ. In the case of the positive dispersion, the growth of the instability leads to a nonlinear deformation of the front's structure, namely, to pushing the field out of the soliton's center and to its (the field) growth on the soliton's "wings" with subsequent formation of one/two collapsing "cavitons" [59,63,64]. For the negative dispersion, the initially observed sub-focusing of the wave field later transfers into the defocusing regime.

The three-dimensional generalization of (0.13), the *3-DNLS equation*, was obtained [65–69] in the form

$$\partial_t h + s\partial_x \left(|h|^2 h\right) - \mathrm{i}\lambda\partial_x^2 h = \kappa \int_{-\infty}^{x} \Delta_\perp h \mathrm{d}x. \tag{0.16}$$

In Refs. [50,70], on the basis of the method of the Hamiltonian's deformations on solutions, the problem of the stability of three-dimensional solutions of (0.16) was investigated. It was demonstrated that the left-polarized as well as right-polarized three-dimensional waves can be stable in some region of values of the Hamiltonian of (0.16). An analysis of numerical results demonstrates that the 3-DNLS equation can also have, in addition to the collapsing and spreading solutions, a three-dimensional solution in the form of *Alfvén soliton.* We stress here that equation (0.16), in contrast to DNLS equation (0.13), is not completely integrable [65], and therefore cannot be solved analytically by the IST method.

It is therefore clear that for three-dimensional problems of the above classes, the application of effective methods such as the IST method and the perturbation theory can be difficult. Moreover, for a number of cases (e.g., 3-DNLS equation) this is merely impossible since the application demands introduction of some limitations to fix classes of the initial and boundary (if effectively acting boundaries appear in the problem) conditions. In this situation, in order to obtain information on nonlinear processes in a wide range of the system parameters, it appears necessary to use powerful methods of computational mathematics developed to solve various problems in plasma physics, as well as in hydro-, gas-, and fluid dynamics (see, e.g., Refs. [71–77] as well as the literature cited therein). In the context of problems studied in this book, we note the works in Refs. [31,58,64,78,79]. Numerical methods also make sense for solution of many practically important applied problems when the use of cumbersome and sometimes overly complicated analytical methods appears unreasonable. Here we note that the first two-dimensional soliton solutions of the KP equation as well as three-dimensional solutions of the 3-DNLS equation were found numerically [31,65], as well as a number of results we discuss in this book (see the corresponding references).

Regarding the *KP equation*, we would like to remark the following. In some cases, the coefficient at the third derivative in equations of the class (0.3) can be close (or even equal) to zero. This can happen, e.g., for the *gravity-capillary waves* in shallow water when $H \to (3\sigma/\rho g)^{1/2}$, for *fast magnetosonic waves* (*FMS waves*) when $\theta \to \arctan(m_i/m_e)^{1/2}$ (see expressions (0.8) and (0.9), respectively). This situation, however, does not mean that the dispersion disappears; indeed, the balance between the nonlinear and dispersive terms can still be maintained in this case by taking into account the next term of the expansion of the full dispersion equation in the wave number k. Thus the Taylor expansion of the *dispersion relation* for waves on the surface of an ideal incompressible fluid $\omega^2 = gk\tanh[(H^2 - 3\sigma/\rho g)^{1/2}k]$ [3] in the limiting case of *shallow water* (namely, $k\delta \ll 1$, where $\delta = (|\beta|/c_0)^{1/2}$) gives us

$$\omega \approx c_0 k\left\{1 + \frac{1}{6}\left(3\tilde{\sigma} - H^2\right)k^2 \right.$$
$$\left. + \frac{1}{6}\left[H^2\left(\frac{2}{5}H^2 - \tilde{\sigma}\right) - \frac{1}{12}\left(3\tilde{\sigma} - H^2\right)^2\right]k^4\right\}, \qquad (0.17)$$

where $\tilde{\sigma} = \sigma/\rho g$. The Fourier transform of (0.17) allows us to find the dispersion correction for an equation of the class (0.3). The correction is proportional to the fifth derivative, it appears in the expression for $\mathcal{R}[u]$, and is given by $-\gamma\partial_x^5 u$, where the coefficient γ is [80]

$$\gamma = \frac{c_0}{6}\left[H^2\left(\frac{2}{5}H^2 - \tilde{\sigma}\right) - \frac{1}{12}\left(3\tilde{\sigma} - H^2\right)^2\right]. \qquad (0.18)$$

For waves propagating in a *magnetized plasma* where $B_0^2 \gg 8\pi nT$ as well as $c_0 = v_A \ll c$ and $\omega \ll \omega_{pi}$, the dispersion law in the linear approximation is

$$\omega_{1,2} = \frac{v_A k}{2\sqrt{1+c^2k^2/\omega_{pe}^2}} \left\{ \left[(1+\cos\theta)^2 + \frac{c^2k^2}{\omega_{pi}^2} \frac{\cos^2\theta}{1+c^2k^2/\omega_{pe}^2} \right]^{1/2} \pm \left[(1-\cos\theta)^2 + \frac{c^2k^2}{\omega_{pi}^2} \frac{\cos^2\theta}{1+c^2k^2/\omega_{pe}^2} \right]^{1/2} \right\}, \quad (0.19)$$

where $\omega_{pe} = (4\pi n_e e^2/m_e)^{1/2}$ is the (unperturbed) *electron plasma frequency.* Here, we consider the magnetosonic mode only, i.e., choose the upper (positive) sign in (0.19). Keeping the terms up to the forth order in k (inclusive) in the expansion of (0.19), we obtain

$$\omega \approx v_A k \left\{ 1 + \frac{c^2k^2}{2\omega_{pi}^2} \left(\cot^2\theta - \frac{m_e}{m_i} \right) + \frac{c^4k^4}{8\omega_{pi}^4} \left[3\left(\frac{m_e}{m_i} - \cot^2\theta \right)^2 - 4\cot^4\theta \left(1+\cot^2\theta \right) \right] \right\}. \quad (0.20)$$

Thus, analogous to the previous case, we find [81–83]

$$\gamma = v_A \frac{c^4}{8\omega_{pi}^4} \left[3\left(\frac{m_e}{m_i} - \cot^2\theta \right)^2 - 4\cot^4\theta \left(1+\cot^2\theta \right) \right]. \quad (0.21)$$

Note that an expansion, analogous to (0.20), can also be done for the Alfvén mode corresponding to the negative sign in the dispersion relation (0.19). This, however, is not physically justified since it does not correspond to reality because the coefficient (in the dimensional units) $\mathrm{i}\lambda r_A^2 \equiv \mathrm{i}\lambda v_A^2/\omega_{Bi}^2$ at the dispersion term in equations (0.13) and (0.16) does not tend to zero in any case. As we can see from (0.17) and (0.20), the dispersion character determined in a general case by a relation between the parameters β and γ can sometimes have different signs for large and small wave numbers k. Thus the situation is becoming more complex as compared to the standard KdV and KP models.

It is often inadmissible to ignore dissipation, and the basic equation should be complemented by the corresponding terms. For example, if we consider ion oscillations in a plasma with frequencies much less than the electron plasma frequency (in this case for $T_e \gg T_i$ their Landau damping is small), then the dissipative effects related to the plasma relaxation lead to the imaginary term $-\mathrm{i}\nu k_x^2$ in the dispersion equation. Correspondingly, the Burgers term $\nu\partial_x^2 u$ [15] appears on the right-hand side of equations of the type (0.3) and (0.16). It was demonstrated [3] that

$$\nu = (\rho_0/2\rho)\left(c_\infty^2 - c_0^2\right)\tau \int_0^\infty \xi\varphi(\xi)\mathrm{d}\xi$$

has the sense of the coefficient of the relaxation damping of the "sound", where c_∞ and c_0 are the velocities of the high-frequency and the low-frequency sound, respectively (note that the latter is equal to $c_s = (T_e/m_i)^{1/2}$), and $\varphi(t,\tau)$ is the function determining the relaxation process. If, on the other hand, for *ion-acoustic waves* in a plasma the *Landau damping* is significant, the dissipation can be taken into account by the corresponding integral term [3] in the equation

$$\mathcal{R}[u] = -\hat{\mathrm{L}}[u] = -\kappa \int_{-\infty}^{\infty} \frac{\mathrm{d}k}{2\pi} |k| \int_{-\infty}^{\infty} u(x')\mathrm{e}^{\mathrm{i}k(x-x')}\mathrm{d}x',$$

where $\kappa = c_0(\pi m_e/8m_i)^{1/2}$. Below in this book, taking into account that the hydrodynamic approximation is considered (and $\omega \ll \omega_{pe}$), we limit ourselves to the investigation of the influence on the structure and evolution of nonlinear waves of only the viscous-type dissipative processes.

Taking into account the above arguments, we thus generalize the *KP equation* using the *higher order dispersion* correction as well as the dissipative term and obtain [81]

$$\partial_x \left(\partial_t u + \alpha u \partial_x u - \nu \partial_x^2 u + \beta \partial_x^3 u + \gamma \partial_x^5 u\right) = \kappa \Delta_\perp u, \tag{0.22}$$

where $\kappa = -c_0/2$. This equation is universal in the same sense as the KdV and KP equations, and corresponds to the following *dispersion law*:

$$\omega \approx c_0 k_x \left(1 + \frac{k_\perp^2}{2k_x^2} - \frac{\mathrm{i}\nu k_x}{c_0} + \frac{-\beta k_x^2 + \gamma k_x^4}{c_0}\right). \tag{0.23}$$

The *3-DNLS equation* (0.16), taking into account the viscous-type dissipative processes, can be written as

$$\partial_t h + s\partial_x \left(|h|^2 h\right) - \mathrm{i}\lambda\partial_x^2 h - \nu\partial_x^2 h = \kappa \int_{-\infty}^{x} \Delta_\perp h \mathrm{d}x. \tag{0.24}$$

For this equation, with $\kappa = -c_0/2$ and $c_0 = v_A$, and with the formal change $\beta k_x = \lambda$, $\gamma = 0$, and $s = c_0$, relation (0.23) also holds.

Finally, various instabilities (determined by the particular character and parameters of the medium where the waves propagate) leading usually to rapid increase of perturbations with eventual formation of developed turbulent structures and transfer of the wave energy into other degrees of freedom, can be taken into account by introducing a term proportional to the fourth derivative, $\delta\partial_x^4$, into the left-hand side of equations (0.22) and (0.24). This term is necessary because of the additional imaginary term $-\mathrm{i}\delta k_x^4/c_0$ appearing in the dispersion relation (0.23).

Regarding the model equations introduced above we would like to note the following. The stationary soliton solutions of (0.22) for $\nu = \kappa = 0$ were first obtained numerically by Kawahara [84]. It was demonstrated that for $\beta > 0$ and $\gamma > 0$, the one-dimensional soliton acquires an oscillating structure. The two-dimensional equation of the type (0.22) with $\beta = \gamma = 0$ and $\Delta_\perp = \partial_y^2$ was first considered by Zabolotskaya and Khokhlov [85] when describing propagation of two-dimensional nonlinear *sound waves* in a medium with dissipation. The equation of the type (0.22), with $\nu = 0$ and $\Delta_\perp = \partial_y^2$ when $\beta \geq 0$, $\gamma > 0$, and $\kappa < 0$, for *gravity-capillary waves* in *shallow water* was numerically investigated by Abramyan and Stepanyants [80] using the *method of stabilizing factor*, suggested in Ref. [31] to find stationary solutions of the two-dimensional KP equation. As a result, for the above parameters in Ref. [80] solutions were obtained in the form of a stationary two-dimensional soliton and a pair of solitons (a *bisoliton*) with oscillating tails. The cross-sections of these solutions along the x-axis (i.e., at $y = 0$) were qualitatively similar to the solutions obtained by Kawahara in the one-dimensional case [84] as well as results found in experiments on modeling of nonlinear processes described by the KdV equation with the fifth-order derivative for $\beta = 0$ in electric lines [51]. However, the conditions and the dynamics of formation of such structures remained unclear, as well as the character of solutions for $\beta < 0$, $|\gamma| > 0$, and $\beta \geq 0$, $\gamma < 0$ (note that for $\gamma < 0$ the method used in ref. [80] diverges), their stability in the whole range of $|\gamma| > 0$ and β, and the influence of dissipation on the structure and evolution of these nonlinear waves. The three-dimensional equation of the type (0.22), having wide applications in the physics of nonlinear dispersive waves, was not investigated by other authors before the works of Belashov and co-authors (see Ref. [83] as well as references therein). The problems outlined above for two-dimensional systems, as well as specific processes of the *wave self-influence*, described by the three-dimensional KP equation [18,59,86,87], are especially important in the three-dimensional setting.

The *DNLS equation* of the type (0.24) for $\nu = 0$ and $\kappa = 0$ was first investigated numerically by Dawson and Fontán [33], and analytically by Kaup and Newell [35]. In the three-dimensional case for $\nu = 0$ it was studied by Petviashvili and Pokhotelov [7]. However, as we already noted above, only one-dimensional solutions without accounting for dissipative processes were considered in these works. The considered one-dimensional geometry did not allow investigation of the effects of the wave self-influence, the stability of non-one-dimensional solutions, and the influence of dissipation on the structure and evolution of nonlinear *Alfvén waves.*

Note that exact analytic solutions of the generalized equations considered in this book are not known. Therefore, we widely use numerical methods for integration of nonlinear systems of the type (0.22) and (0.24). The analytical approaches to investigations of such systems are limited by qualitative, asymptotic, and stability analyses of their solutions, as well as by studies of some particular cases (important for applications) of external influence of a medium on the structure and dynamics of multidimensional nonlinear waves and solitons [83]. The above approaches to investigation of non-one-dimensional nonlinear systems of KP and DNLS classes constitute the main content of the corresponding sections of this book.

The book consists of an Introduction, four Chapters, subdivided into Sections and subsections, and two Appendixes.

In Chap. 1, Sect. 1.1, we derive the *Korteweg–de Vries equation* (*KdV equation*) from the full set of *hydrodynamic equations*. We realize the general approach, Sect. 1.1.1, i.e., do not consider any particular type of a medium, and work further with the general equations. We demonstrate that the KdV model is universal in the sense that it describes the nonlinear wave dynamics in any medium where the wave dispersion is of a certain sort, and consider the *scale transforms* and the *similarity principle* for KdV equation, Sect. 1.1.2. We also briefly consider some other one-dimensional equations of the KdV class (see Sect. 1.1.3). In Sect. 1.2, we present fundamentals of the theory of the *inverse scattering transform* (Sect. 1.2.1). The powerful *IST method* enables us to analytically obtain general solutions of some classes of canonical nonlinear equations. On this basis, we approach the problem of (exact) analytical integration of the KdV equation (Sect. 1.2.2). In Sect. 1.2.3, we consider generalization of the integral *Gelfand–Levitan–Marchenko equation* (*GLM equation*) for the purpose of possible separation of the study of the soliton and non-soliton parts of the solution. Then we consider the *variational principle* which practically enables us to separate these types of solutions (Sect. 1.2.4). The inverse scattering problem for the multidimensional cases is considered later in Chap. 3 (see Sect. 3.2).

However, even development of such powerful and effective analytical apparatus as the IST method does not remove the problem of numerical integration of the KdV equation as well as other equations of KdV-class from the agenda. It is because, first, for arbitrary initial conditions it is not possible to obtain an analytical solution in its closed form using the IST method. The second reason is that there are models not integrated analytically among the equations of KdV-class (for example, the *Korteweg–de Vries–Burgers equation* (*KdVB equation*) or KdV equations with additional terms describing, e.g., instability of some type in the medium). Therefore, the problems of developing numerical codes and setting up numerical experiments for this class of problems are highly important. In Sect. 1.3, by example of the KdV equation, we consider some difference schemes used for the numerical analysis and present numerical solutions obtained with their help. These schemes are also used to obtain solutions of other one-dimensional equations of KdV-class, and some elements of these schemes will be further used (see Chap. 4) when studying numerical methods of integration of the (1+2)- and (1+3)-dimensional problems. Finally, in Sect. 1.4 we consider applications of the results obtained for the KdV equation to description of the structure and dynamics of one-dimensional waves in a plasma. Specifically, we study *ion-acoustic waves* in an *unmagnetized plasma*, including weakly *relativistic effects* as well.

In Chap. 2, Sect. 2.1, we present some generalizations of the KdV equation taking into account *dissipative effects*, *higher order dispersion* corrections and instability (Sect. 2.1.1 and 2.1.2), as well as considering the *modified KdV equation* (*MKdV equation*, Sect. 2.1.3). In Sect. 2.1.4, by example of *surface waves* in a plasma, we present a KdV-type equation with higher order dispersive nonlinearity and discuss some types of solitary surface plasma waves. Other problems related to evolution of generalized KdV solitons and classification of solutions of the *generalized KdV equation*, using methods of the qualitative and asymptotic analyses, are discussed in Sect. 2.2. In particular, we consider evolution of solitons of the KdV equation generalized by the terms accounting for dissipation, higher order dispersion correction, and instability, Sect. 2.2.1). We also consider the soliton evolution in a medium with *stochastic fluctuations of the wave field* within the limits of the *stochastic KdV equation* (Sect. 2.2.2). Section 2.2.3 is devoted to the qualitative and asymptotic analysis of all possible classes of solutions of the generalized KdV equation.

In Sect. 2.3, we consider effects related to *modulational interactions*, where change of the *wave envelope* occurs as a result of the *modulational instability*, an instability of a wave with respect to its modulations. The final stage of this instability, in the one-dimensional case, leads to the formation of *envelope solitons.* The canonic equation describing this type of processes is the *nonlinear Schrödinger equation* (*NLS equation*), with solution in the form of an envelope *NLS soliton.* An important generalization of the NLS equation including interactions via propagating low-frequency perturbations is called the Zakharov system of equations or *Zakharov equations*, with numerous applications in plasma physics and nonlinear optics. Thus we first, in Sect. 2.3.1, derive the NLS equation for the simplest case of slow modulations of *Langmuir waves* in an *unmagnetized plasma.* The inverse scattering problem for the NLS equation is outlined in Sect. 2.3.2. Then in Sect. 2.3.3 we generalize the NLS equation for faster modulations of the Langmuir wave in a plasma, derive the Zakharov system of equations, and, in Sect. 2.3.4, demonstrate its solution in the form of an envelope *Langmuir soliton.* In Sect. 2.3.5, we consider an interesting case of a Langmuir *near-sonic soliton* and study the influence of dissipative processes on its propagation. To conclude Chap. 2, in Sect. 2.4 we introduce the *derivative nonlinear Schrödinger equation* (*DNLS equation*), derive it from the plasma *magnetohydrodynamic equations* (*MHD equations*) (Sect. 2.4.1), consider the DNLS equation as an integrability condition for two linear differential equations (Sect. 2.4.2) and discuss stability of such DNLS solutions as the *DNLS soliton*, Sect. 2.4.3. Finally, we present some numerical approaches to study the dynamics of *Alfvén soliton* (Sect. 2.4.4) and consider results of numerical simulations of the soliton's evolution (Sect. 2.4.5).

In Chap. 3, we discuss generalization of the KdV equation on a weakly non-one-dimensional case when the *Kadomtsev–Petviashvili equation* (*KP equation*) appears (Sect. 3.1.1). Furthermore, we consider various classes of solutions of the KP equation and investigate the problem of their stability in more detail (Sects. 3.1.2 and 3.1.3). Finally, we present some effective methods of numerical integration of the KP equation (Sect. 3.1.4) used in particular for study of the *wave self-focusing* phenomenon and the *wave collapse* in the classic KP model. In Sect. 3.2, we present the *dressing method*, an analytical integration of the KP equation by "dressing" of L–A pairs on an example of the two-dimensional KP equation (Sect. 3.2.1). Furthermore, in Sect. 3.2.2 we consider the method of three-dimensional inverse scattering problem, *3-ISP*, and investigate problems related to the *wave self-influence* phenomena, namely, the *wave collapse* and *wave self-focusing* in the three-dimensional KP model, Sect. 3.2.3.

In Chap. 4, we generalize the classic KP equation by introducing the *higher order dispersion* correction, the terms describing dissipation of the viscous type as well as an instability and stochastic fluctuations of the wave field (Sect. 4.1). Thus we derive the *generalized KP equation* (*GKP equation*). We then reduce this equation to a simplified form allowing its subsequent analysis (Sect. 4.1.1). Furthermore, in Sect. 4.1.2, we derive the three-dimensional derivative nonlinear Schrödinger equation (*3-DNLS equation*) from the full set of the plasma one-fluid *MHD equations*, and then, using the *scale transforms*, reduce it to the dimensionless form convenient for further analysis. Also, a generalization of 3-DNLS equation in the presence of dissipation in a medium is presented. Then, in Sect. 4.1.3, we study in detail the stability of two- and three-dimensional solutions of the GKP and 3-DNLS equations. In Sect. 4.2, on the basis of the results of Sect. 2.2, we study the structure of (possible) multidimensional solutions of the GKP equation with an arbitrary nonlinearity exponent. We employ an approach taking into account the asymptotics of the solutions along the direction of the wave propagation. The study of the asymptotic behavior of solutions of 3-DNLS equation along the direction of the wave propagation is a simpler problem because we can explicitly obtain exact solutions in the one-dimensional approximation [32,33]. We also present some considerations on the construction of the 8-dimensional *phase portraits* for the systems described by the GKP equation on the basis of the results of qualitative analysis of the generalized KdV-class equations.

In Sect. 4.3, we consider a few relatively simple methods of numerical integration of the GKP- and 3-DNLS-class equations used in Sects. 4.4–4.6 to study dynamics of multidimensional solitons and non-stationary wave packets. The methods are based on the explicit and implicit finite-difference schemes (Sects. 4.3.1 and 4.3.2) with $O(\tau^2, h^2_{x,y})$ and $O(\tau^2, h^4_{x,y})$ approximations. We consider also the *dynamic spectral method* (Sect. 4.3.3) including first the Fourier transform of the basic equations in the space variables and then the subsequent solution of the resulting first-order differential equations by the Runge–Kutta method. For every algorithm, we formulate the stability conditions. Unlike the schemes considered in Sect. 3.1, the methods presented here not only enable us to control the evolution of the solutions as well as the dynamics of the solitons and their interactions, they, exhibiting the high accuracy characteristics, are less cumbersome than the *iterative splitting* method and the *hop-scotch method* sometimes used for numerical integration of KP equation. We consider all the methods on an example of the standard KP equation aiming, first of all, to avoid the inconveniences of those considered earlier as well as to compare all of them. When applied to the 3-DNLS equation, these methods, unlike the method based on the *Ablowitz–Ladik scheme* for example, take into account the multidimensionality of the problem. To conclude this section, we discuss comparative characteristics of the schemes of different types obtained when testing the basic equations on the exact solutions.

In Sect. 4.4, we consider numerical solutions of the two-dimensional *GKP equation* written in the differential form and describing formation and interaction of solitons. We also present the evolution of nonstationary wave packets. Numerical integration of GKP equation is performed by using both the *dynamic spectral method* and the *implicit scheme with* $O(\tau^2, h^4_{x,y})$ *approximation*. The initial conditions are assumed in the form of the soliton solutions of the KP equation (see Sect. 3.1.2) with various values of parameters defining the amplitudes, phases, velocities and other soliton characteristics. The numerical integration is controlled by the conservation of the momentum and Hamiltonian of the soliton solutions. We consider the structure of two-dimensional numerical solutions estimating their stability (Sect. 4.4.1), the interaction of two-dimensional solitons (Sect. 4.4.2), the influence of the viscous-type dissipation on their evolution (Sect. 4.4.3), as well as the evolution of two-dimensional solitons in a dispersive medium with *stochastic fluctuations of the wave field* (Sect. 4.4.4) and the dynamics of solitons in a medium with variable dispersion (Sect. 4.4.5). In Sect. 4.5, we numerically investigate the structure and evolution of three-dimensional solutions of the GKP equation (Sects. 4.5.1 and 4.5.3). We also study the *3-DNLS equation* (Sects. 4.5.2 and 4.5.3) in the axially-symmetric geometry. For numerical integration, we use the methods presented in Sect. 4.3. In Sects. 4.5.1 and 4.5.2 we present results of the numerical study of the structure of solutions, estimate their stability, and consider the dynamics of the evolution of the three-dimensional axially-symmetric pulses in the GKP and 3-DNLS models, respectively. Finally, in Sect. 4.5.3 we present numerical results of the investigation of the influence of a viscous-type dissipation for the GKP and 3-DNLS models on the evolution of the three-dimensional solutions obtained.

As we already mentioned above, equations (0.10) and (0.22) are universal in the sense that they describe a wide class of nonlinear waves in dispersive media in the absence and in the presence of dissipation, respectively. Such situations occur, e.g., for the wave perturbations of the acoustic type such as the *ion-acoustic waves* and *magnetosonic waves* in a plasma, *surface waves* in *shallow water*, and *internal gravity waves* (*IGW*) in the Earth's upper atmosphere and ionosphere (applications in the one-dimensional case are given in Sect. 1.4). Equations of the DNLS class (0.13), (0.16), and (0.24) directly describe the nonlinear evolution of *Alfvén waves* in a *magnetized plasma* without and with dissipation in the medium, respectively. In Sect. 4.6, we consider applications of the results obtained for multidimensional cases to the investigation of: (a) the propagation of nonlinear ion-acoustic waves in an *unmagnetized plasma*, including the *relativistic effects*, Sect. 4.6.1; (b) the dynamics of three-dimensional *fast magnetosonic waves* (*FMS waves*) propagating in a *magnetized plasma*, Sect. 4.6.2; (c) the dynamics of two-dimensional solitary nonlinear internal gravity waves, generated in the *F-layer of the Earth's ionosphere* by fronts of the *solar terminator* and the *solar eclipse* as well as by seismic sources, and excitation by them of the traveling perturbations of the plasma electron density (so-called *traveling ionospheric disturbances*, *TID*), Sect. 4.6.3; and (d) the evolution of two-dimensional solitons of the *gravity waves* and *gravity-capillary waves* on the surface of a *shallow water* with the bottom relief varying in time and space, Sect. 4.6.4. The main results here are obtained by the analytical and numerical methods detailed above in the previous sections of the book.

Finally, in Appendixes we elaborate and present two technical problems. In the first one, we investigate expansion of four-dimensional dynamic systems linearized in the vicinity of singular points (and the corresponding canonical systems) into two sub-system. Similarly, we consider expansion of three-dimensional dynamic systems into a two-dimensional system and one equation that used in Sect. 2.2.3 when constructing the *phase portraits* of solutions in four-dimensional and three-dimensional phase spaces, respectively. In the second part, we investigate an algebraic equation of the fourth order appearing when analyzing possible extrema of the Hamiltonian of the GKP equation in Sect. 4.1.

1. KdV-Class Solitons

1.1 Korteweg–de Vries Equation and KdV-Class Equations

In this section, we derive the KdV equation from the full set of hydrodynamic equations. Here, we realize the general approach, i.e., we do not consider any particular type of a medium (Sect. 1.1.1), and work further with the general equations. We demonstrate that the KdV model is universal in the sense that it describes the nonlinear wave dynamics in any medium where the wave dispersion is of a certain class, and consider the scale transformations and the similarity principle for the KdV equation (Sect. 1.1.2). We also briefly consider some other one-dimensional equations of the KdV class.

1.1.1 Derivation of the KdV Equation

The fundamental equation describing the propagation of nonlinear waves in the one-dimensional case in a medium with weak dispersion is the *Korteweg–de Vries equation* (*KdV equation*) [12], with solutions as stable solitary wave structures, i.e., *solitons*. Let us show how this equation can be derived.

Introduce, first of all, the generalized "density" ρ and generalized "sound" velocity $c(\rho)$, neglecting the medium's dispersion for the moment.

• For the *surface waves* in the water the above are: $\rho = H$ is the depth and $c(\rho)$ is the phase velocity of the waves; $c(\rho) = c_0 = \sqrt{gH}$ for small-amplitude waves.

• For the *ion-acoustic waves* in a collisionless plasma: ρ is the plasma (gas) density and $c(\rho)$ is the phase velocity of the ion sound; $c(\rho) = c_s = \sqrt{T_e/m_i}$ for the long-wavelength linear waves, where T_e is the electron temperature in energy units (Boltzmann constant equals unity), and m_i is the ion mass.

• For the *magnetosonic waves* in a magnetized plasma: $\rho = |\mathbf{B}_0|$ is (the strength of) the external magnetic field and $c(\rho) = v_A = |\mathbf{B}_0|/\sqrt{4\pi nm}$ is the Alfvén velocity, in this case usually the plasma density $nm \approx n_i m_i$ where n_i is the ion density.

In the following, we work with equations involving these generalized functions. We assume the dissipation is negligible and choose the following as the basic the set of gas dynamics (or hydrodynamics) equations:

$$\begin{aligned} &\partial_t \mathbf{v} + (\mathbf{v} \cdot \boldsymbol{\nabla})\, \mathbf{v} + \left(c^2/\rho\right) \boldsymbol{\nabla} \rho = 0, \\ &\partial_t \rho + \boldsymbol{\nabla} \cdot (\rho \mathbf{v}) = 0. \end{aligned} \tag{1.1}$$

These are the equation of motion and the continuity equation for the generalized velocity and density, respectively. For the *surface waves* in shallow water, $\mathbf{v}$ is the hydrodynamic velocity in the wave (the "mass" velocity); for the *ion-acoustic waves* it is the ion velocity, for the *magnetosonic waves* $\mathbf{v} = \mathbf{h} \equiv \mathbf{B}_{\sim}/|\mathbf{B}_0|$ is the wave magnetic field normalized to the external magnetic field.

Assuming that the gas motion is a potential one, i.e., $\mathbf{v} = \boldsymbol{\nabla}\Phi$, we integrate the first equation of the set (1.1) and obtain

$$\partial_t \Phi + \frac{1}{2} (\boldsymbol{\nabla}\Phi)^2 + \frac{c^2(\rho - \rho_0)}{2\rho} + \frac{c^2 z}{\rho} = 0, \tag{1.2}$$

where ρ_0 =const. The continuity equation for the velocity, $\boldsymbol{\nabla} \cdot \mathbf{v} = 0$, gives the Laplace equation

$$\Delta\Phi = 0. \tag{1.3}$$

We supplement this set by the boundary conditions

$$\begin{aligned} &\partial_t \eta + \partial_x \eta \partial_x \Phi + \partial_y \eta \partial_y \Phi - \partial_z \Phi = 0, \\ &\partial_t \Phi + (\boldsymbol{\nabla}\Phi)^2 /2 + \left(c^2/\rho\right) \eta = 0, \\ &z = \eta\,(x, y, t)\,, \\ &\partial_z \Phi|_{z=-\rho_0} = 0, \end{aligned} \tag{1.4}$$

where for a fluid, for example, the third and fourth conditions can be interpreted as the equation of the fluid surface and the boundary condition on the bottom, i.e., at $z = -H$.

We have now the full set of equations. We then introduce small parameters $\mu = v_0/c \ll 1$ and $\varepsilon = \rho_0/l \ll 1$ (here, l is the characteristic linear scale of perturbations and v_0 is the amplitude of the particle velocity in the wave, note also that $\varepsilon^2 \sim \mu$), expand the equations into the series of μ and ε, and convert the obtained expressions to dimensionless variables:

$$x' = x/l, \quad y' = y/l, \quad z' = z/\rho_0, \quad t' = c_0 t/l,$$

$$\Phi' = \Phi/v_0 l, \quad \eta' = c_0 \eta/v_0 \rho_0, \quad c_0 = \sqrt{g\rho_0}.$$

In the linear approximation in μ and ε^2 we obtain

$$\begin{gathered} \partial^2_{z'} \Phi' + \varepsilon^2 \left(\partial^2_{x'} \Phi' + \partial^2_{y'} \Phi'\right) = 0, \\ \partial_{t'} \Phi' + \frac{\mu}{2} \left[(\partial_{x'} \Phi')^2 + (\partial_{y'} \Phi')^2\right] + \frac{\mu}{2\varepsilon^2} (\partial_{z'} \Phi')^2 + \eta'|_{z'=\mu\eta'} = 0, \\ \Phi'|_{z'=-1} = 0. \end{gathered} \tag{1.5}$$

The first of these equations is obtained from the Laplace equation (1.3), the second from the equation of motion for the potential Φ, and the third from the last boundary condition of set (1.4). Expanding the potential Φ in z', we obtain from Eqs. (1.5)

$$\partial_{t'}\varphi' + \frac{\mu}{2}\left(\nabla\varphi'\right)^2 + \eta' = O\left(\varepsilon^4,\ \mu^2,\ \varepsilon^2\mu\right)$$

and

$$\partial_{t'}\eta' + \mu\nabla\varphi'\cdot\nabla\eta' + \mu\eta'\Delta\varphi' + \Delta\varphi' + \frac{\varepsilon^2}{3}\Delta^2\varphi' = O\left(\varepsilon^4,\ \mu^2,\ \varepsilon^2\mu\right),$$

where $\varphi' = \Phi'|_{z'=0}$. Now return back to the old variables, introduce the new potential $\psi = \varphi + \rho_0^2\Delta\varphi/3$, where $\varphi = v_0 l\varphi' - c_0 t$, and write with the accepted accuracy

$$\begin{aligned}&\partial_t\psi + \left(\nabla\psi\right)^2/2 + c^2 + 2c_0\beta\Delta\rho/\rho_0 = 0,\\ &\partial_t\rho + \nabla\cdot\left(\rho\nabla\psi\right) = 0,\end{aligned} \tag{1.6}$$

where $\beta = c_0\rho_0^2/6$ and $\rho = \rho_0 + \eta(t, x, y)$. Eqs. (1.6) are the equation of motion for the potential ψ and the continuity equation for the generalized density ρ. Here, β is the dispersion parameter, and the last equation, for example, for waves on the fluid surface can be interpreted in the following way: the height of the fluid level is a sum of the depth of the "tank" (where the fluid is contained) and the height of the perturbation set by the equation of the surface.

Now, if instead of the potential ψ we introduce the velocity

$$\mathbf{v}(t, x, y) = \nabla\psi(t, x, y)$$

and apply the gradient to the first equation of (1.6), we obtain

$$\begin{aligned}&\partial_t\mathbf{v} + \left(\mathbf{v}\nabla\right)\mathbf{v} + c^2\nabla\rho/\rho_0 + 2c_0\beta\nabla\Delta\rho/\rho_0 = 0,\\ &\partial_t\rho + \nabla\left(\rho\mathbf{v}\right) = 0.\end{aligned} \tag{1.7}$$

Set (1.7) represents the *Boussinesq equations* describing waves on the surface of shallow water as well as waves in plasmas and other dispersive media. These are three-dimensional nonlinear evolution equations, and it is generally very difficult to solve them.

Assume that v_0/c, $(\rho - \rho_0)/\rho_0$, and δ/λ are small (of the first order in μ), ε^2 (here, $\delta = (|\beta|/c_0)^{1/2}$ is the effective dispersion length, and λ is the characteristic wavelength). Further we suppose (as in standard hydrodynamics) that

$$c^2(\rho) = c_0^2\left(\rho/\rho_0\right)^{\gamma-1}, \quad \text{and} \quad \gamma = c_\mathrm{p}/c_\mathrm{v},$$

and look for a solution of (1.7) written as

$$\rho(t, x) = \rho(v) + \varphi(t, x), \tag{1.8}$$

where

$$\begin{aligned} \rho(v) &= \pm c(v)\mathrm{d}_v\rho, \\ c(v) &= c_0 + (\gamma - 1)v/2, \end{aligned} \tag{1.9}$$

and $\varphi(t, x)$ is small (of the second order in μ, ε^2). Then consider a wave propagating in the positive direction of the x-axis. Neglecting the terms of order higher than the second, we find that the function $\varphi(t, x)$ satisfies the equation

$$\partial_t\varphi + c_0\partial_x\varphi = 0.$$

Substituting (1.8) into (1.7), using expressions (1.9), and excluding the potential φ from the obtained expressions, we have

$$\partial_t v + \left(c_0 + \frac{\gamma + 1}{2}v\right)\partial_x v + \beta\partial_x^3 v = 0. \tag{1.10}$$

Converting now to new variables

$$\xi = x - c_0 t \quad \text{and} \quad u = \frac{\gamma + 1}{2}v,$$

we finally obtain from (1.10)

$$\partial_t u + u\partial_\xi u + \beta\partial_\xi^3 u = 0. \tag{1.11}$$

This is the *Korteweg–de Vries equation* (*KdV equation*) written in the reference frame moving with the velocity c_0 along the x-axis. It is obvious that this equation is much simpler than the initial three-dimensional set of hydrodynamic equations or Boussinesq equations.

We thus realized the marvelous ideas of Korteweg and de Vries that: to investigate complex nonlinear equation(s), we can simplify them while preserving their basic qualitative features; it is necessary, with the same order of accuracy, to keep the terms leading to the opposite effects, in our case - to maintain the balance of the nonlinear and dispersive terms.

It is very important that the KdV equation has an infinite number of the integrals of motion; this determines its complete integrability. Consider briefly this problem and rewrite the *KdV equation* (1.11) as

$$\partial_t u + \partial_x(u^2/2 + \beta\partial_x^2 u) = 0. \tag{1.12}$$

Equation (1.12) has the form of the conservation law for the "momentum" of the system,

$$\mathcal{J}_1 = \int_{-\infty}^{\infty} u(t, x)\mathrm{d}x = \text{const}. \tag{1.13}$$

Analogously we can obtain the next integrals of motion for the higher orders of u [3]:

$$\partial_t(u^2/2) + \partial_x\left\{u^3/3 + \beta\left[u\partial_x^2 u - (\partial_x u)^2\right]\right\} = 0 \tag{1.14}$$

and

$$\partial_t(u^3/3 - \beta\partial_x^2 u) + \partial_x \left\{u^4/4 + \beta\left[u^2\partial_x^2 u + 2\partial_t u \partial_x u\right] \right.$$
$$\left. +\beta^2(\partial_x^2 u)^2\right\} = 0. \qquad (1.15)$$

Equation (1.14) can be interpreted as the conservation law for the "energy" of the system [3], and equation (1.15), first obtained by Whitham [88], does not have any obvious physical interpretation. Other conservation laws were obtained by Kruskal, Zabusky and Miura [89] who also proved [90,91] that the KdV equation has an infinite number of conserved integrals, i.e., invariants given by

$$\mathcal{J}_m = \int_{-\infty}^{\infty} Q_m(t,x)\mathrm{d}x. \qquad (1.16)$$

For example, the first six invariants are [89–91][1]

$$Q_1 = u, \qquad Q_2 = u^2/2, \qquad Q_3 = u^3/3 - \beta(\partial_x u)^2,$$
$$Q_4 = u^4/4 - 3\beta u(\partial_x u)^2 + 2\beta^2(\partial_x^2 u)^2/5,$$
$$Q_5 = u^5/5 - 6\beta u^2(\partial_x u)^2 + 36\beta^2 u(\partial_x^2 u)^2/5 - 108\beta^3(\partial_x^3 u)^2/35,$$

and

$$Q_6 = u^6/6 - 10\beta u^3(\partial_x u)^2 + \beta^2\left[-5(\partial_x u)^4 + 18u^2(\partial_x^2 u)^2\right]$$
$$+\beta^3\left[-108u(\partial_x^3 u)^2/7 + 120(\partial_x^2 u)^3/7\right] + 36\beta^4\partial_x^4 u.$$

As we already noted, the infinite number of the *conservation laws* proves that the KdV equation represents a completely integrable system; we will also see that in the next section.

1.1.2 Universality of the KdV Model. Scaling Transformations and Similarity Principle

It is important that the KdV equation is universal in the sense that it describes the propagation of nonlinear waves when the linearized wave dispersion law is given by

$$\omega = c_0 k\left(1 - \frac{\beta k^2}{c_0}\right). \qquad (1.17)$$

For example, this is the case for:
(a) *Surface waves* in shallow water where u is the amplitude of the hydrodynamic velocity (in some sense – the amplitude of the wave), $\beta = c_0(H^2 - 3\sigma/\rho g)/6$ for the gravity-capillary waves, and $\beta = c_0 H^2/6$ for the gravity waves.

[1] Detailed information on this problem including a general algorithm to calculate the numerical factors in Q_m can be found in Refs. [90,92].

(b) *Ion-acoustic waves* in an unmagnetized plasma where u is the velocity of the ion sound wave, $\beta = c_s r_D^2/2$, $r_D = (T_e/4\pi n_e e^2)^{1/2}$ is the electron Debye length, and n_e is the unperturbed number density of plasma electrons.
(c) *Magnetosonic waves* in a magnetized plasma where u is the normalized wave magnetic field perturbation,

$$\beta = v_A \frac{c^2}{2\omega_{pi}^2}\left(\frac{m_e}{m_i} - \cot^2\theta\right),$$

c is the speed of light, $\omega_{pi} = (4\pi n_i e^2/m_i)^{1/2}$ is the ion plasma frequency, m_e is the electron mass, and θ is the angle between the direction of the wave propagation (vector $\mathbf{k}$) and the external magnetic field $\mathbf{B}_0$.[2]

The *KdV equation* is often written in a slightly different form, e.g.,

$$\partial_t u + \alpha u \partial_x u + \beta \partial_x^3 u = 0, \tag{1.18}$$

which can be easily obtained from (1.11) via the scale transformations $u \to \alpha u$, $t \to t$, and $\xi \to x$. For the analytical (exact) integration using the *inverse scattering transform* method (*IST method*) (see Sect. 1.2) the KdV equation (1.11) is transformed by $u \to -6u$, $t \to t$, and $\xi \to x$. In this case, the equation for $\beta = 1$ is given by

$$\partial_t u - 6u\partial_x u + \partial_x^3 u = 0. \tag{1.19}$$

The negative sign of the dispersion term can be obtained by the change $u \to -u$ and $x \to -x$. The sign of α in (1.18) determines the "polarity" of the solution (which can be positive, $u > 0$, or negative, $u < 0$), while the sign of β determines the direction of the wave propagation. Thus the KdV equation can be equally applied to a medium with the negative dispersion (when the phase velocity of linear waves decreases with the increasing wavenumber) as well as to a medium with the positive dispersion – the difference is only in the direction of the x-axis. Furthermore, for the convenience of an ad hoc study, we use various forms of the KdV equation.

The first solutions of the KdV equation were obtained numerically [1,93]. The studies showed that the equation can have two kinds of locally stationary solutions:
(a) In the form of moving solitons; and
(b) Of the type of periodic *cnoidal waves* (see below).
Later in 1967 [19] it was shown that the KdV equation is a completely integrable Hamiltonian system (see Sect. 1.2) and its soliton solution, found analytically for equation in the form (1.19), is given by [3]

$$u(t,x) = \frac{a^2}{2}\cosh^{-2}\left[\frac{a}{2}\left(x - x_0 - a^2 t\right)\right],$$

[2] Note that in the case $\theta = \pi/2$ and $n_e = n_i = n_0$ (perpendicular propagation of the wave in a two-component plasma), we have $\beta = v_A c^2/2\omega_{pe}^2$ since $\omega_{pe} = \omega_{pi}\sqrt{m_i/m_e}$.

where a^2 is the amplitude of the soliton whose velocity is proportional to a^2 and whose width is inversely proportional to the square root of the amplitude, $1/a$. For the periodic cnoidal wave, the solution $u(t,x)$ can be written using Jacoby elliptic functions, namely, the elliptic cosine $\mathrm{cn}[f(x)]$ (note that the name "cnoidal" originates here). In the following, we do not consider the periodic cnoidal wave solutions.

Note that the KdV equation can have N-soliton solutions describing the dynamics of interactions (collisions) of N solitons. We assume that

$$u(t,x) = 2\partial_x^2 \ln F \tag{1.20}$$

(obtained as an exact solution using the IST method, see Sect. 1.2.1) and (following [94]) choose the function F as

$$F_N = \sum_{\bar{\mu}} \exp\left[\sum_{i=1}^{N} \mu_i \xi_i + \sum_{1\leq i<j}^{N} \mu_i \mu_j A_{ij}\right], \tag{1.21}$$

where the sum on $\bar{\mu}$ is for all sets μ_i (the element μ_i can take a value 0 or 1, $i = 1, \ldots, N$). Furthermore, $\xi_i = k_i(x - x_{0i} - k_i^2 t)$ and the factors A_{ij} are defined by

$$e^{A_{ij}} = \left(\frac{k_i - k_j}{k_i + k_j}\right)^2 . \tag{1.22}$$

Thus we obtain the N-soliton solution of the KdV equation. For example, for $N = 1$ we have $F_1 = 1 + \mathrm{e}^{\xi_1}$ from (1.21), and the solution is as that above with $a = k_1$; for $N = 2$ we have

$$F_2 = 1 + \mathrm{e}^{\xi_1} + \mathrm{e}\xi_2 + \mathrm{e}^{\xi_1+\xi_2+A_{12}}, \tag{1.23}$$

where $\xi_2 = -(k_2^2 - k_1^2)t + \xi_1$, and substituting (1.23) into (1.20), we obtain the 2-soliton solution where the phase delay appearing as a result of the solitons' interaction is defined by the factor A_{12}.

To conclude this brief remark on the construction of the soliton solutions, we note that the choice of values of the constants x_{0i} is quite arbitrary. Thus, using the above direct method, we can also construct other classes of solutions. For example, choosing $\exp k_1 x_{01} = -1$ in (1.21) for $N = 1$, we obtain a *singular solution*

$$u(t,x) = -\frac{k_1^2}{2} \sinh^{-2} \frac{k_1}{2}\left(x - k_1^2 t\right) .$$

In the long-wavelength limit $k_1 \to 0$, we obtain the solution $u = -2/x^2$ [26] which is the first representative of the class of the rational solutions of the KdV equation. Generally, by choosing proper phase constants in (1.21) we can obtain a non-trivial limit for any function F_N.

To obtain the criterion enabling us to distinguish solutions of the KdV equation into two classes (the soliton and non-soliton types of waves), Karpman formulated the *similarity principle* in 1966 [93]. Here, we consider it on the basis of (1.18) with $\alpha = 1$. We set up the initial value problem, i.e., write the initial condition for the KdV equation as given by

$$u(0,x) = u_0 \varphi(x/l), \tag{1.24}$$

where u_0 is the initial amplitude and l is the linear size of the initial disturbance. Changing u_0 and l, we obtain a family of similar initial conditions characterized by the dimensionless function $\varphi(x/l)$. Introducing new variables

$$\eta = u/u_0, \quad \xi = x/l, \quad \text{and} \quad \tau = u_0 t/l, \tag{1.25}$$

we obtain from the KdV equation (1.18) with $\alpha = 1$ and initial condition (1.24) the equation

$$\partial_\tau \eta + \eta \partial_\xi \eta + \sigma^{-2} \partial_\xi^3 \eta = 0, \tag{1.26}$$

where $\eta(0,\xi) = \varphi(\xi)$ and

$$\sigma = l \left(u_0/\beta\right)^{1/2}. \tag{1.27}$$

It follows from (1.26) that solutions corresponding to the same value of σ and the same initial function $\varphi(\xi)$ are similar. For solitons,

$$u(x) = u_0 \cosh^{-2}\left[\left(u_0/12\beta\right)^{1/2} x\right], \tag{1.28}$$

we have

$$\sigma = \sigma_s = \sqrt{12}.$$

The parameter σ is, in fact, the index of nonlinearity of the problem, and σ_s is in some sense a characteristic value: as numerical simulations [93] demonstrated, for the same form of the initial function $\varphi(\xi)$ in the cases $\sigma \gg \sigma_s$ and $\sigma \ll \sigma_s$, qualitatively different solutions are observed. We study them in detail in the next sections (especially in Sect. 1.3.5) after we briefly consider other equations of the KdV-class.

1.1.3 Other (1+1)-Dimensional KdV-Class Equations

Other (1+1)-dimensional[3] KdV-class equations to be considered here are the linearized KdV equation, the Burgers equation and the Korteweg–de Vries–Burgers (KdVB) equation, as well as the generalized KdV equation with the higher order dispersion correction (Kawahara equation [84]) and terms taking into account dissipation and instability in a medium. Numerical solutions of these equations are given below in Sect. 1.3 after considering the analytical and numerical methods of integration of KdV-class equations.

[3] This terminology, often used in the mathematical literature, refers to the number of independent variables, namely, the time and the space coordinate in the present case.

The linearized KdV Equation. Linear approximation to (1.18) is given by the *linearized KdV equation*

$$\partial_t u + c_0 \partial_x u + \beta \partial_x^3 u = 0. \tag{1.29}$$

In the reference frame moving with the velocity c_0 we obtain

$$\partial_t u + \beta \partial_x^3 u = 0. \tag{1.30}$$

The general solution of this equation expressed via the Airy function is [3]

$$u(t,x) = \pi^{-1/2}(3\beta t)^{-1/3} \int_{-\infty}^{\infty} \mathrm{Ai}\left[\frac{x-x'}{(3\beta t)^{1/3}}\right] u(0,x')\mathrm{d}x', \tag{1.31}$$

where

$$\mathrm{Ai}(z) = \frac{1}{\sqrt{\pi}} \int_0^{\infty} \cos\left(\frac{\upsilon^3}{3} + \upsilon z\right) \mathrm{d}\upsilon \tag{1.32}$$

has the following asymptotics

$$\mathrm{Ai}(z) = \begin{cases} (z^{-1/4}/2)\exp\left(-2z^{3/2}/3\right), & z \to \infty, \\ |z|^{-1/4}\cos\left((2\,|z|^{3/2}/3) - \pi/4\right), & z \to -\infty. \end{cases}$$

Although we are able to conclude on the solution's behavior from this general solution, it is difficult to see the change of dynamics of an initial disturbance (in the region of its localization) as a function of the dispersion parameter β since it is necessary to construct the corresponding Airy function in this case. Therefore numerical methods are better for the analysis of the structure and evolution of the solution of the linearized KdV equation (1.30). The results obtained numerically are given below in Sect. 1.3.

The Burgers equation. Consider the equation describing nonlinear waves in a medium with the "viscous" type of damping,

$$\partial_t u + u\partial_x u = \nu \partial_x^2 u. \tag{1.33}$$

This equation was obtained and analyzed by Burgers in 1940 [15] and is now called the *Burgers equation.* The general solution of this equation can also be obtained analytically. Namely, if

$$u = -2\nu \partial_x \ln \varphi(t,x), \tag{1.34}$$

for the function φ then one can obtain the heat conductivity (diffusion) equation

$$\partial_t \varphi = \nu \partial_x^2 \varphi.$$

In this case, the solution of the Burgers equation with the initial condition $u(0,x) = \psi(x)$ can be written as (1.34), where [3]

$$\varphi(t,x) = \frac{1}{\sqrt{4\pi\nu t}} \int\limits_{-\infty}^{\infty} \exp\left[-\frac{(x-\eta)^2}{4\nu t} - \frac{1}{2\nu}\int\limits_{0}^{\eta} \psi(\eta')\mathrm{d}\eta'\right]\mathrm{d}\eta.$$

However, some additional condition(s) need to be imposed for convergence of the integral on the right-hand side of this expression. In general, the structure of solution (1.34) is not that transparent and it is difficult to estimate its dynamics for various ν. In this regard, it is helpful to integrate the Burgers equation numerically, and the corresponding results are also presented in Sect. 1.3.

1.2 Inverse Scattering Transform and Analytical Integration

In this section, we consider fundamentals of the theory of the inverse scattering transform (Sect. 1.2.1), the powerful method that enables us to obtain analytically general solutions of some classes of canonical nonlinear equations. On this basis, we approach the problem of (exact) analytical integration of the KdV equation in Sect. 1.2.2. In Sect. 1.2.3, we consider generalization of the Gelfand–Levitan–Marchenko integral equation for the purpose of possible separation of the study of the soliton and non-soliton parts of the solution. Finally, we consider the variational principle which practically enables us to separate the solutions (Sect. 1.2.4). The inverse scattering problem for the multidimensional cases is considered later in Chap. 3 (see Sect. 3.2).

1.2.1 Fundamentals of the Inverse Scattering Theory

The inverse scattering transform (IST) method in its classic form is a very convenient tool for solving the initial value (Cauchy) problem for nonlinear evolution equations. Note that when the Cauchy problem is solved on a class of functions decreasing sufficiently fast at infinity ($u(\mathbf{r}) \to 0$ when $|\mathbf{r}| \to \infty$), the *IST method* is as effective as the Fourier method for integration of linear partial differential equations with the constant coefficients (when the Fourier transform reduces the partial differential equation to an infinite set of ordinary linear differential equations for Fourier harmonics). When the equation's coefficients do not depend on the coordinates, the equations for the Fourier harmonics are independent and can be trivially integrated. The situation is similar when the IST method is applied. The mapping of the coefficient functions of a linear differential operator to the set of so-called "scattering data" plays the role of the Fourier transform here.

For example, the KdV equation is integrated with the help of the transition from the potential of the one-dimensional *Schrödinger equation* ,

$$-\mathrm{d}_x^2\psi + u(x)\psi = k^2\psi,$$

to the *reflection coefficient* $r(k)$ on this potential. As the potential $u(x)$ evolves according to the KdV equation, the dependence of the reflection coefficient on time is trivial:

$$r(k,t) = r(k,0)\exp\left(-8\mathrm{i}k^3t\right).$$

Thus, the problem of integrating the KdV equation is converted to the problem of reconstruction of the potential $u(x)$ for the given reflection coefficient, i.e., the inverse of the (quantum) scattering problem.

Consider fundamental statements of the scattering theory for the one-dimensional *Sturm–Liouville operator* (*Schrödinger operator*),

$$\hat{\mathrm{L}} = -\mathrm{d}_x^2 + u(x),$$

on the (whole) real axis $-\infty < x < \infty$. We assume that the real potential $u(x)$ is a sufficiently smooth function of x that turns to zero as $|x| \to \infty$. We consider the eigenvalue problem

$$\hat{\mathrm{L}}\psi \equiv -\mathrm{d}_x^2\psi + u(x)\psi = \lambda\psi, \tag{1.35}$$

where the eigenvalues $\lambda = k^2$ is the (full) energy and k is the momentum of the system, on the class of (limited on the whole x-axis) functions $\psi(x)$. The spectrum of the operator $\hat{\mathrm{L}}$ consists of two parts, discrete and continuous, such that the *continuous spectrum* occupies the real half-axis $\lambda > 0$ (real k), and the eigenvalues of the *discrete spectrum* are negative (corresponding to the points of the imaginary axis $k = \mathrm{i}\kappa_n$, $n = 1, 2, \ldots, N$, $\kappa_n > 0$).

Changing λ on the right-hand side of (1.35) to k^2, we write the equation defining the eigenfunctions as

$$-\mathrm{d}_x^2\psi + u(x)\psi = k^2\psi. \tag{1.36}$$

First, consider the characteristics of the continuous spectrum. For every real $k \neq 0$, the set of solutions of (1.36) constitutes the two-dimensional linear space $\mathbb{G}_k$. We fix two bases in $\mathbb{G}_k$.

The first basis consists of solutions $\psi_{1,2}(x,k)$ of (1.36) determined by the asymptotic conditions at $+\infty$ in x:

$$\begin{cases} \psi_1(x,k) = \mathrm{e}^{-\mathrm{i}kx} + O(1), \\ \psi_2(x,k) = \mathrm{e}^{\mathrm{i}kx} + O(1), \quad x \to +\infty. \end{cases}$$

The second basis consists of solutions $\varphi_{1,2}(x,k)$ determined by the analogous conditions at $-\infty$:

$$\begin{cases} \varphi_1(x,k) = \mathrm{e}^{-\mathrm{i}kx} + O(1), \\ \varphi_2(x,k) = \mathrm{e}^{\mathrm{i}kx} + O(1), \quad x \to -\infty. \end{cases}$$

Due to the real character (zero imaginary part) of the potential, we have

$$\varphi_1(x,k) = \varphi_2^*(x,k) \quad \text{and} \quad \psi_1(x,k) = \psi_2^*(x,k). \tag{1.37}$$

It is also obvious that

$$\varphi_1(x,k) = \varphi_2(x,-k) \quad \text{and} \quad \psi_1(x,k) = \psi_2(x,-k). \tag{1.38}$$

Vectors of either basis are the linear combination of vectors of the other:

$$\varphi_i(x,k) = \sum_{l=1,2} T_{il}(k)\psi_l(x,k), \quad i = 1,2.$$

The matrix $T(k)$, introduced above, is called the *transition matrix.* [4] We have

$$T(k) = \begin{bmatrix} a(k) & b(k) \\ b^*(k) & a^*(k) \end{bmatrix}.$$

Omitting the indices 1 and 2 of the functions φ and ψ, we write

$$\varphi(x,k) = a(k)\psi(x,k) + b(k)\psi^*(x,k). \tag{1.39}$$

The Wronskian $W(f_1, f_2) = f_1 d_x f_2 - f_2 d_x f_1$ of any pair of the solutions f_1 and f_2 of (1.36) does not depend on x. It is clear that $W(\varphi, \varphi^*) = W(\psi, \psi^*) = 2ik$. This relation together with (1.39) gives us

$$|a(k)|^2 - |b(k)|^2 = 1, \tag{1.40}$$

i.e., we see that the transition matrix is a unimodular one: $\det T(k) = 1$.

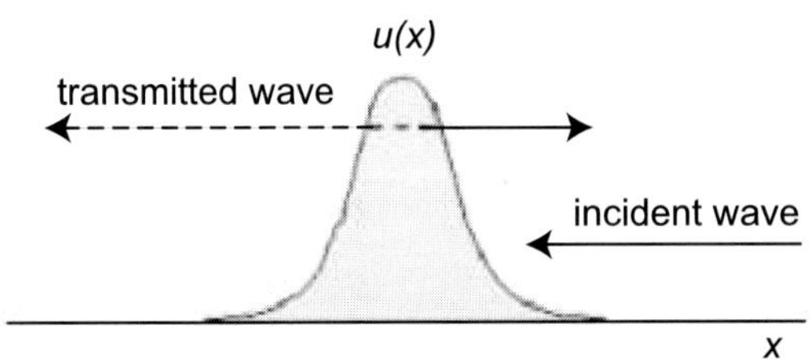

Fig. 1.1. Sketch of the scattering problem for a wave incident on the potential $u(x)$

The functions $a^{-1}(k)$ and $b(k)a^{-1}(k)$ are the *transmission coefficient* and the *reflection coefficient*, respectively, of the wave incident to the potential $u(x)$ from the right (see Fig. 1.1). In reality, the asymptote of the eigenfunction $\varphi(x,k)/a(k)$, see (1.39), for $x \to +\infty$ is given by

$$a^{-1}\varphi(x,k) = \mathrm{e}^{-ikx} + ba^{-1}\mathrm{e}^{ikx} + O(1),$$

i.e., it is a superposition of the incident and reflected waves. At the other end of the x-axis we have

[4] Written as $T(k) \equiv S(k) = \begin{bmatrix} s_{11}(k) & s_{12}(k) \\ s_{21}(k) & s_{22}(k) \end{bmatrix}$, this matrix is often called the S-matrix or the scattering matrix.

$$a^{-1}\varphi(x,k) = a^{-1}\mathrm{e}^{-ikx} + O(1),$$

i.e. only the transmitted wave propagates there. In other words, $t(k) = a^{-1}(k)$ is the amplitude of the forward scattering and $r(k) = b(k)a^{-1}(k)$ is the amplitude of the backward scattering. It follows from (1.40) that the scattering is unitary, i.e.,

$$|t(k)|^2 + |r(k)|^2 = 1.$$

An analysis of the analytic properties of the amplitude of the forward scattering on the physical sheet of the Riemann surface (we omit details here to save space) shows that (see Fig. 1.2) $a(k)$ is an analytic function in the upper semi-plane of complex k and has simple zeros at the points $k_n = \mathrm{i}\kappa_n$, $\kappa_n^2 = -\lambda_n$; in addition, $a(k) \to 1$ for $|k| \to \infty$, $\mathrm{Im}\ k \geq 0$. These analytic properties of the diagonal elements of the transition matrix are principally important and, to a certain degree, universal. Analogous statements can be made in the scattering theory for other differential operators.

The *transition matrix* $T(k)$ gives us comprehensive information on the *continuous spectrum* of the Schrödinger operator. The information on $T(k)$ is fully contained in the reflection coefficient

$$r(k) = \frac{b(k)}{a(k)}, \tag{1.41}$$

that, in view of relations (1.37) and (1.38), can be determined only on the semi-axis $k > 0$ since $r(-k) = r^*(k)$. From (1.40) we can easily obtain

$$|a(k)|^2 = \frac{1}{1 - |r(k)|^2}, \tag{1.42}$$

i.e., the modulus of the reflection coefficients uniquely defines $|a(k)|$. Knowing zeros of the analytic (in the upper semi-plane) function $a(k)$, it is possible to find the unique argument $\arg[a(k)]$ by its modulus. Thus $a(k)$ is reconstructed

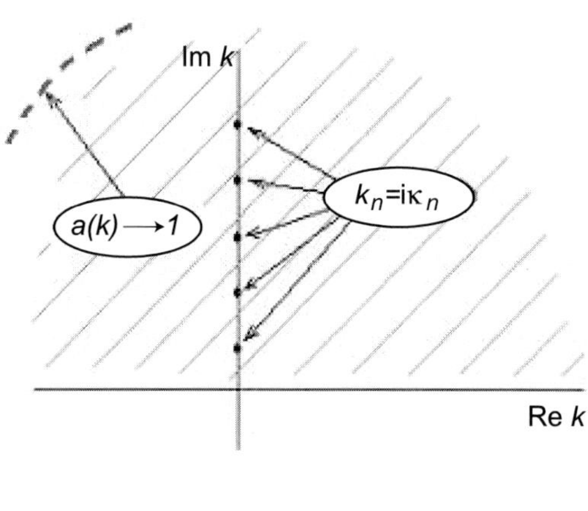

Fig. 1.2. Analytic properties of the amplitude of the forward scattering $a(k)$

according to the modulus of the reflection coefficient, and the function $b(k)$ is simply $r(k)a(k)$.

Consider now characteristics of the discrete spectrum of the *Sturm–Liouville operator* (*Schrödinger operator*) matching naturally the scattering characteristics. The eigenfunctions corresponding to the eigenvalue $\lambda_n = -\kappa_n^2$ satisfy the equation

$$-\mathrm{d}_x^2\psi + u(x)\psi = -\kappa_n^2\psi.$$

The discrete spectrum of the Schrödinger operator, as it is known, is simple: all solutions of this equation can be obtained from any one by multiplying it by a constant. At infinity, the eigenfunctions have the asymptotes

$$\psi \to c_\pm \exp\left(\mp\kappa_n x\right). \quad x \to \pm\infty.$$

Fix the eigenfunction $\varphi^{(n)}(x)$ at $-\infty$ in x by its asymptote

$$\varphi^{(n)}(x) = \mathrm{e}^{\kappa_n x} + O(\mathrm{e}^{\kappa_n x}).$$

When $x \to +\infty$ the eigenfunctions are given by

$$\varphi^{(n)}(x) = b_n \mathrm{e}^{-\kappa_n x} + O(\mathrm{e}^{-\kappa_n x}); \tag{1.43}$$

obviously, they are real, and therefore the factors b_n are also real. If we arrange the eigenvalues λ_n in a usual way in order of increasing λ_n (i.e., decreasing κ_n), namely,

$$\lambda_1 < \lambda_2 < \ldots < \lambda_N < 0$$

(here, λ_1 is the energy of the ground (basic) state of a quantum system and the corresponding function $\varphi^{(1)}$ is the wave function of this state), then $\varphi^{(1)}$ has no zeros and $\varphi^{(n)}$ crosses zero exactly $(n-1)$ times. Thus

$$b_n = (-1)^{n-1}\left|b_n\right|.$$

The positive quantities $|b_n|$ consist additional (to the eigenvalues λ_n) characteristics of the discrete spectrum. These characteristics together with the reflection coefficient $r(k)$ and the set of eigenvalues $\lambda_1, \lambda_2, \ldots, \lambda_N$ fully determine the spectrum of problem (1.36). The set

$$\mathsf{S} = \{r(k), \kappa_n, \left|b_n\right|, \quad n = 1, 2, \ldots, N\}$$

is called the *scattering data*. Mapping $u(x) \to \mathsf{S}$ of potentials $u(x)$ to the scattering data is uniquely reversible. The procedure of reconstruction of the potential $u(x)$ on S is the subject of the inverse problem of the scattering theory.

The analytic properties of the special solutions of the Schrödinger equation introduced via the scattering data enable us to write the integral equations equivalent to (1.36) on the functions

$$\chi_+(x,k) = \varphi(x,k)\mathrm{e}^{\mathrm{i}kx} \quad \text{and} \quad \chi_-(x,k) = \psi(x,k)\mathrm{e}^{-\mathrm{i}kx}$$

using the Green's function. The function $\chi_+(x,k)$ can be analytically continued to the upper semi-plane k (i.e. to Im $k > 0$), and the function $\chi_-(x,k)$ can be analytically continued to the lower semi-plane k. Thus for $|k| \to \infty$ we have

$$\chi_+(x,k) = 1 + \frac{1}{2ik}\int_{-\infty}^{x} u(x')\mathrm{d}x' + O\left(\frac{1}{k^2}\right),$$

$$\chi_-(x,k) = 1 + \frac{1}{2ik}\int_{x}^{\infty} u(x')\mathrm{d}x' + O\left(\frac{1}{k^2}\right). \tag{1.44}$$

For the Schrödinger operator, the most known form of the IST equation, i.e., of the equation describing the transition from the scattering data S to the potential $u(x)$, is the *Gelfand–Levitan–Marchenko equation* (*GLM equation*). Now we are ready to obtain this equation. The procedure involves the following steps:

1. First of all, we note that Fourier transform of the function $\chi_-(x,k)$ (analytic in the lower semi-plane k) is cut off. Thus the function $\chi_-(x,k)$ can always be written as

$$\chi_-(x,k) = 1 + \int_0^{\infty} \mathbf{A}(x,y)\mathrm{e}^{-iky}\mathrm{d}y.$$

2. For the function $\psi(x,k)$, this means that there is a function $\mathbf{K}(x,y)$ such that

$$\psi(x,k) = \mathrm{e}^{-ikx} + \int_x^{\infty} \mathbf{K}(x,y)\mathrm{e}^{-iky}\mathrm{d}y. \tag{1.45}$$

   Obviously, $\mathbf{K}(x,y) = \mathbf{A}(x, y-x)$.
3. It follows from (1.45) that there exists a linear operator transforming the solution of Schrödinger equation with the zero potential (the plane wave solution e^{-ikx}) to the solution of this equation with the potential $u(x)$. The function $\mathbf{K}(x,y)$ is called the kernel of the transformation operator and it is real because $\psi(x,-k) = \psi(x,k)$ for Im $k = 0$.
4. Consider now equality (1.39). We multiply it by ratio $\mathrm{e}^{-iky}/a(k)$ and integrate in all real k:

$$\int_{-\infty}^{\infty}\left[\frac{\varphi(x,k)}{a(k)} - \underline{\mathrm{e}^{-ikx}}\right]\mathrm{e}^{iky}\mathrm{d}k$$

$$= \int_{-\infty}^{\infty}\left[\psi(x,k) - \underline{\mathrm{e}^{-ikx}} + r(k)\psi^*(x,k)\right]\mathrm{e}^{iky}\mathrm{d}k. \tag{1.46}$$

(Note that the underlined terms sum up to zero and they are introduced because of convenience of further transformations – in fact, we just added to the both sides of (1.39) the term $a(k)\mathrm{e}^{-\mathrm{i}kx}$).
Calculation of the integral on the left-hand side of (1.46), taking into account that the function under the integral has, in the upper semi-plane, only a finite number of simple poles and decreases for $|k| \to \infty$, gives us

$$2\pi\mathrm{i}\sum_{n=1}^{N}\frac{\varphi(x,\mathrm{i}\kappa_n)\mathrm{e}^{-\kappa_n y}}{a'(\mathrm{i}\kappa_n)},$$

where a' is the k-derivative of the function $a(k)$ for $k = \mathrm{i}\kappa_n$. Taking into account (1.43), taking place for the discrete spectrum, which can be written as

$$\varphi(x,\mathrm{i}\kappa_n) = b_n\psi^*(x,-\mathrm{i}\kappa_n) = b_n\psi(x,-\mathrm{i}\kappa_n) \tag{1.47}$$

(the functions φ, ψ and ψ^* on the imaginary axis are obviously equivalent), and (1.45) for $k = \mathrm{i}\kappa_n$, $n = 1,\ldots,N$, we obtain for the integral on the left-hand side of (1.46) the expression

$$2\pi\mathrm{i}\sum_{n=1}^{N}\frac{b_n\mathrm{e}^{-\kappa_n(x+y)}}{a'(\mathrm{i}\kappa_n)} + 2\pi\mathrm{i}\int_x^{\infty}\mathbf{K}(x,z)\sum_{n=1}^{N}\frac{b_n\mathrm{e}^{-\kappa_n(z+y)}}{a'(\mathrm{i}\kappa_n)}\mathrm{d}z.$$

Substituting now formula (1.45) into the right-hand side of (1.46) and introducing the new notation,

$$\mathbf{F}(x) \equiv \sum_{n=1}^{N}\frac{b_n\mathrm{e}^{-\kappa_n x}}{\mathrm{i}a'(\mathrm{i}\kappa_n)} + \frac{1}{2\pi}\int_{-\infty}^{\infty}r(k)\mathrm{e}^{\mathrm{i}kx}\mathrm{d}k, \tag{1.48}$$

we obtain finally the *GLM equation* given by

$$\mathbf{K}(x,y) + \mathbf{F}(x+y) + \int_x^{\infty}\mathbf{K}(x,z)\mathbf{F}(z+y)\mathrm{d}z = 0. \tag{1.49}$$

5. It follows from expression (1.45) that the asymptotic expansion for the function $\chi_-(x,k)$ gives us

$$\chi_-(x,k) = 1 + \frac{1}{\mathrm{i}k}\mathbf{K}(x,x) + O\left(\frac{1}{k^2}\right), \quad |k| \to \infty.$$

Comparing that with (1.44) we can express the potential $u(x)$ by the kernel of the transformation operator

$$u(x) = -2\mathrm{d}_x\mathbf{K}(x,x). \tag{1.50}$$

Thus, to determine the potential $u(x)$, it is necessary to solve the integral GLM equation, i.e., to obtain the kernel of the transformation operator via the scattering data fully contained in the function $\mathbf{F}(x)$, see (1.48). Now, having considered the fundamentals of the theory of scattering, we are able to proceed to the integration of the KdV equation using the IST method.

1.2.2 Integration of the KdV Equation Using the IST Method

Generally speaking, the *IST method* is the result of the outstanding observation by Gardner, Green, Kruskal and Miura who found [19] that solutions of the KdV equation can be associated with the potential of the Schrödinger equation.[5] Consider therefore the *Schrödinger equation* of the form

$$-\,\mathrm{d}_x^2 f + u(x,t) f = k^2 f, \tag{1.51}$$

with the potential $u(x,t)$ decreasing at x-infinity. We note that the function f can be a special solution of problem (1.51), e.g., with the asymptote

$$f(x,k) = \mathrm{e}^{-\mathrm{i}kx} + O(1), \quad x \to -\infty,$$

i.e., the solution $\varphi(x,k)$ defined in the previous section.

If the potential $u(x,t)$ depends on t, the function $\varphi(x,k)$ also depends on t and then the scattering data become functions of t, too. The asymptote of $\varphi(x,k)$ at $x \to +\infty$ is then given by

$$\varphi(x,k,t) = a(k,t)\mathrm{e}^{-\mathrm{i}kx} + b(k,t)\mathrm{e}^{\mathrm{i}kx} + O(1). \tag{1.52}$$

In the case of an arbitrary dependence of $u(x,t)$ on t, it is obviously not possible to find the general dependence of the scattering data on t. However, if $u(x,t)$ changes in time as a solution of the KdV equation, $\partial_t u - 6u\partial_x u + \partial_x^3 u = 0$, then the "coefficient functions" $a(k,t)$ and $b(k,t)$ satisfy the Gardner–Green–Kruskal–Miura (GGKM) equations [19]

$$\dot{a}(k) = 0 \quad \text{and} \quad \dot{b}(k,t) = 8\mathrm{i}k^3 b(k,t) \tag{1.53}$$

(here, the dot stands for the time derivative), and the dependence of $\varphi(x,k,t)$ on t is given by the equation

$$\dot{\varphi}(x,k,t) = -\hat{\mathrm{A}}\varphi(x,k,t) + 4\mathrm{i}k^3\varphi(x,k,t), \tag{1.54}$$

where the operator $\hat{\mathrm{A}}$ is

$$\hat{\mathrm{A}} = 4\mathrm{d}_x^3 - 3\left(u\mathrm{d}_x + \mathrm{d}_x u\right). \tag{1.55}$$

The inverse is also true: if the scattering data change in time as the GGKM-equation solutions then the potential $u(x,t)$ (uniquely defined by them) satisfies the KdV equation. The simplest way to prove that is to note that the KdV equation is identical to the equation for the operators $\hat{\mathrm{L}} = -\mathrm{d}_x^2 + u$ and $\hat{\mathrm{A}}$:

$$\dot{\hat{\mathrm{L}}} = \left[\hat{\mathrm{L}}, \hat{\mathrm{A}}\right] = \hat{\mathrm{L}}\hat{\mathrm{A}} - \hat{\mathrm{A}}\hat{\mathrm{L}}. \tag{1.56}$$

[5] Later this method was naturally extended to other evolution equations, e.g., the nonlinear Schrödinger (NLS) equation and the sin-Gordon equation.

Indeed, $\dot{\hat{\text{L}}}$ is just the multiplication operator $\partial_t u$, and the commutator $[\hat{\text{L}}, \hat{\text{A}}]$, as simple calculations demonstrate, is the multiplication operator $6u\partial_x u - \partial_x^3 u$, so that (1.56) coincides with the KdV equation, $\partial_t u = 6u\partial_x u - \partial_x^3 u$.

The representation of the evolution equations by (1.56) is called the *Lax representation* or Lax $\hat{\text{L}}$–$\hat{\text{A}}$ *pair* representation. With the help of the Lax representation it is not difficult to obtain for $\varphi(x,k,t)$, instead of (1.54), the following expression:

$$\dot{\varphi}(x,k,t) = \dot{a}\mathrm{e}^{-\mathrm{i}kx} + \dot{b}\mathrm{e}^{\mathrm{i}kx} = \left(-4d_x^3 + 4\mathrm{i}k^3\right)\left(a\mathrm{e}^{-\mathrm{i}kx} + b\mathrm{e}^{\mathrm{i}kx}\right),$$

which is equivalent to (1.53). Note that the functions a and b fully define the time dependence of the amplitude of the backward scattering

$$r(k,t) = r(k,0)\mathrm{e}^{8\mathrm{i}k^3 t} = \frac{b(k,t)}{a(k,t)},$$

characterizing the continuous spectrum. Analogously we can solve the problem of the time evolution of the *scattering data* $\kappa_n(t)$ and $b_n(t)$ $(n = 1, \ldots, N)$ for the discrete spectrum:

1. It follows from the first equation of (1.53) that $\mathrm{i}\kappa_n$ are the zeros of an analytic function $a(k)$ independent of t, i.e., $\dot{\kappa}_n = 0$;
2. The dependence of $b_n(t)$ can be obtained from (1.54) because by definition b_n is a factor in the asymptotic expansion of the function $\varphi(x, \mathrm{i}\kappa_n)$, where

$$\varphi(x, \mathrm{i}\kappa_n) = b_n(t)\mathrm{e}^{-\kappa_n x} + O\left(\mathrm{e}^{-\kappa_n x}\right), \quad x \to +\infty.$$

Substituting this asymptotic equation into (1.54), we obtain for $k = \mathrm{i}\kappa_n$ that $\dot{b}_n = 8\kappa_n^3 b_n$. Thus the time evolution of the *scattering data* is given by

Table 1.1. Scheme of solution of the initial value problem for the KdV equation

$u(x,0)$	**Ist stage**	Consists of calculation of the scattering data $\mathsf{S}(0)$, with the initial condition $u(x,t)\vert_{t=0} = u(x,0)$, by finding the eigenfunctions of the Schrödinger operator with the potential $u(x,0)$.
$\Downarrow$		
$\mathsf{S}(0)$	**IInd stage**	Involves solving the initial value problem in terms of the scattering data. The problem is trivial and its solution is given by (1.57).
$\Downarrow$		
$\mathsf{S}(t)$	**IIIrd stage**	Takes place when using the GLM equation, the inverse problem is solved, i.e., the potential $u(x,t)$ in the Schrödinger operator is determined with $\mathsf{S}(t)$ as the scattering data.
$\Downarrow$		
$u(x,t)$		Each stage implies solution of a **linear problem** only.

$$\mathsf{S}(t) = \left\{ r(k,0)\mathrm{e}^{8\mathrm{i}k^3 t}, \kappa_n, b_n \mathrm{e}^{8\kappa_n^3 t}, n = 1, \ldots, N \right\}. \tag{1.57}$$

We in fact integrated the KdV equation by means of the (implicit) change of the variables $u(x) \to \mathsf{S}$. The inverse change $\mathsf{S}(t) \to u(x,t)$ gives us the solution of the KdV equation. Overall, the scheme of solution of the initial value problem for the KdV equation is given by Table 1.1. This scheme, despite the absence of a general analytical solution for both direct and inverse problems, enables us to find very important exact solutions of the KdV equation analytically, in particular, the one-soliton solution, and, in a more general case, the N-soliton solutions describing interactions (collisions) of KdV solitons.

1.2.3 Generalization of the GLM Equation

Consider the *Sturm–Liouville operator* (*Schrödinger operator*)

$$\hat{\mathrm{H}}_j = -\mathrm{d}_x^2 + u_j(x), \quad -\infty < x < \infty, \tag{1.58}$$

where $u_j(x)$ is an element in the class of the scattering potentials. We suppose that $u_j(x) \to 0$ (when $|x| \to \infty$) sufficiently fast and, therefore, the standard theory of scattering is applicable. Then the spectrum of the operator $\hat{\mathrm{H}}_j$ consists of the continuous and discrete parts. The continuous spectrum occupies the real semi-axis $\lambda = k^2 \geq 0$, and the discrete spectrum consists of the negative $\lambda = k^2 < 0$ non-degenerated point-like eigenvalues (we assume that their number is finite). We name the eigenfunctions of the continuous spectrum describing the scattering as $\varphi_j(x,k)$. They satisfy the *Schrödinger equation*

$$\hat{\mathrm{H}}_j \varphi_j(x,k) = k^2 \varphi_j(x,k) \tag{1.59}$$

with the boundary conditions

$$\begin{cases} \varphi_j(x,k) = \mathrm{e}^{-\mathrm{i}kx} + ba^{-1}\mathrm{e}^{\mathrm{i}kx}, & x \to +\infty, \\ \varphi_j(x,k) = a^{-1}\mathrm{e}^{-\mathrm{i}kx}, & x \to -\infty. \end{cases} \tag{1.60}$$

Recall that $t(k) = 1/a(k)$ is the amplitude of the forward scattering, and $r(k) = b(k)/a(k)$ is the amplitude of the backward scattering for the potential u_j.

For the direct and inverse scattering problems, another set of eigenfunctions, namely, in the form of the *Jost function* satisfying the Schrödinger equation (1.59), but with another boundary condition,

$$\psi_j(x,k) = \mathrm{e}^{\mathrm{i}kx}, \quad x \to -\infty, \tag{1.61}$$

is also important (see Sect. 1.2.1). They are in fact pseudo-eigenfunctions because they are not quadratically integrated in general. The functions $r(k)$, $a^{-1}(k)$, and $\psi_j(x,k)$ satisfy the relations

$$\begin{cases} r(-k) = r^*(k), \\ a^{-1}(-k) = \left(a^{-1}\right)^* (k), \\ \psi_j(x,k) = \psi_j^*(x,-k). \end{cases}$$

Note that the functions ψ_j and ψ_j^* are linearly independent. It is obvious that (see Sect. 1.2.1)

$$\varphi_j(x,k) = \psi_j(x,k) + r_j(k)\psi_j^*(x,k).$$

The discrete spectrum that the operator $\hat{\mathrm{H}}_j$ can have consists of a finite number of the point-like eigenvalues represented by

$$\lambda_{jn} = -\kappa_{jn}^2, \quad \kappa_{jn} > 0. \tag{1.62}$$

The corresponding eigenfunctions (note that they are true eigenfunctions in contrast to ψ_j which, as noted above, are "pseudo-eigenfunctions") satisfy the Schrödinger equation

$$\hat{\mathrm{H}}_j \varphi_{jn}(x) = \lambda_{jn}\varphi_{jn}(x) \tag{1.63}$$

together with the asymptotic condition

$$\lim_{x\to\pm\infty} \varphi_{jn}(x) = c_\pm \mathrm{e}^{\mp\kappa_{jn}x}, \quad c_+ = b_n, \quad c_- = 1, \tag{1.64}$$

where the normalization factors are

$$|b_n| = \int_{-\infty}^{\infty} [\varphi_{jn}(x)]^2 \,\mathrm{d}x. \tag{1.65}$$

In the inverse problem, we begin with the spectral data

$$\mathsf{S} = \{r(k), \kappa_n, |b_n|, n = 1, 2, \ldots, N\}$$

to obtain the scattering potential $u_j(x)$ by using the *GLM equation* which was written in the previous section as

$$\mathbf{K}(x,y) + \mathbf{F}(x+y) + \int_x^{\infty} \mathbf{K}(x,z)\mathbf{F}(z+y)\mathrm{d}z = 0, \tag{1.66}$$

where $\mathbf{F}(x)$ is fully defined by the scattering data and the problem is, therefore, to determine the transform kernel $\mathbf{K}(x,y)$. The purpose of generalization of the GLM equation (1.66) is to introduce the "basic potential" $u_m(x)$. To do that we represent the potential $u_j(x)$ as [95,96]

$$u_j(x) = u_m(x) + u_{jm}(x). \tag{1.67}$$

We assume that the spectral data S are known. The number of the point-like eigenvalues for $u_m(x)$ is generally not equal to that of $u_j(x)$, i.e., the number of values of n in κ_{mn} is in general not equal to that in κ_{jn}. Using the GLM equation (1.66) it is possible to obtain $u_m(x)$ as well as the eigenfunctions $\psi_m(x,k)$ and $\varphi_{mn}(x)$. Let $\mathbf{K}_m(x,y)$ be the kernel of the GLM equation for $u_m(x)$. Then we can represent the "pseudo-eigenfunctions" of the *discrete spectrum* $\varphi_{jn}(x)$ satisfying the *Schrödinger equation*

$$\hat{\mathrm{H}}_m \ddot{\varphi}_{jn}(x) = -\kappa_{jn}^2 \ddot{\varphi}_{jn}(x)$$

as those given by

$$\ddot{\varphi}_{jn}(x) = \mathrm{e}^{\kappa_{jn}x} + \int_{-\infty}^{x} \mathbf{K}_m(x,y)\mathrm{e}^{\kappa_{jn}y}\mathrm{d}y \tag{1.68}$$

(compare with (1.45) of the previous section). If the basic potential $u_{m(x)}$ is a sufficiently short-range one then we can show that

$$\ddot{\varphi}_{jn}(x) = \psi_m(x, -\mathrm{i}\kappa_{jn}), \tag{1.69}$$

where the right-hand side is obtained from $\psi_m(x,k)$ through analytic continuation. The functions $\ddot{\varphi}_{jn}(x)$, as we noted above, are the "pseudo-eigenfunctions" of the operator $\hat{\mathrm{H}}_m$ since they are not quadratically integrated in general (although they satisfy the equation on the eigenfunctions of the operator $\hat{\mathrm{H}}_m$ with the discrete eigenvalue $-\kappa_{jn}^2$). An exception is the case when $\kappa_{jn} = \kappa_{ml}$ for some n and l. Then

$$\ddot{\varphi}_{jn}(x) = \varphi_{ml}(x) \tag{1.70}$$

and the pseudo-eigenfunction becomes the true eigenfunction for the operator $\hat{\mathrm{H}}_m$.

Now we can write the generalized GLM equation. Introduce (instead of $\mathbf{F}(x,y)$ in the GLM equation (1.49)) the function [95]

$$\begin{aligned}\Omega_{jm}(x,y) = {} & \frac{1}{2\pi}\int_{-\infty}^{\infty} \psi_m^*(x,k)\left[r_j(k) - r_m(k)\right]\psi_m^*(y,k)\mathbf{d}k \\ & + \sum_n \frac{\ddot{\varphi}_{jn}(x)\ddot{\varphi}_{jn}(y)}{|b_{jn}|} - \sum_n \frac{\varphi_{mn}(x)\varphi_{mn}(y)}{|b_{jn}|}\end{aligned} \tag{1.71}$$

and require that the GLM kernel $\mathbf{K}_{jm}(x,y)$ satisfies the one-dimensional *generalized GLM equation*, i.e.,

$$\mathbf{K}_{jm}(x,y) = -\Omega_{jm}(x,y) - \int_{-\infty}^{x} \mathbf{K}_{jm}(x,z)\Omega_{jm}(z,y)\mathbf{d}z. \tag{1.72}$$

Then, as was shown in Sect. 1.2.1, we have

$$u_{jm}(x) = -2\mathrm{d}_x \mathbf{K}_{jm}(x,x) \tag{1.73}$$

(compare the latter with (1.50)) and the functions $\psi_j(x,k)$ and $\varphi_{jn}(x)$ are determined by [95]

$$\psi_j(x,k) = \psi_m(x,k) + \int_{-\infty}^{x} \mathbf{K}_{jm}(x,y)\psi_m(y,k)\mathrm{d}y, \tag{1.74}$$

and

$$\varphi_{jn}(x) = \ddot{\varphi}_{jn}(x) + \int_{-\infty}^{x} \mathbf{K}_{jm}(x,y)\ddot{\varphi}_{jn}(y)\mathrm{d}y. \tag{1.75}$$

The function $\Omega_{jm}(x,y)$, as we see from (1.71), is fully determined by the scattering data. It is not difficult to see from the obtained expressions that if we take $r_m(k) \equiv 0$ (where $r_m(k)$ is the reflection coefficient for the basic potential) in the spectral data for $u_m(x)$ and assume the absence of bound states (note that solitons appear as the result of the bound states), then we obtain $u_m(x) \equiv 0$ and the generalized GLM equation (1.72) together with Eqs. (1.73)–(1.75) converts to the initial one-dimensional GLM equation with the properly defined functions $\psi_j(x,k)$ and $\varphi_{jn}(x)$.

In the case of the KdV equation it is necessary to introduce the parameter t into the scattering data. We have

$$r_m(k,t) = r_m(k)\mathrm{e}^{-\mathrm{i}8k^3t}, \quad |b_{mn}(t)| = |b_{mn}|\,\mathrm{e}^{8\kappa_{mn}^3 t},$$

and

$$r_j(k,t) = r_j(k)\mathrm{e}^{-\mathrm{i}8k^3t}, \quad |b_{jn}(t)| = |b_{jn}|\,\mathrm{e}^{8\kappa_{jn}^3 t}.$$

Then the parameter t is included in the function $\Omega_{jm}(x,y,t)$ (1.71) as well as the generalized GLM equation (1.72). In this case, instead of (1.73), we have

$$u_{jm}(x,t) = -2\mathrm{d}_x\mathbf{K}_{jm}(x,x,t).$$

Thus, t is included in the potentials $u_m(x,t)$, $u_j(x,t)$, and $u_{jm}(x,t)$, where the first two satisfy the *KdV equation*

$$\partial_t u + 6u\partial_x u + \partial_x^3 u = 0.$$

Instead of (1.67) where the basic potential $u_m(x)$ is introduced, we now have

$$u_j(x,t) = u_m(x,t) + u_{jm}(x,t). \tag{1.76}$$

Thus we see that the generalized GLM equation allows us to separate $u_m(x,t)$, the soliton part of the solution of KdV equation, from $u_{jm}(x,t)$, the part corresponding to the *continuous spectrum*. When $t \to \infty$, because of the effect of dispersion, $u_{jm}(x,t) \to 0$, and the potential $u(x,t)$ disintegrates into a (finite) number of solitons [96].

To be able to explain numerical results obtained for solitons of the KdV equation in the "non-stationary stage" [96] (see Sect. 1.3.5), consider here correlations of the potentials $u_{jm}(x,t)$ and $u_m(x,t)$ in the region of small x for $t \ll \infty$, i.e., in the initial stage of dispersive spreading of the potential $u_{jm}(x,t)$. Choose in $\Omega_{jm}(x,y,t)$ (1.71)

$$r_m(k) = 0, \quad |b_{mn}| = |b_{jn}|, \quad \text{and} \quad \kappa_{mn} = \kappa_{jn} \tag{1.77}$$

for all j (thereby we choose the necessary spectral data). Then $u_m(x,t)$ is the soliton part of the solution (the non-reflective potential), and $u_{jm}(x,t)$ is its non-soliton part corresponding to the continuous spectrum. The particular form of the solution is defined by the right-hand side of (1.71). The *Jost function* $\psi_m(x,k)$ for the potential $u_m(x,k)$ can be found explicitly but detailed analysis of (1.71) and (1.72) is difficult. Nevertheless, proceeding from the known asymptotic relations for $\psi_m(x,k)$ for $x \geq 0$ and accounting for the dependence of the reflection coefficient $r_j(k,t)$ on t, we can conclude from the right-hand side of (1.71) (for the scattering data (1.77)) [96] the following:

1. A part of the solution determined by the potential $u_{jm}(x,t)$ is an oscillating wave packet where the amplitude of the oscillations decreases in time due to dispersive spreading.
2. Since the "momentum" $\int_{-\infty}^{\infty} u_j(x,t)\mathrm{d}x$ conservation takes place for the KdV equation, we can write

$$\partial_t\left(u_j\right) + \partial_x\left[3u_j^2 + \partial_x^2\left(u_j\right)\right] = 0.$$

Substituting here the expression for $u_j(x,t)$ from (1.76) and taking into account that u_j and u_m satisfy the KdV equation, we obtain [96]

$$u_m(x,t) \cdot u_{jm}(x,t) = \text{const.} \tag{1.78}$$

In the region of small $x > 0$ when $t \ll \infty$, we see that the decrease of the potential $u_{jm}(x,t)$ with time leads to the increase of the potential $u_m(x,t)$, corresponding to the soliton part (i.e., growth of the amplitudes of the solitons is observed) until the oscillating tail shifts as a whole to the region of negative x. With u_{jm} decreasing to zero in the region $x > 0$, the soliton parameters determined by the potential u_m tend to be constant values. For more detailed investigation of the behavior of the potential $u_{jm}(x,t)$ corresponding to the non-soliton part of the solution (the non-reflective potential), one can use the variational principle first proposed for this purpose by Moses [95], as shown below.

1.2.4 The Variational Principle

Following Refs. [95,96] we introduce the functional

$$\mathbf{G}(\mathbf{N},x) = -2\Bigg\{\int_{-\infty}^{x} \mathbf{N}(x,y)\mathrm{d}y\Bigg[2\Omega(x,y)+\mathbf{N}(x,y) + \int_{-\infty}^{x} \Omega(y,z)\mathbf{N}(x,z)\mathrm{d}z\Bigg] + \Omega(x,x)\Bigg\}, \tag{1.79}$$

where $\mathbf{N}(x,y)$ is the test function for the transformation kernel $\mathbf{K}(x,y)$. The *variational principle* states [95]:

Theorem. *The functional* $\mathbf{G}(\mathbf{N},x)$ *has an absolute maximum for all* x *if and only if* $\mathbf{N}(x,y)=\mathbf{K}(x,y)$*, i.e., the test function* $\mathbf{N}(x,y)$ *is a solution of the generalized GLM equation, namely:*

$$\max_{x\in]-\infty,\infty[}\{\mathbf{G}(\mathbf{N},x)\} = \mathbf{G}(\mathbf{K},x), \quad \mathbf{N}(x,y)=\mathbf{K}(x,y).$$

Besides, the absolute maximum of $\mathbf{G}(\mathbf{K},x)$ *is equal to the area bound by the curve* $u_{jm}(x)$*:*

$$\int_{-\infty}^{x} u_{jm}(x')\mathrm{d}x' = \mathbf{G}(\mathbf{K},x).$$

This theorem allows us to calculate the potential $u_{jm}(x)$ with increased accuracy. For such calculations, it is necessary: (a) to have an algorithm providing the test functions; and (b) to foresee the process allowing only increasing $\mathbf{G}(\mathbf{N},x)$ when testing the possible (test) functions in the computational program.

For example, for fixed x we can construct the histogram on y for $\mathbf{N}(x,y)$, which is varied in such a way that the functional $\mathbf{G}(\mathbf{N},x)$ is increased[6] (see Fig. 1.3). The same operation is repeated for the set of x-values for $\mathbf{G}(\mathbf{N},x)$ increasing from set to set. Furthermore, it is necessary to introduce time so that $u_{jm}(x,t)$ is a part of the solution of the KdV equation corresponding to the continuous spectrum, and $u_m(x,t)$ is its purely soliton part. Then one can obviously use the variational principle considered above. The functional now depends on x and t: $\mathbf{G}=\mathbf{G}(\mathbf{N},x,t)$, and this corresponds to the continuous part of the spectrum, i.e., there should be the integral

[6] For example, one can construct an algorithm changing the height of each column of the histogram according to some procedure, then changing the widths of the columns using the best from the former histograms as the initial approximation.

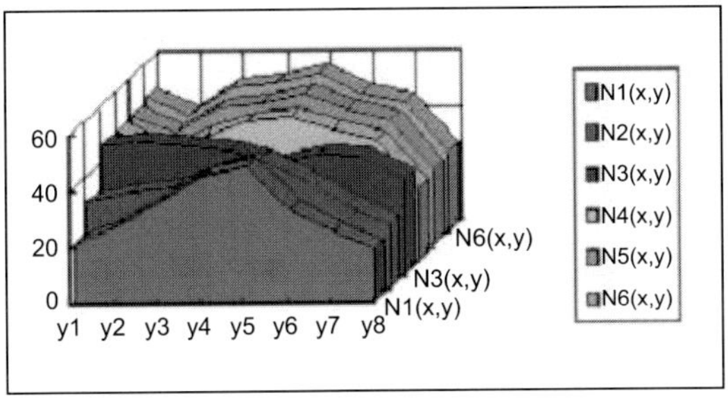

Fig. 1.3. Example of the histogram on y for $\mathbf{N}(x, y)$ (x =const)

$$\int_{-\infty}^{x} u_{jm}(x', t)\mathrm{d}x = \mathbf{G}(\mathbf{K}, x, t)$$

and there must be the maximum for all x and t, namely,

$$\max_{x \in]-\infty,\infty[, 0 \le t < \infty} \{\mathbf{G}(\mathbf{N}, x, t)\} = \mathbf{G}(\mathbf{K}, x, t), \quad \mathbf{N}(x, y, t) = \mathbf{K}(x, y, t).$$

Thus the variational principle provides new methods for computation of the continuous part of the wave solution, whereas the "classic" GLM equation is unable to separate the non-soliton part of the solution.

1.3 Numerical Integration of (1+1)-Dimensional KdV-Class Equations

As we already noted above, solutions of the KdV equation were first obtained numerically [1,93]. Later in 1967, Gardner, Green, Kruskal, and Miura [19] found the method to solve the KdV equation analytically using the IST method (see the previous section), and obtained exact solutions in the form of solitons.

However, even development of such powerful and effective analytical apparatus as the IST method does not remove the problem of numerical integration of the KdV equation as well as other equations in the KdV-class from the agenda because, first, it is not possible to obtain an analytical solution in its closed form using the IST method with arbitrary initial conditions, and, second, among the equations in the KdV-class there are models not integrated analytically (for example, KdV–Burgers equation (KdVB equation) or KdV equations with additional terms describing, for example, instability of some type in the medium). Therefore, developing numerical codes as well as setting up numerical experiments for this class of problems is highly important.

In this section, using the example of the KdV equation (1.18) we consider some difference schemes used for the numerical analysis and present numerical solutions obtained with their help. These schemes are also used to obtain solutions of other one-dimensional equations in the KdV-class, and some elements of these schemes will be further used (see Chap. 4) when studying numerical methods of integration of the (1+2)- and (1+3)-dimensional problems.

1.3.1 Explicit Difference Schemes

For the KdV equation written in the form (1.18), we first consider the three-layer *explicit scheme with* $O(\tau^2, h^2)$ *approximation:*

$$u_i^{n+1} = u_i^{n-1} - \frac{\alpha\tau}{h} u_i^n \left(u_{i+1}^n - u_{i-1}^n\right) - \frac{\beta\tau}{h^3}\left(u_{i+2}^n - 2u_{i+1}^n + 2u_{i-1}^n - u_{i-2}^n\right). \tag{1.80}$$

As we can see from the above difference equation, the scheme is realized on the 5-point template (Fig. 1.4). This scheme was used to obtain some of the first numerical solutions of the KdV equation in 1965–1968 [1,3,93]. Investigation of the stability of scheme (1.80) using Fourier analysis gives the

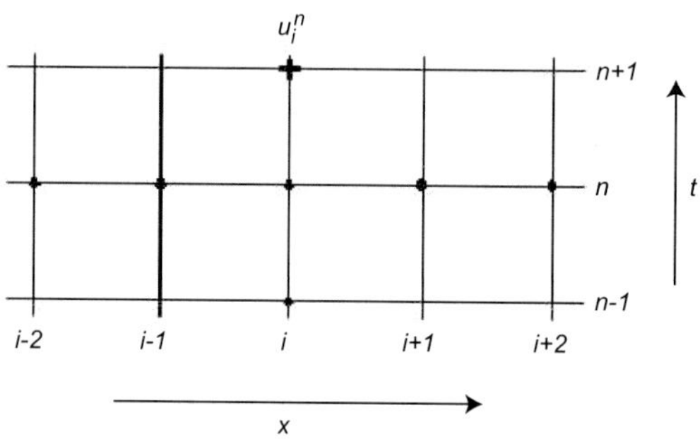

Fig. 1.4. Template for the difference scheme (1.80): n is the number of the time layer, and i is the number of the space layer

condition

$$\frac{\tau}{h}\max\left|\sin kh\left(\alpha u - \frac{4\beta}{h^2}\sin^2\frac{kh}{2}\right)\right| \le 1,$$

that is,

$$\frac{\tau}{h}\left(\alpha\left|u\right| + \frac{3\sqrt{3}\beta}{2h^2}\right) \le 1,$$

or, for sufficiently small steps

$$\tau \le \frac{2h^3}{3\sqrt{3}\beta} \cong 0.384\frac{h^3}{\beta}. \tag{1.81}$$

Calculations demonstrate that this condition is quite accurate. For example, for $\beta = 2\times 10^{-4}$, $\alpha = 1$, and $h = 0.01$, the stability of the scheme is maintained for the time step $\tau = 1.91\times 10^{-3}$, and the scheme is already unstable when $\tau = 1.93 \times 10^{-3}$. Inequality (1.81) gives $\tau = 1.92 \times 10^{-3}$ in this case.

Consider now the three-layer *explicit scheme with $O(\tau^2, h^4)$ approximation*. Since here the approximation of the x-derivatives is used by finite differences of higher order than in the previous scheme, we introduce the 7-point template (Fig. 1.5). The scheme is given by

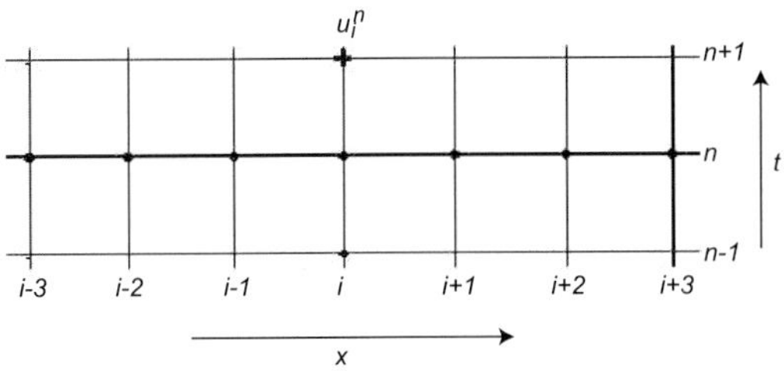

Fig. 1.5. Template for the difference scheme (1.82)

$$u_i^{n+1} = u_i^{n-1} + \frac{\alpha\tau}{6h} u_i^n \left(u_{i+2}^n - 8u_{i+1}^n + 8u_{i-1}^n - u_{i-2}^n\right)$$
$$+ \frac{\beta\tau}{4h^3}\left(u_{i+3}^n - 8u_{i+2}^n + 13u_{i+1}^n - 13u_{i-1}^n + 8u_{i-2}^n - u_{i-3}^n\right). \quad (1.82)$$

According to Fourier analysis, the scheme is stable when the condition on the steps τ and h,

$$\frac{\tau}{h}\max\left|\left[\frac{4\beta}{h^2}\left(1+\sin^2\frac{kh}{2}\right)\sin^2\frac{kh}{2} - \left(1+\frac{2}{3}\sin^2\frac{kh}{2}\right)\alpha u\right]\sin kh\right| \le 1,$$

are satisfied. For sufficiently small steps h, we then obtain

$$\tau \le \frac{108h^3}{(43+7\sqrt{73})\sqrt{10\sqrt{73}-62}\beta} \cong 0.216\frac{h^3}{\beta}. \quad (1.83)$$

A comparison of conditions (1.83) and (1.81) shows that the stability condition is more strict for the scheme with the higher approximation order.

1.3.2 Implicit Difference Schemes

For the example of an implicit difference scheme, consider first the *implicit scheme with $O(\tau^2, h^4)$ approximation* [81,83]:

$$
\begin{aligned}
\frac{u_i^{n+1}-u_i^n}{\tau} &= \frac{\alpha}{24h}\left[u_i^n\left(u_{i+2}^{n+1}-8u_{i+1}^{n+1}+8u_{i-1}^{n+1}-u_{i-2}^{n+1}\right)\right. \\
&\quad \left. +u_i^{n+1}\left(u_{i+2}^n-8u_{i+1}^n+8u_{i-1}^n-u_{i-2}^n\right)\right] \\
&+\frac{\beta}{16h^3}\left(u_{i+3}^{n+1}-8u_{i+2}^{n+1}+13u_{i+1}^{n+1}-13u_{i-1}^{n+1}+8u_{i-2}^{n+1}-u_{i-3}^{n+1}\right. \\
&\quad \left. +u_{i+3}^n-8u_{i+2}^n+13u_{i+1}^n-13u_{i-1}^n+8u_{i-2}^n-u_{i-3}^n\right). \qquad (1.84)
\end{aligned}
$$

Note that if we approximate the derivatives using the finite differences of another order, we obtain an implicit scheme of that approximation order. We see that in scheme (1.84), it is impossible to explicitly express the value of the function on the knot in the given time layer using the values of the function in the previous time layer (whence follows the name "implicit scheme"). Implicit schemes for sufficiently small steps h are absolutely stable and can be realized using the *sweep method.* In particular, for scheme (1.84), the method of the monotonous 7-point sweep (discussed below) is highly effective.

Let us represent (1.84) as

$$
\begin{aligned}
-a_0^1u_3^n+a_0^2u_2^n-a_0^3u_1^n+a_0^4u_0^n &= h_0^n, && i=0, \\
-a_1^1u_4^n+a_1^2u_3^n-a_1^3u_2^n+a_1^4u_1^n-a_1^5u_0^n &= h_1^n, && i=1, \\
-a_2^1u_5^n+a_2^2u_4^n-a_2^3u_3^n+a_2^4u_2^n & && \\
-a_2^5u_1^n+a_2^6u_0^n &= h_2^n, && i=2, \\
-a_i^1u_{i+3}^n+a_i^2u_{i+2}^n-a_i^3u_{i+1}^n+a_i^4u_i^n & && \\
-a_i^5u_{i-1}^n+a_i^6u_{i-2}^n-a_i^7u_{i-3}^n &= h_i^n, && 3\le i\le N-3, \quad (1.85) \\
a_{N-2}^2u_N^n-a_{N-2}^3u_{N-1}^n+a_{N-2}^4u_{N-2}^n-a_{N-2}^5u_{N-3}^n & && \\
+a_{N-2}^6u_{N-4}^n-a_{N-2}^7u_{N-5}^n &= h_{N-2}^n, && i=N-2, \\
-a_{N-1}^3u_N^n+a_{N-1}^4u_{N-1}^n-a_{N-1}^5u_{N-2}^n & && \\
+a_{N-1}^6u_{N-3}^n-a_{N-1}^7u_{N-4}^n &= h_{N-1}^n, && i=N-1, \\
a_N^4u_N^n-a_N^5u_{N-1}^n+a_N^6u_{N-2}^n-a_N^7u_{N-3}^n &= h_N^n, && i=N.
\end{aligned}
$$

Here, for $3\le i\le N-3$,

$$
h_i^n=-b_i\left(u_{i+3}^{n-1}-8u_{i+2}^{n-1}+13u_{i+1}^{n-1}-13u_{i-1}^{n-1}+8u_{i-2}^{n-1}-u_{i-3}^{n-1}\right)+u_i^{n-1},
$$

and

$$
\begin{aligned}
n=1,2,\ldots,N1, &\quad b_i=a_i^7, \\
a_i^1=-\tau\beta/16h^3, &\quad a_i^7=-a_i^1, \\
a_i^2=(\tau/2h)\left(\beta/h^2-\alpha u_i^{n-1}/12\right), &\quad a_i^6=-a_i^2, \qquad (1.86) \\
a_i^3=(\tau/h)\left(\alpha u_i^{n-1}/3-13\beta/16h^2\right), &\quad a_i^5=-a_i^3, \\
a_i^4=1+(\alpha\tau/24h)\left(u_{i+2}^{n-1}-8u_{i+1}^{n-1}+8u_{i-1}^{n-1}-u_{i-2}^{n-1}\right), &
\end{aligned}
$$

where $N1$ stands for the number of the time layers. The values h_i^n for $i=0,1,2$ and $i=N-2,N-1,N$ are defined by

$$
\begin{aligned}
h_0^n &= -a_0^1 u_3^{n-1} + a_0^2 u_2^{n-1} - a_0^3 u_1^{n-1} + a_0^4 u_0^{n-1}, \\
h_1^n &= -a_1^1 u_4^{n-1} + a_1^2 u_3^{n-1} - a_1^3 u_2^{n-1} + a_1^4 u_1^{n-1} - a_1^5 u_0^{n-1}, \\
h_2^n &= -a_2^1 u_5^{n-1} + a_2^2 u_4^{n-1} - a_2^3 u_3^{n-1} + a_2^4 u_2^{n-1} - a_2^5 u_1^{n-1} + a_2^6 u_0^{n-1}, \\
h_{N-2}^n &= a_{N-2}^2 u_N^{n-1} - a_{N-2}^3 u_{N-1}^{n-1} + a_{N-2}^4 u_{N-2}^{n-1} - a_{N-2}^5 u_{N-3}^{n-1} \\
&\quad + a_{N-2}^6 u_{N-4}^{n-1} - a_{N-2}^7 u_{N-5}^{n-1}, \qquad (1.87) \\
h_{N-1}^n &= -a_{N-1}^3 u_N^{n-1} + a_{N-1}^4 u_{N-1}^{n-1} - a_{N-1}^5 u_{N-2}^{n-1} \\
&\quad + a_{N-1}^6 u_{N-3}^{n-1} - a_{N-1}^7 u_{N-4}^{n-1}, \\
h_N^n &= a_N^4 u_N^{n-1} - a_N^5 u_{N-1}^{n-1} + a_N^6 u_{N-2}^{n-1} - a_N^7 u_{N-3}^{n-1}.
\end{aligned}
$$

The coefficients $a_i^1, \ldots, a_i^7$ for $i = 0, 1, 2$ and $i = N-2, N-1, N$ are obtained from the boundary conditions of the problem (see below in this section). The set (1.86) is solved using the monotonous 7-point sweep method; the procedure of developing the corresponding equations is analogous to the case of the 5-point sweep method considered in detail in numerous monographs and textbooks (see, for example, Ref. [97]). Therefore, here we present the algorithm in its final form [83,98].

The algorithm of monotonous 7-point *sweep method* for scheme (1.84) is based on the expressions

$$
\begin{aligned}
&\alpha_{i+1} = (1/\Delta_i)\left\{a_i^3 - a_i^5\beta_i + a_i^6(\beta_i\alpha_{i-1} - \delta_{i-1}) - a_i^7\right. \\
&\left.\times\left[\alpha_{i-2}(\beta_i\alpha_{i-1} - \delta_{i-1}) - \beta_{i-2}\beta_i\right]\right\}, \qquad \text{for} \quad i = 3, 4, \ldots, N-1; \\
&\qquad \alpha_1 = a_0^3/a_0^4, \qquad \alpha_2 = (1/\Delta_1)\left(a_1^3 - a_1^5\beta_1\right), \\
&\qquad \alpha_3 = (1/\Delta_2)\left[a_2^3 - a_2^5\beta_2 + a_2^6(\beta_2\alpha_1 - \delta_1)\right]; \\
&\beta_{i+1} = \frac{1}{\Delta_i}\left[a_i^2 - a_i^5\delta_i + a_i^6\alpha_{i-1}\delta_i - a_i^7(\alpha_{i-2}\alpha_{i-1}\delta_i - \beta_{i-2}\delta_i)\right], \\
&\qquad\qquad \text{for} \quad i = 3, 4, \ldots, N-2; \\
&\qquad \beta_1 = a_0^2/a_0^4, \qquad \beta_2 = (1/\Delta_1)\left(a_1^2 - a_1^5\delta_1\right), \\
&\qquad \beta_3 = (1/\Delta_2)\left(a_2^2 - a_2^5\delta_2 + a_2^6\alpha_1\delta_2\right); \\
&\qquad \delta_{i+1} = a_i^1/\Delta_i, \qquad \text{for} \quad i = 3, 4, \ldots, N-3; \qquad (1.88) \\
&\qquad \delta_1 = a_0^1/a_0^4, \qquad \delta_2 = a_1^1/\Delta_1, \qquad \delta_3 = a_2^1/\Delta_2; \\
&\gamma_{i+1} = \frac{1}{\Delta_i}\left\{a_i^5\gamma_i - a_i^6(\alpha_{i-1}\gamma_i + \gamma_{i-1}) + a_i^7\left[\alpha_{i-2}(\alpha_{i-1}\gamma_i\right.\right. \\
&\left.\left.+\gamma_{i-1}) - \beta_{i-2}\gamma_i + \gamma_{i-2}\right] - h_i\right\}, \qquad \text{for} \quad i = 3, 4, \ldots, N; \\
&\qquad \gamma_1 = -h_0/a_0^4, \qquad \gamma_2 = (1/\Delta_1)\left(-h_1 + a_1^5\gamma_1\right), \\
&\qquad \gamma_3 = (1/\Delta_2)\left[-h_2 - a_2^6(\alpha_1\gamma_2 + \gamma_1) + a_2^5\gamma_2\right]; \\
&\Delta_i = \left\{a_i^4 - a_i^5\alpha_i + a_i^6(\alpha_i\alpha_{i-1} - \beta_{i-1}) - a_i^7\left[\alpha_{i-2}(\alpha_i\alpha_{i-1}\right.\right. \\
&\left.\left.-\beta_{i-1}) - \beta_{i-2}\alpha_i + \delta_{i-2}\right]\right\}, \qquad \text{for} \quad i = 3, 4, \ldots, N; \\
&\qquad \Delta_1 = a_1^4 - a_1^5\alpha_1, \qquad \Delta_2 = a_2^4 - a_2^5\alpha_2 + a_2^6(\alpha_2\alpha_1 - \beta_1)
\end{aligned}
$$

for the sweep coefficients α_i, β_i, γ_i, and δ_i. Then, using the expressions

$$u_i = \alpha_{i+1}u_{i+1} - \beta_{i+1}u_{i+2} + \delta_{i+1}u_{i+3} - \gamma_{i+1},$$
$$\text{for} \quad i = N-2, N-3, \ldots, 0, \tag{1.89}$$
$$u_{N-1} = \alpha_N u_N - \gamma_N, \qquad u_N = -\gamma_{N+1},$$

we obtain the unknown quantities u_i one after another. Note that the sweep coefficients α_i and β_i in (1.89) and (1.89) should not be mixed up with the coefficients α and β of the original differential problem (1.18).

It is well known that the implicit difference schemes, theoretically, are absolutely stable. In our case of the nonlinear equation, however, the function u_i is included into the expressions for a_i (1.87). Therefore there are limitations on the correctness of the algorithm. In particular, the algorithm of the monotonous sweep method is correct when the following conditions are fulfilled [83,98]:

$$\begin{aligned}
&|a_i^7| > 0, \quad 3 \le i \le N; \qquad |a_i^6| > 0, \quad 2 \le i \le N;\\
&|a_i^5| > 0, \quad 1 \le i \le N; \qquad |a_i^3| > 0, \quad 0 \le i \le N-1;\\
&|a_i^2| > 0, \quad 0 \le i \le N-2; \qquad |a_i^1| > 0, \quad 0 \le i \le N-3;
\end{aligned} \tag{1.90}$$

and

$$\begin{aligned}
|a_0^4| \ge |a_0^3| + |a_0^2| + |a_0^1|, \quad |a_1^4| &\ge |a_1^5| + |a_1^3| + |a_1^2| + |a_1^1|,\\
|a_2^4| &\ge |a_2^5| + |a_2^6| + |a_2^3| + |a_2^2| + |a_2^1|,\\
|a_N^4| &\ge |a_N^5| + |a_N^6| + |a_N^7|,\\
|a_{N-1}^4| &\ge |a_{N-1}^3| + |a_{N-1}^5| + |a_{N-1}^6| + |a_{N-1}^7|,\\
|a_{N-2}^4| &\ge |a_{N-2}^2| + |a_{N-2}^3| + |a_{N-2}^5| + |a_{N-2}^6| + |a_{N-2}^7|,\\
|a_i^4| \ge |a_i^1| + |a_i^2| + |a_i^3| + |a_i^5| &+ |a_i^6| + |a_i^7|, \quad 3 \le i \le N-3.
\end{aligned} \tag{1.91}$$

These conditions impose restrictions on KdV equation (1.18). Indeed, for $3 \le i \le N-3$, we have to satisfy the conditions given by

$$u_i^{n-1} \ne \frac{12\beta}{\alpha h^2} \quad \text{and} \quad u_i^{n-1} \ne \frac{39}{16\alpha h^2}. \tag{1.92}$$

These inequalities are satisfied for a sufficiently small step $h \le 0.2$. Also, the last inequality of (1.92) should be satisfied for the coefficients (1.87). If the wave amplitude $u_i^{n-1} \le 30$ (i.e., it changes within reasonable limits), then

$$|a_i^4| \ge \frac{\tau}{4h}\left(\frac{11\beta}{h^2} - 3\alpha u_i^{n-1}\right). \tag{1.93}$$

Furthermore, if the following inequality is satisfied for a sufficiently smooth function u^{n-1}:

$$\frac{\alpha\tau}{24h}\left(u_{i-2}^{n-1} - 8u_{i-1}^{n-1} + 8u_{i+1}^{n-1} - u_{i+2}^{n-1}\right) \le 1, \tag{1.94}$$

then follows the restriction [81,83,98]

$$\tau \le 4h\left[\frac{\alpha}{6}\left(u_{i-2}^{n-1} - 8u_{i-1}^{n-1} - 18u_i^{n-1} + 8u_{i+1}^{n-1} - u_{i+2}^{n-1}\right) + \frac{11\beta}{h^2}\right]^{-1} \cong \frac{4h}{3|\alpha u|}, \tag{1.95}$$

where $u = \max_{i,n} |u_i^n|$, that should be taken into account in calculations. In reality, as numerical simulations demonstrate, the given restriction on the time step is too strict since the adequate accuracy of the solution has been observed already for $h = 0.1$ and $\tau = 0.0025$. For $i = 0, 1, 2$ and $i = N - 2, N - 1, N$, in order to satisfy the conditions in (1.91), it is sufficient to choose proper boundary conditions, as described in the next section.

Consider now another *implicit scheme with $O(\tau^2, h^2)$ approximation*:

$$\begin{aligned}\frac{u_i^{n+1} - u_i^n}{\tau} &+ \frac{\alpha}{4h}\left[u_i^n\left(u_{i+1}^{n+1} - u_{i-1}^{n+1}\right) + u_i^{n+1}\left(u_{i+1}^n - u_{i-1}^n\right)\right]\\ &+\frac{\beta}{4h^3}\left(u_{i+2}^{n+1} - 2u_{i+1}^{n+1} + 2u_{i-1}^{n+1} - u_{i-2}^{n+1}\right.\\ &\left.+u_{i+2}^n - 2u_{i+1}^n + 2u_{i-1}^n - u_{i-2}^n\right) = 0.\end{aligned} \tag{1.96}$$

Analogously to what was done for the scheme (1.84), we represent (1.96) as

$$\begin{aligned} a_0^1 u_2^n - a_0^2 u_1^n + a_0^3 u_0^n &= h_0^n, && i = 0,\\ a_1^1 u_3^n - a_1^2 u_2^n + a_1^3 u_1^n - a_1^4 u_0^n &= h_1^n, && i = 1,\\ a_i^1 u_{i+2}^n - a_i^2 u_{i+1}^n + a_i^3 u_i^n - a_i^4 u_{i-1}^n &\\ +a_i^5 u_{i-2}^n &= h_i^n, && 2 \le i \le N-2,\\ -a_{N-1}^2 u_N^n + a_{N-1}^3 u_{N-1}^n - a_{N-1}^4 u_{N-2}^n &\\ +a_{N-1}^5 u_{N-3}^n &= h_{N-1}^n, && i = N-1,\\ a_N^3 u_N^n - a_N^4 u_{N-1}^n + a_N^5 u_{N-2}^n &= h_N^n, && i = N.\end{aligned} \tag{1.97}$$

Here, for $2 \le i \le N - 2$, we have

$$\begin{aligned} h_i^n &= (\tau/4h^3)\left(u_{i+2}^{n-1} - 2u_{i+1}^{n-1} + 2u_{i-1}^{n-1} - u_{i-2}^{n-1}\right) + u_i^{n-1},\\ a_i^1 &= -\tau\beta/4h^3, \qquad\qquad a_i^5 = -a_i^1,\\ a_i^2 &= (\tau/4h)\left(\alpha u_i^{n-1} - 2\beta/h^2\right), \qquad a_i^4 = -a_i^2,\\ a_i^3 &= 1 + (\alpha\tau/4h)\left(u_{i+1}^{n-1} - u_{i-1}^{n-1}\right).\end{aligned} \tag{1.98}$$

For $i = 0, 1$ and $i = N - 1, N$ in (1.98) we have

$$\begin{aligned} h_0^n &= a_0^1 u_2^{n-1} - a_0^2 u_1^{n-1} + a_0^3 u_0^{n-1},\\ h_1^n &= a_1^1 u_3^{n-1} - a_1^2 u_2^{n-1} + a_1^3 u_1^{n-1} - a_1^4 u_0^{n-1},\\ h_{N-1}^n &= -a_{N-1}^2 u_N^{n-1} + a_{N-1}^3 u_{N-1}^{n-1} - a_{N-1}^4 u_{N-2}^{n-1} + a_{N-1}^5 u_{N-3}^{n-1},\\ h_N^n &= a_N^3 u_N^{n-1} - a_N^4 u_{N-1}^{n-1} + a_N^5 u_{N-2}^{n-1}.\end{aligned}$$

The coefficients $a_i^1, \ldots, a_i^5$ are determined by boundary conditions of the problem (see the next section). The set (1.98) can be effectively solved using the non-monotonous *sweep method*. We consider here neither the non-monotonous 5-point sweep method nor the final calculation expressions since there are plenty of monographs and textbooks devoted to realization of various sorts of non-monotonous sweep methods. Here, we only note that the scheme (1.96) is correct under the condition that the matrix A of the set (1.98) is non-degenerated, i.e., $\det A \neq 0$.

Thus we considered here some examples of explicit and implicit difference schemes constructed directly for the KdV equation. Some remarks regarding their adaptation to other KdV-class equations as well as the problem of boundary conditions are given below in the next section.

1.3.3 Remarks on Numerical Integration

1. In the case when the study of KdV-class equations involves some sort of an additional term, this term must be included into the difference schemes considered above with the appropriate order of approximation of the derivative. For example, if we investigate the KdVB equation with the term (on the right-hand side of the equation) describing wave damping as a result of a dissipative process in the medium, $\nu\partial_x^2 u$, it is necessary to include that term into the difference scheme with approximation of the appropriate order. For the scheme (1.80) this term is given by
$$\frac{2\nu\tau}{h^2}\left(u_{i+1}^n - 2u_i^n + u_{i-1}^n\right). \tag{1.99}$$
2. If some term is absent in the considered equation (as compared to the KdV equation) then it is sufficient to assume in the difference scheme that the corresponding coefficient equals zero (e.g., in the classic Burgers equation when $\beta = 0$; of course, in this case it is still necessary to include the term (1.99)) in the right-hand side of the difference equation. Thus, difference schemes (1.80), (1.82), (1.84) and (1.96) are general (in a certain sense) for the whole class of (1+1)-dimensional equations of the KdV type. Below, when presenting numerical methods for multidimensional equations (Sects. 3.1 and 4.3) we will also see that the schemes considered above are used there as their inalienable elements.
3. This remark is related to the boundary conditions of the problem. We note that although we solve the initial value problem for all presented cases, we should still impose constraints at the boundaries of the one-dimensional grid because of the limits of the region of numerical integration, i.e., the problem acquires features of the initial-boundary problem. Naturally, the difference derivatives in the schemes must be defined for $i = 0, 1, 2$ and $i = N-2, N-1, N$. In this case, terms approximating the function u at the points corresponding to the limits of the integration region (where the function u is not defined) on the x-axis, appear in these schemes; accordingly, it is necessary to impose some conditions defining the function at these points. Here, different variants are possible, and the following discusses two possibilities most often used.

In the case when the initial condition $u(0,x) = \psi(x)$ has the asymptotics $|\psi(x)| \to 0$ for $|x| \to \infty$ (tending to zero sufficiently fast with the increasing modulus x, e.g., exponentially where $\psi(x) \sim \exp(-x^2)$ for $|x| \to \infty$), we can use on the boundaries of the integration region the so-called zero boundary conditions

$$u(t,x) = \partial_x u(t,x) = \partial_x^2 u(t,x) = \partial_x^3 u(t,x) = 0.$$

If the condition on the (space) localization of the function $\psi(x)$ defining the initial condition does not take place, the boundary conditions become more complex. For example, when the initial condition is $\psi(x) = a\cos(mx + \varphi)$, the periodic boundary conditions are usually used (they are given here for the grid $i = 0, 1, \ldots, N$):

$$\begin{cases} u_{N+1} = u_1, & u_{N+2} = u_2, & u_{N+3} = u_3, \\ u_{-1} = u_{N-1}, & u_{-2} = u_{N-2}, & u_{-3} = u_{N-3}. \end{cases}$$

In other cases, especially when solutions have more complex asymptotics with a slow tendency of the function u approaching zero at infinity, we can also use other boundary conditions. For example, the condition of the "total absorption" at the boundary or the impedance boundary condition of the Leontovich type is often used for simulations of more complex multidimensional evolution equations (see Sect. 4.3.2 for details).

Having considered various methods and schemes of numerical approach to the integration of (1+1)-dimensional KdV-class equations based on the finite-difference approximation of the derivatives, we now discuss their characteristics related to the calculation's accuracy and productivity, in terms of the time expenses and demands on the computer memory. After that we can consider numerical solutions obtained by use of these schemes.

1.3.4 Test of Numerical Methods, Their Comparative Characteristics, and Use

To test the above difference schemes (1.80), (1.82), (1.84), and (1.96), we investigate their characteristics related to integration in the space coordinate x. As an initial condition we use the exact solution of the KdV equation

$$u_0(x) = \frac{3v}{\alpha}\cosh^{-2}\left[\frac{\sqrt{v}}{2\beta}(x - x_0)\right],$$

where we choose $v = \alpha = 6$ and $\beta = 1$. The control of the accuracy for all time layers is fulfilled by a comparison of the numerical solution with the above exact (analytical) solution. At each time step τ, we calculate the relative mean deviation ε,

$$\varepsilon = \frac{|u_\tau^{\mathrm{num}} - u_\tau^{\mathrm{exact}}|}{u_\tau^{\mathrm{exact}}},$$

as well as the mean-square-root deviation of the numerical solution from the exact one,

$$s = \left[\frac{1}{N} \sum_{i=1}^{N} \left| \left(u_i^{\mathrm{num}}\right)^2 - \left(u_i^{\mathrm{exact}}\right)^2 \right| \right]^{1/2} .$$

For example, for $t = 0.4$ (the soliton is near the boundary of the integration region) the scheme (1.84) with $h = 0.1$ and $\tau = 0.0025$ gives us

$$\varepsilon = 6.38775 \times 10^{-3} \quad \text{and} \quad s = 1.74663 \times 10^{-4}.$$

This result is quite acceptable and is approximately on an order of magnitude better than the corresponding results for schemes proposed for the solution of the KdV equation in Refs. [79,93].

Comparative analysis of the results of numerical calculation employing the difference schemes (1.80), (1.82), (1.84), and (1.96) allows us to make the following conclusions. The schemes (1.82) and (1.84) naturally demonstrate the best accuracy characteristics. Conditions on the time grid step for these schemes are approximately the same and are less restrictive than those for scheme (1.80). Better possibilities to choose the time step in scheme (1.96) allow us to considerably decrease the necessary computer time as compared to other schemes. Obviously, the implicit schemes (1.84) and (1.96) have another advantage over the explicit schemes (1.80) and (1.82), viz., their two-layer structure ensures less requirements to the computer memory resources. For the same time steps τ as in the implicit schemes, explicit schemes (1.80) and (1.82) are more preferable because of their lesser time expenses. Their three-layer structure, however, imposes severe requirements on the computer memory resources. These schemes can be effectively used for calculations on small time scales in the investigation of evolution of the initial condition at the "non-stationary stage" when there is a "birth" and the subsequent formation of locally stationary objects – solitons (see the next section).

1.3.5 Numerical Solutions of Some KdV-Class Equations

Consider now numerically obtained solutions of the KdV equation as well as some other equations in the KdV class. We begin with the linearized KdV equation (1.30) and Burgers equation (1.33) considered in Sect. 1.1. Although exact analytical solutions can be found for these equations, see (1.31) and (1.34), they are not transparent and are not convenient for detailed analysis of the dynamics of an initial disturbance.

For the fast decreasing initial disturbance for $|x| \to \infty$, numerical integration of *linearized KdV equation* (1.30) results in the following. The local wave number is given by

$$k(x,t) = \left| \frac{x}{3\beta t} \right|^{1/2},$$

and the local frequency is defined by $\omega = -\beta k^3$, as can be seen from the dispersion law for the linearized KdV equation. Evolution of the single initial disturbance

$$u(0,x) = u_0 \exp(-x^2/l^2) \tag{1.100}$$

(here, u_0 and l are arbitrary parameters determined by the convenience of numerical calculation for a particular size of the region of the numerical integration) leads to formation of the oscillating wave packet shown in Fig. 1.6. In the case where $\beta > 0$, the "fast" oscillations are in the region $x < 0$,

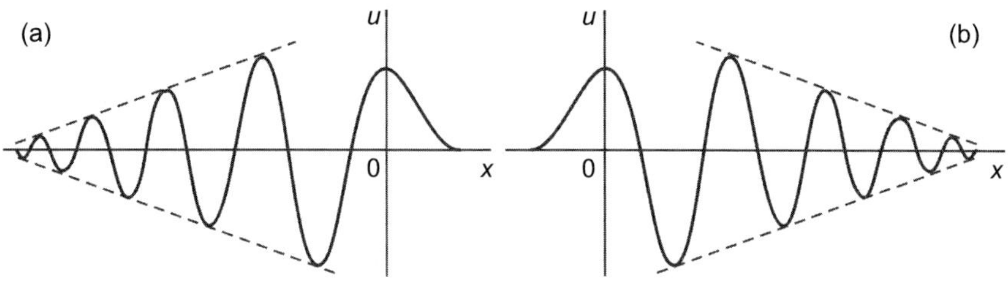

Fig. 1.6. Numerical solution of the linearized KdV equation. **a.** $\beta > 0$ **b.** $\beta < 0$

i.e., the short wavelength oscillations fall behind the large wavelength ones. If $\beta < 0$ (note that it is generally easy to convert equation (1.30) with $\beta < 0$ to the equation with $\beta > 0$ by changing $x \to -x$), the fast oscillations are in the region $x > 0$, i.e. the short wavelength waves are propagating forward. Thus the change $x \to -x$ is equivalent to the change of the dispersion sign.

For numerical integration of the *Burgers equation* (1.33), we consider evolution of the initial disturbance $u(0,x) = \psi(x)$ decreasing at $|x| \to \infty$. Following Ref. [3] we assume that

$$\int_{-\infty}^{\infty} \psi(x)\mathrm{d}x = C < \infty, \tag{1.101}$$

and the profile of the initial condition can be arbitrary in other respects. We note only that for any t [3]

$$\int_{-\infty}^{\infty} u(t,x)\mathrm{d}x = \int_{-\infty}^{\infty} u(0,x)\mathrm{d}x = C;$$

this is easy to prove by writing the Burgers equation in the divergence form,

$$\partial_t u + \partial_x \left(\frac{1}{2} u^2 - \nu \partial_x u \right) = 0,$$

and then integrating both parts in x from $-\infty$ to $+\infty$. Thus, the area bound by the function $\psi(x)$ does not vary with time, i.e., it is the integral of motion.

Consider now the results of numerical simulations of the *Burgers equation.* Choosing for simplicity of numerical realization of the algorithm (the simplicity of the boundary conditions), the initial condition in the form of the single disturbance (1.100) as in the case of the linearized KdV equation, we obtain for large t the result presented in Fig. 1.7. For $\nu \to 0$ (see curve 1

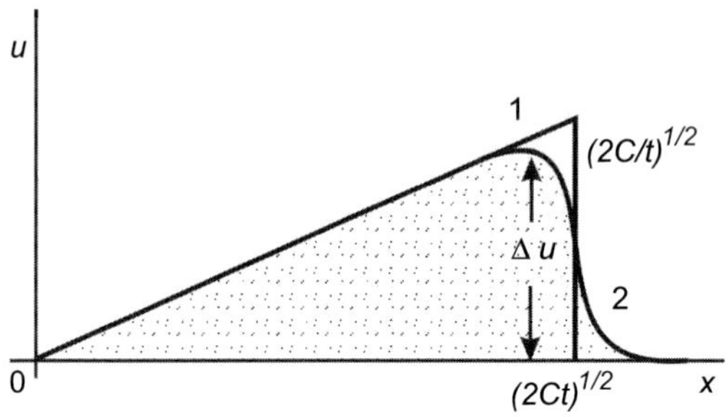

Fig. 1.7. Numerical solution of Burgers equation. Curve 1 corresponds to $\nu = 0.0001$, curve 2 corresponds to $\nu = 0.1$

in Fig. 1.7), the profile of the solution is a triangle with the shock wave structure at its front. The jump of u at the shock wave is $\sqrt{2C/t}$, i.e., it decreases as the (inverse) square root of time; the basis (width) of the profile, on the contrary, increases as the square root of time, so the area of the profile is a constant equal to C.

For finite ν (see curve 2 in Fig. 1.7), we obtain another solution, namely, the stationary wave traveling (without deformation) with a constant velocity w:

$$u = f(x - wt).$$

The jump of the wave, Δu, and the width of the transition region, $\delta = 2\nu/\Delta u$, are determined by the parameter ν of the problem, i.e., by the dissipation factor defining the level of damping. We note that for $\nu \to 0$ the parameter δ also tends to zero, and in this limit we come again to the solution of type (a) of Fig. 1.7.

The results shown in Fig. 1.7 correspond to the case $C > 0$, i.e., when the area of the profile of the initial perturbation is positive. If $C < 0$ then the change (in the Burgers equation) $u \to -u$, $x \to -x$, and $t \to -t$ allows us to return to the solutions considered above.

We consider now numerical solutions of the KdV equation itself, and also discuss, in general, the influence on the structure and dynamics of its solutions of the terms taking into account dissipation and instability[7]. For the analysis of numerical solutions we use the similarity principle formulated above in Sect. 1.1. Consider solutions of (1.11) corresponding to the most typical initial disturbances decreasing for $|x| \to \infty$. The evolution of disturbances of a

[7] Sections 2.1 and 2.2 below are specially devoted to detailed investigation of this problem.

similar type within the limits of the KdV model was studied numerically for the first time by Berezin and Karpman in 1966 [93].

Choosing the initial disturbance as a single pulse (1.100), we consider the dependence (for this initial condition) of the character of solutions on the parameter

$$\sigma = l\left(\frac{u_0}{\beta}\right)^{1/2}$$

(see (1.18) of Sect. 1.1). Numerical experiments of Ref. [93] (also repeated by us when testing the numerical schemes for more complicated generalized equations, see Chap. 4) show that for the sufficiently large $\sigma \gg \sigma_s = \sqrt{12}$, the initial wave pulse practically fully disintegrates with its evolution into individual solitons – see Fig. 1.8 (in addition to the solitons, an oscillating tail is also formed as the wave packet of a small amplitude). We note that solutions of the same type was obtained by Zabusky and Kruskal in 1965 [1] for the periodical initial condition $u(0,x) = \cos(\pi x)$. It follows from numerical

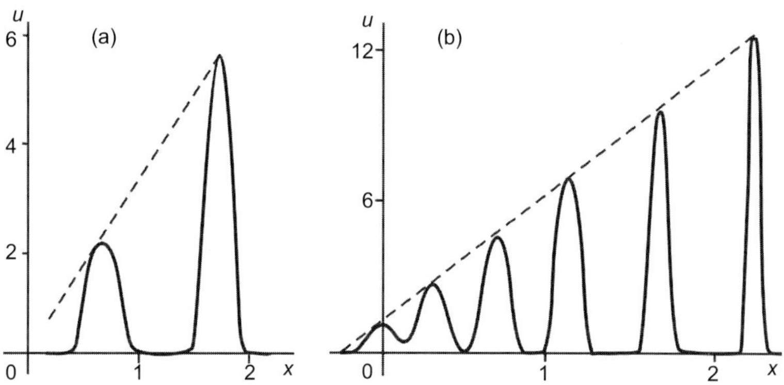

Fig. 1.8. Numerical solution of KdV equation for $\beta > 0$. **a.** $\sigma = 5.9$; **b.** $\sigma = 16.5$

calculations that the initial condition (1.100):

for $4 < \sigma < 7$, decays into 2 solitons,

for $7 < \sigma < 11$, decays into 3 solitons,

for $\sigma \sim 11$, decays into 4 solitons,

for $\sigma \sim 16$, decays into 6 solitons,

i.e., with increasing σ (decreasing β) the corresponding initial disturbance decays into a growing number of solitons. This is clear because the decrease of the dispersion parameter β corresponds to the increasing role of nonlinear effects as related to the dispersion effects. We note that in the evolution of the solution, the integral

$$\int_{-\infty}^{\infty} u(t,x)\mathrm{d}x$$

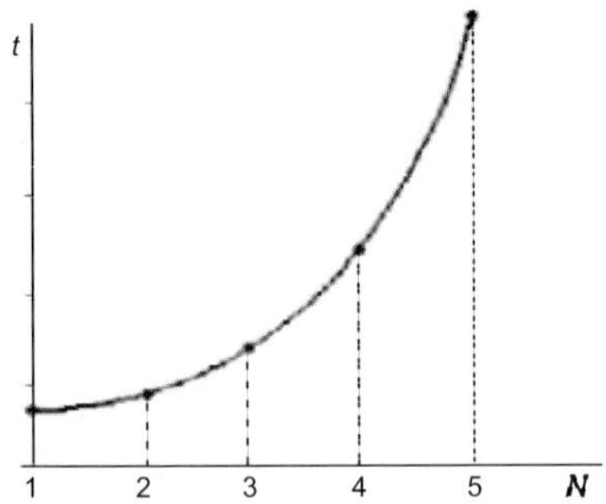

Fig. 1.9. Duration of the stage of the "non-stationary" evolution as a function of the number of the KdV soliton

is conserved, i.e. it is the integral of motion.

The evolution of an initial disturbance in the case when $\sigma \gg \sigma_s$ has two stages which can be conditionally defined as:

1. The "non-stationary stage" corresponding to decay of the initial disturbance and to formation of individual solitons, i.e., solitary structures with the unchanging shape propagating with the constant velocities and amplitudes.
2. The stage of the stationary evolution, i.e. propagation of the solitons as stable solitary wave structures.

Up till 1985 practically all analytical and numerical investigations (constructions of exact solutions, investigations of collisional dynamics of solitons, etc.) were primarily limited to studies of the solutions on the second stage of the evolution, i.e., studies of the evolution and dynamics of already formed soliton structures. In 1984, Belashov [96] succeeded in analytical and numerical investigation of the evolution of solitons of the KdV equation on the non-stationary stage, i.e., he studied dynamics of the formation of solitons as (subsequently) stationary wave objects. The analytical study was based on a generalization of the IST method for the KdV equation considered in Sect. 1.2.3. The numerical studies were done on the basis of scheme (1.82) of numerical integration of the KdV equation. The results of these investigations demonstrated that:

1. The duration of the stage of the "non-stationary" evolution for every soliton of the solution $u(t, x)$ is different and increases exponentially with the increasing "number" of the soliton (see Fig. 1.9).
2. The velocities of the solitons on the "non-stationary" stage decrease to their stationary values determined by the asymptotic expression

$$v_n(t, x) = \lim_{t \to \infty} \left[\frac{u_n(t, x)}{3} \right],$$

where u_n is the amplitude of the nth soliton (Fig. 1.10).

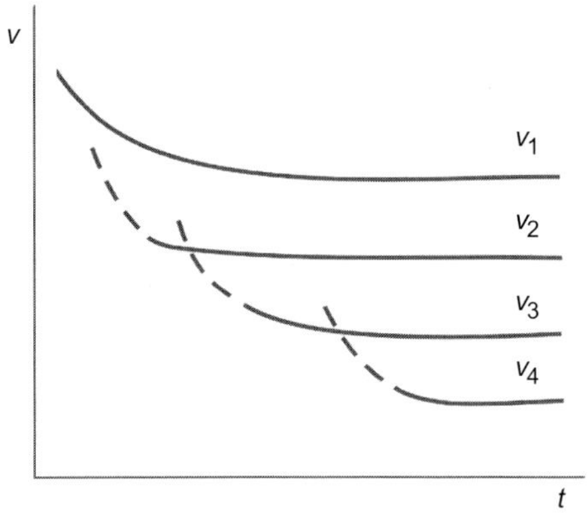

Fig. 1.10. Change of the soliton velocities on the stage of the "non-stationary" evolution

3. The growth dynamics of the soliton amplitudes also reveals the exponential character, with the exponent depending on the number of the soliton (Fig. 1.11).

It is interesting to note that the changes of the amplitudes and velocities (to their stationary asymptotic values) are inversely proportional to each other at this stage, while in the limit $t \to \infty$ the amplitudes and velocities are directly proportional:

$$v_n = \frac{u_n}{3}.$$

We note again that all numerical results obtained for the "non-stationary" stage can be naturally interpreted within the limits of the (properly modified) IST theory.

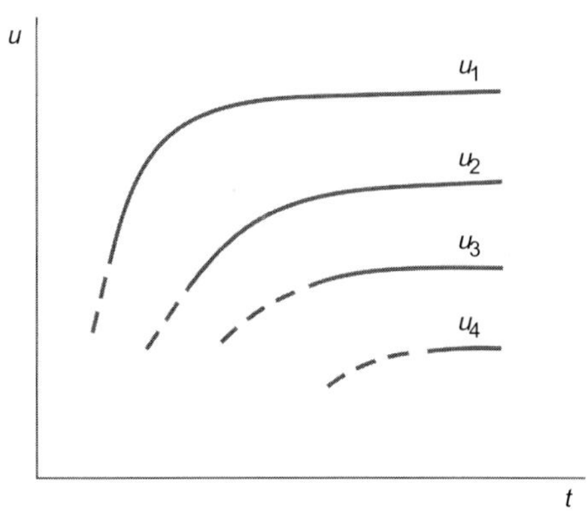

Fig. 1.11. Evolution of the soliton amplitudes on the stage of the "non-stationary" evolution

Consider now solutions of the KdV equation in the opposite limiting case when $\sigma \ll \sigma_s$. As demonstrated in Ref. [93], the "non-soliton" solutions corresponding to disturbances not decaying into solitons are observed in this case. These solutions are fast-oscillating wave packets as shown in Fig. 1.12, looking qualitatively like the *self-similar solution* [3] (see below), although such a packet can differ quantitatively from the self-similar solution by the law with which the amplitude decreases in time and space. It is clear that

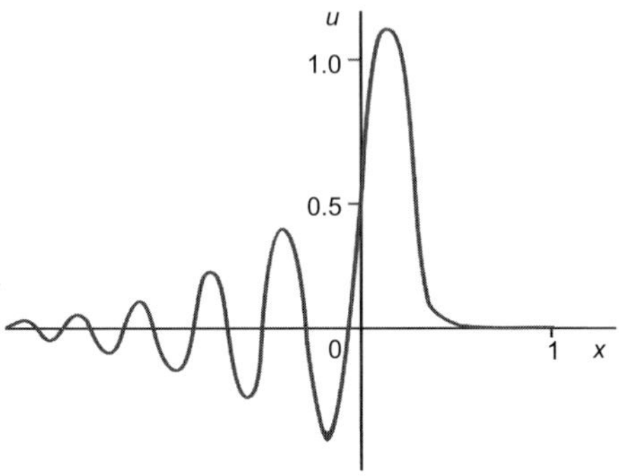

Fig. 1.12. Solution of the KdV equation with $\sigma = 1.9$ ($\beta > 0$)

for $\beta < 0$, the picture is qualitatively opposite – a "train" of fast oscillations goes forward from the main maximum since with the change $x \to -x$ we obtain the same KdV equation (1.11) but with $\beta < 0$. We also note that for some initial conditions, one can observe a solution of the clearly pronounced mixed type having (alongside with the solitons, propagating as the locally stationary objects) an oscillating "tail" falling behind from the solitons of the form similar to that shown in Fig. 1.12.[8]

Now, a few words about the *self-similar solution* of the KdV equation. If we consider a series of initial disturbances with $l \to 0$ in (1.24) (see Sect. 1.1) but take into account that $u_0 l^2$ =const, then solutions for the same β appear to be similar since the parameter σ remains constant. Such a limit solution is given by [3]

$$u(t,x) = \frac{\beta}{(3\beta t)^{2/3}} \psi\left[\frac{x}{(3\beta t)^{1/3}}\right]. \tag{1.102}$$

Substituting this solution into the KdV equation (1.11), limiting our study by the solutions $\psi(z)$ exponentially decreasing at $z \to \infty$, and also assuming that $\psi(z) = f'(z)$, we obtain the Airy equation (for the function $f(z)$),

$$f''(z) - zf(z) = 0.$$

Its solution for $z \to \infty$ is the Airy function $\mathrm{Ai}(z)$ (1.32). Thus the asymptotics of the considered solutions for the function $\psi(z)$ is given by [3]

[8] It is necessary, however, to state that, as was shown in Ref. [96] where evolution of the solitons on the "non-stationary" stage was studied in detail, in reality the non-soliton oscillating part of the solution is always present. The difference consists only in the comparative amplitudes of the solitons and the "tail."

$$\psi(z) = c\frac{d\mathrm{Ai}(z)}{dz} \approx -\frac{c}{2}z^{1/4}\exp\left(-\frac{2}{3}z^{3/2}\right) \quad \text{as} \quad z \to +\infty.$$

Behavior of such solutions for not too large $z > 0$ and $z < 0$ was investi-

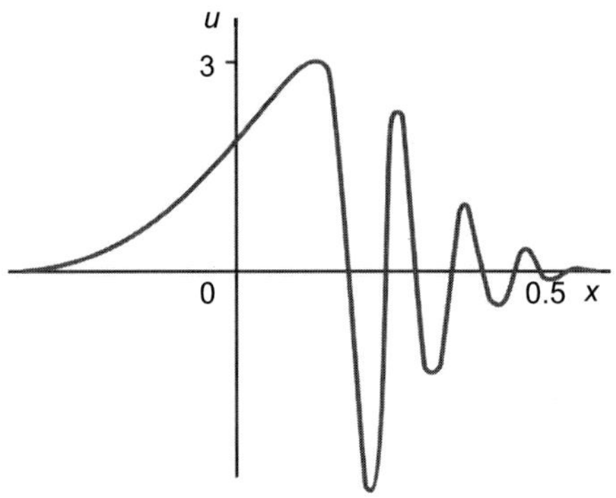

Fig. 1.13. Self-similar solution of the KdV equation with $\beta < 0$ for $\sigma = 10$ $(z < 0)$

gated numerically [93]. The studies showed (see Fig. 1.13) that the solution is qualitatively similar to the solutions of the linearized KdV equation (see Sect. 1.1) as well as the KdV equation with finite l in the initial condition and $\sigma \ll \sigma_s$. Physically, the self-similar solution (1.102) describes the evolution of initial disturbances given by

$$\partial_x\left[\left(\frac{C}{\pi^{1/2}}\right)\exp\left(-x^2/l^2\right)\right]$$

for $x \gg l$ and $t^{1/3} \gg l/\beta^{1/3}$, where $C = u_0 l^2$ =const, i.e., $u_0 = C/l^2$ is the characteristic velocity.

Consider now the character of the solution of the KdV equation with the initial condition taken as a "smoothed step,"

$$u(0,x) = \frac{c}{1+\exp(x/l)}, \tag{1.103}$$

where l gives the width of the front of the initial disturbance. As a result, the step (1.103) evolves into the wave "train" (Fig. 1.14). The amplitude of the first oscillation achieves its stationary value proportional to c. This result was obtained in Ref. [93] by employing the implicit scheme represented by (1.84).

We see that various initial disturbances lead, within the limits of the KdV model, to different sorts of soultions. The form of a particular solution is defined (for the same initial condition) by the dispersion parameter β (or σ in the KdV equation (1.26)). Thus we demonstrated in the above examples that numerical methods presented here can be effectively used for numerical

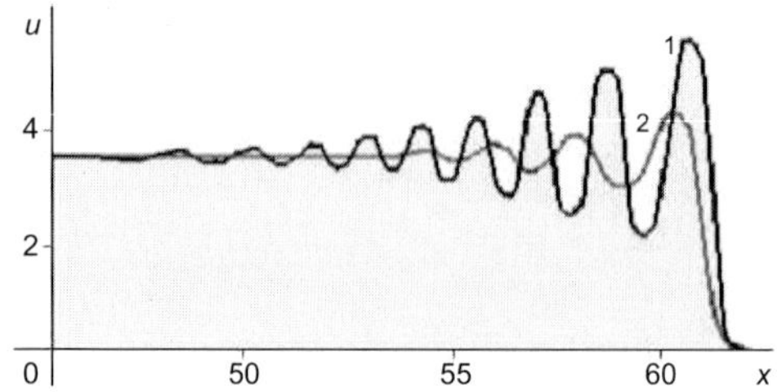

Fig. 1.14. Solution of the KdV equation for the initial condition (1.103) with $c = 4$ at $t = 0.9$. Curve 1 – without dissipation, curve 2 – with dissipation, see (2.2) in Sect. 2.1

integration of the KdV-class equations. More KdV-class equations generalizing the KdV model and taking into account the dissipation, instability, and higher order dispersion correction are considered below in specially dedicated sections of Chap. 2.

1.4 Ion-Acoustic Waves in Plasmas

In this section we consider applications of the results obtained for the KdV equation in the previous sections to description of the structure and dynamics of one-dimensional waves in a plasma. We consider ion-acoustic waves in an unmagnetized plasma and we also include discussions on weakly-relativistic effects.

1.4.1 The Ion-Acoustic Waves

In the Introduction section, we already mentioned that both the KdV-class and KP-class equations are universal in the sense that they describe a wide class of nonlinear wave motions in dispersive media. In Sect. 1.1 we demonstrated the above for the KdV equation (1.11) when the dispersion law in the linear approximation can be written as

$$\omega = c_0 k \left(1 - \frac{\beta k^2}{c_0}\right), \tag{1.104}$$

where c_0 is the phase velocity of the wave and the factor β is determined by the particular type of the medium considered. In this section, we consider one of numerous applications of the KdV-class equations and study *ion-acoustic waves* in an unmagnetized plasma, when the dispersion parameter and the phase velocity are given by

$$\beta = c_0 r_D^2/2 \quad \text{and} \quad c_0 = c_s = (T/m_i)^{1/2}. \tag{1.105}$$

Here, $r_D = (T_e/4\pi n_e e^2)^{1/2}$ is the electron Debye length, T_e is the electron temperature (in energy units such that the Boltzmann constant is unity), n_e

is the unperturbed electron density, c_s is the speed of the ion sound, T is the effective temperature (e.g., equal to T_e when $T_i \ll T_e$), and m_i is the ion mass.

The ion-acoustic wave is a mode commonly occurring in both collisionless and collisional plasmas. Physically, in a collisionless and non-isothermal plasma where the electron temperature is much larger than the ion temperature ($T_e \gg T_i$), these waves are driven by the electron pressure and ion inertia, the coupling between the species being achieved by the electrostatic forces. Although the dispersion relation remains similar to that of the collisionless case, the physics of the ion-acoustic waves in a *collision-dominated plasma* is more complicated, since both electrostatic and collisional effects enter into play [99]. For example, collisions between the dissimilar particles can also couple the dynamics of the ions and the electrons. Thus, the collisional ion-acoustic waves can involve both plasma and neutral-fluid properties. Furthermore, collision-driven resistive and dissipative instabilities can occur if external free-energy sources, such as external currents, density and velocity inhomogeneities, etc., are present [100–103], and the waves can become nonlinear and/or turbulent.

1.4.2 Nonrelativistic Approximation

Assume that a plasma has two fully ionized components (electron and ion) and can be approximated by two-fluid hydrodynamics when the electron and ion components are described by the equation of motion and the continuity equations of type (1.1):

$$\begin{gathered}\partial_t \mathbf{v}_{e,i} + (\mathbf{v}_{e,i} \cdot \boldsymbol{\nabla})\mathbf{v}_{e,i} = -(nm_{e,i})^{-1}\boldsymbol{\nabla} p_{e,i} - (e_{e,i}/m_{e,i})\boldsymbol{\nabla}\varphi, \\ \partial_t n_{e,i} + \boldsymbol{\nabla} \cdot (n_{e,i}\mathbf{v}_{e,i}) = 0.\end{gathered} \tag{1.106}$$

In addition, instead of the Laplace's equation (1.3), we have the Poisson's equation for the electric potential φ:

$$\Delta\varphi = -4\pi e(n_i - n_e). \tag{1.107}$$

In these equations, the subscripts e, i stand for the particle type (electron or ion, respectively) and for simplicity we assume that $-e_e = e_i \equiv e$. Furthermore, we suppose that the plasma is non-isothermal, i.e., its (electron and ion) components have distinctively different temperatures, $T_i \ll T_e$ (and, correspondingly, the different pressures, $p_{e,i} \sim n_{e,i}T_{e,i}$), and consider the low-frequency branch of the oscillations when the condition

$$\tau^{-1} \ll \omega_{pe} = \left(4\pi n_e e^2/m_e\right)^{1/2}$$

is satisfied. Assume also that the electrons are Boltzmann-distributed, i.e., their relaxation time is very short compared to the period of the ion plasma oscillations [3]. In the limit $T_i \ll T_e$ the low-frequency oscillations are weakly damped, and one can further assume that $T_i = 0$ (and therefore in the following section, T can stand for the electron temperature T_e).

Thus omitting the index i for the ion velocities and number densities, we can rewrite Eqs. (1.106) and (1.107) as

$$\begin{aligned} \partial_t \mathbf{v} + (\mathbf{v} \cdot \boldsymbol{\nabla})\,\mathbf{v} &= -(e/m_i)\boldsymbol{\nabla}\varphi, \\ \partial_t n + \boldsymbol{\nabla} \cdot (n\mathbf{v}) &= 0, \\ \Delta\varphi &= 4\pi e\left[n_0 \exp(e\varphi/T) - n\right]. \end{aligned} \tag{1.108}$$

For the long wavelength *ion-acoustic waves* when $kr_D \ll 1$ (the condition of weak dispersion), the dispersion relation (1.12) is valid. Now, using the technique used above in Sect. 1.1 to derive the KdV equation, we can easily obtain the *Boussinesq equations* for the ion-acoustic wave,

$$\begin{aligned} \partial_t \mathbf{v} + (\mathbf{v} \cdot \boldsymbol{\nabla})\mathbf{v} &= -c_s^2 \boldsymbol{\nabla} \ln n - (2c_s\beta/n_0)\boldsymbol{\nabla}\Delta n, \\ \partial_t n + \boldsymbol{\nabla}(n\mathbf{v}) &= 0 \end{aligned} \tag{1.109}$$

(compare with (1.7) in Sect. 1.1), where β and c_s are defined by (1.105).

Consider the wave propagating along the x-axis when the x-component of the ion velocity is much smaller than the phase velocity c_s. In this case, following the results of Sect. 1.1, we obtain the KdV equation for propagation of the ion-acoustic wave in the x-direction:

$$\partial_t v + c_s \partial_x v - c_s \delta^2 \partial_x^3 v + v \partial_x v = 0, \tag{1.110}$$

where $\delta^2 = r_D^2/2 = \beta/c_s$, which is similar to (1.10). After the homothetic transformation and transition to the reference frame moving along the x-axis with the velocity c_s, we obtain the *KdV equation* in its standard form

$$\partial_t u + \alpha u \partial_x u + \beta \partial_x^3 u = 0, \tag{1.111}$$

where u is the velocity of the ion "sound," and the factor at the nonlinear term is $\alpha = 3c_s/2n$ [3].

Turning to (1.110), we note that the term $c_s\partial_x v$ describes the wave propagating in the x-direction with the velocity c_s, while the dispersion and nonlinear terms are responsible for slow changes of the sound wave field on the background of the fast wave motion with the velocity c_s. Such type of sound waves is mostly characteristic of isotropic media (e.g., plasma without magnetic field), but sometimes it takes place in an anisotropic medium as well. For example, if the characteristic frequencies of the ion-acoustic wave packet are much larger than the ion-cyclotron frequency, $\omega_{Bi} = eB_0/m_i c$, appearing in a magnetized plasma, the plasma anisotropy can be neglected and one can still reduce (1.110) to the KdV equation (1.111). In the opposite case, when

$\omega \ll \omega_{Bi}$, the effects of the anisotropy cannot be neglected and it is necessary to consider a non-one-dimensional model[9] (see, for example, Sect. 4.6).

Finally, we note that it is not necessary to specially integrate (1.111), since all results of the previous sections can be applied directly to this case of the ion-acoustic waves in a collisionless unmagnetized plasma. What is necessary is merely to interpret them properly, taking into account the physics behind the terms and factors of this equation.

1.4.3 Weakly-Relativistic Effects

As we demonstrated above, ion-acoustic waves in a plasma can be described by the KdV equation (1.111). If velocities of plasma particles approach the speed of light, the relativistic effects should be taken into account when considering propagation of the one-dimensional solitary ion-acoustic wave. It appears that the *relativistic effects* can strongly influence the phase velocity, the amplitude, and the characteristic length of the wave.

Using the reduced perturbation method [104] for the one-dimensional ion-acoustic solitary waves in a weakly relativistic collisional plasma, we can obtain the KdV equation of type (1.111) by taking into account the relativistic factor u/c:

$$\partial_\tau \Phi_1 + \alpha(\vartheta_1)\,\Phi_1 \partial_\xi \Phi_1 + \frac{1}{2}\beta(\vartheta_1)\,\partial_\xi^3 \Phi_1 = 0, \tag{1.112}$$

where $\Phi_1 = \vartheta_1^{1/2} u_1$ is the first-order perturbation of the electrostatic potential $\Phi = \varepsilon\Phi_1 + \varepsilon^2\Phi_2 + \ldots$ (ε is the small expansion parameter), u_1 is the first-order perturbation of the particle velocity ($u = u_0 + \varepsilon u_1 + \varepsilon^2 u_2 + \ldots$), and

$$\begin{aligned} \alpha(\vartheta_1) &= \beta(\vartheta_1)\left(1 - \vartheta_2/\vartheta_1^{3/2}\right), & \beta(\vartheta_1) &= \vartheta_1^{-1/2}, \\ \vartheta_1 &= 1 + 3u_0^2/2c^2, & \vartheta_2 &= 3u_0/2c^2. \end{aligned} \tag{1.113}$$

Equation (1.112) is written for the reference frame moving along the x-axis, $\xi = \varepsilon^{1/2}(x - \lambda t)$ and $\tau = \varepsilon^{3/2} t$, where λ is the phase velocity. Note that the factor at the nonlinear term is positive, $\alpha > 0$, because of $\vartheta_1 \gg \vartheta_2$. In this case we can obtain a stationary solution in the form of a solitary wave. Introducing a new variable $\zeta = k\xi - \omega\tau$ and substituting it into (1.112), we write the solution for the one-dimensional wave in the form of the *ion-acoustic soliton*:

$$\Phi_1 = \Phi_0 \cosh^{-2}\left(\frac{\zeta}{kW}\right). \tag{1.114}$$

Here, the amplitude Φ_0 and the characteristic scale W are given by

---

[9] On the right-hand side of the equation of motion (1.107), an additional term proportional to $\omega_{Bi}\hat{\mathbf{x}} \times \mathbf{v}$ (where $\hat{\mathbf{x}}$ is the unit vector of the x-axis) appears and it is necessary to include an additional term that is proportional to $k_\perp^2/2k_x^2$ into dispersion relation (1.104) (see also comment in Sect. 4.6.1).

$$\Phi_0 = \frac{3\delta}{\alpha(\vartheta_1)} \quad \text{and} \quad W = \left[\frac{2\beta(\vartheta_1)}{\delta}\right]^{1/2}, \tag{1.115}$$

where $\delta = \omega/k$ and the boundary conditions are $\Phi_1 \to 0$, $\partial_\zeta^n \Phi_1 \to 0$ for $n = 1, 2$ and $|\zeta| \to \infty$. The dispersion law for the described waves has the form $\omega = 2\beta(\vartheta_1)k^3$.

Table 1.2. Comparison of results (1.113)–(1.115) with those of Refs. [105,106]

Parameter	Eqs. (1.113)–(1.115)	$u_0/c = 0$ [105,106]	$u_0/c \neq 0$ [105,106]
λ	$u_0 + \vartheta_1^{-1/2}$	1	$u_0 + \vartheta_1^{-1/2}$
α	$(1 - \vartheta_2/\vartheta_1^{3/2})/\vartheta_1^{1/2}$	1	$(1 - \vartheta_2/\vartheta_1^{3/2})/\vartheta_1^{1/2}$
β	$\vartheta_1^{-1/2}$	1	$\vartheta_1^{-1/2}$
Φ_0	$3\delta\vartheta_1^{1/2}/(1 - \vartheta_2/\vartheta_1^{3/2})$	$3s$	$3s\vartheta_1^{1/2}/(1 - \vartheta_2/\vartheta_1^{3/2})$
W	$\vartheta_1^{-1/4}(2/\delta)^{1/2}$	$(2/s)^{1/2}$	$\vartheta_1^{-1/4}(2/s)^{1/2}$

We can see from (1.113) that factors at the nonlinear term as well as at the dispersion term are determined by the relativistic factor ϑ_1, and relations (1.115) show the dependence of the amplitude and the characteristic scale of the KdV ion-acoustic soliton on the (weakly) relativistic effects. Comparison of results following from (1.113)–(1.115) with those for the two extreme cases $u_0/c = 0$ and $u_0/c \neq 0$ considered in Refs. [105,106] is given in Table 1.2. Here,

$$s = \omega/k \cong v_0 + \vartheta_1^{-1/2}\left(1 - k^2/2\right),$$

where v_0 is the ion velocity (if $v_0 \sim 0$ and the relativistic effects are absent then $s \cong 1 - k^2/2$). We see from Table 1.2 that the results obtained above also include the cases considered in Refs. [105,106].

Consideration of effects in a weakly-collisional and weakly-relativistic plasma is justified by a number of phenomena in a plasma where high energy flows of a particles should be taken into account. In particular, when the kinetic energy of ions, $Mu_0^2/2$, reaches 4.7MeV for $u_0/c \approx 0.1$, a weakly-relativistic ion-acoustic solitary wave starts to form, thus describing the motion of high energy protons with the velocity approaching the speed of light, as observed in the Earth's *magnetospheric plasma* [107]. Investigation of the relativistic nonlinear waves also has application in laser plasma physics [108] and astrophysics [109].

2. Generalized KdV Equations. NLS and DNLS Equations

2.1 Generalized KdV Equations

In this section we consider some generalizations of the KdV equation, taking into account dissipation processes, higher order dispersion corrections and instability (Sect. 2.1.1 and 2.1.2), as well as consider modified KdV equations (Sect. 2.1.3). In Sect. 2.1.4, on example of surface waves in a plasma, we present a KdV-type equation with higher order dispersive nonlinearity and discuss some types of solitary surface plasma waves. Other problems related to evolution of the generalized KdV solitons and classification of solutions of the generalized KdV equation, using methods of the qualitative and asymptotic analyses, are discussed in Sect. 2.2.

2.1.1 The KdV–Burgers Equation. Some Applications

Consider the influence of terms related to dissipation (as well as possible instability in a medium) on the structure of the KdV equation.

Dissipation. The KdV–Burgers (KdVB) Equation. The influence of dissipation in a medium leads to appearance of the new term $-\mathrm{i}\mu k^2 c_0$ in the dispersion relation (1.17). Thus the dispersion law takes the following form:

$$\omega = c_0 k \left(1 - \mathrm{i}\mu k - \beta k^2 / c_0\right). \tag{2.1}$$

The KdV equation with such type of dissipative term,

$$\partial_t u + u\partial_x u + \beta\partial_x^3 u = \nu\partial_x^2 u, \tag{2.2}$$

where $\nu = c_0\mu$, is called the *Korteweg–de Vries–Burgers equation* (*KdVB equation*). This equation describes propagation of nonlinear waves and solitons in media with the viscous-type dissipation, in this case $\nu > 0$. The KdVB equation can be obtained, for example, by the same way as in Ref. [3] (see also below).

As was shown in Refs. [3,83], the dissipative factor ν for ion-acoustic waves in an unmagnetized plasma is given by

$$\nu = (\rho_0/2\rho)\left(c_\infty^2 - c_0^2\right)\tau \int_0^\infty \xi\varphi(\xi)\mathrm{d}\xi,$$

and describes the relaxation damping of "sound" where c_∞ and c_0 are the velocities of the high- and low-frequency waves, respectively (the latter coincides with $c_s = (T_e/m_i)^{1/2}$); here, $\varphi(t, \tau)$ is the function determining the relaxation process. If, on the contrary, the collisionless *Landau damping* is significant for the considered ion-acoustic waves in a plasma, the dissipation can be accounted by introducing (into the right-hand side of the KdV equation) the integral [3,83]

$$\hat{\mathrm{L}}[u] = \sigma \int_{-\infty}^{\infty} \frac{\mathrm{d}k}{2\pi} |k| \int_{-\infty}^{\infty} u(x') \mathrm{e}^{\mathrm{i}k(x-x')} \mathrm{d}x', \tag{2.3}$$

where $\sigma = c_0(\pi m_e/8m_i)^{1/2}$. In this case the dispersion law is given by

$$\omega = c_0 k \left(1 - \frac{\mathrm{i}\sigma|k|}{c_0} - \frac{\beta k^2}{c_0}\right), \tag{2.4}$$

and we obtain the following generalization of the KdV equation:

$$\partial_t u + u\partial_x u + \beta\partial_x^3 u = -\hat{\mathrm{L}}[u]. \tag{2.5}$$

Using the relation [3]

$$\frac{\mathrm{i}\pi|k|}{k} = P \int_{-\infty}^{\infty} \frac{\mathrm{d}z}{z} \mathrm{e}^{\mathrm{i}kz},$$

where P stands for the principal value, we can write [92]

$$\hat{\mathrm{L}}[u] = \frac{\sigma}{\pi} P \int_{-\infty}^{\infty} \frac{\mathrm{d}x'}{x - x'} \partial_{x'} u(x'). \tag{2.6}$$

We note here that the damping coefficient σ in relations (2.3)–(2.6), similar to the factors ν and μ in (2.1) and (2.2), can only be positive; that is natural according to its physics. Furthermore in our study of the influence of dissipation on the structure and evolution of nonlinear waves, we limit ourselves, unless it is specifically mentioned, to the hydrodynamic approximation when, for example, for plasma $\omega \ll \omega_{pe}$, in other words, the inverse time of the (ion) oscillations is much less than the electron plasma frequency, i.e., $\tau^{-1} \ll (4\pi n_0 e^2/m_e)^{1/2}$. In this case for $T_e \gg T_i$ the Landau damping is small.

The presence of dissipation/absorption in a medium leads to the exponential wave damping, with the damping rate proportional to the factor ν in the KdVB equation. In this case, the tail oscillations of a soliton are smoothed. Thus, when simulating evolution of the initial "step" (see above), for example, the oscillation region at the front of the shock wave is narrowed and becomes finite (line 2 in Fig. 1.14, Sect. 1.3).

Instability. When introducing the term describing an instability in the medium, for example, in the form αu into the right-hand side of the KdVB equation (2.2), we can in principle achieve stabilization of solitons since the instability leading to the exponential increase of the wave amplitude is suppressed by the dissipation effects. An instability can also be included into the KdVB (or KdV) equation with the help of a small additional differential term, for example, in the form of $-\delta\partial_x^4 u$, into the right-hand side of the equation. The effect of such type of the instability on the structure and evolution of solutions is discussed in detail in Sect. 2.2, where evolution of solitons of the generalized KdV equation is studied.

Strongly Dissipative Waves in a Plasma. Consider nonlinear propagation of *ion-acoustic waves* in a *collision-dominated plasma*, taking into account variations of the particle densities, fluid velocities, as well as temperatures in the wave field [99]. It was found that the propagation is governed by the KdVB equation [99], similar to that for the weakly collisional plasmas [110]. The scaling and therefore the physics are completely different from the collisional case, however. Also, in contrast to the latter, where the nonlinearity originates mainly from ion convection and electron pressure, it is dominated here by the thermal forces and inter-particle heat transfer. As a result, there is no regime of weak dissipation, and KdV solitons cannot propagate in such a system.

Unlike the case for a nearly *collisionless plasma*, where the electrons are in thermal equilibrium and are governed by the Boltzmann distribution, the full dynamics of both the ions and electrons must be considered here. Accordingly, we start with the equations for the fluid velocities, $\mathbf{v}_e$ and $\mathbf{v}_i$, of the electrons and ions [111]:

$$m_e n_e(\partial_t + \mathbf{v}_e \cdot \boldsymbol{\nabla})v_{e;j} = -\nabla_j n_e T_e - \nabla_l \pi_{lj}^{(e)} - e n_e E_j + R_j, \tag{2.7}$$

and

$$m_i n_i(\partial_t + \mathbf{v}_i \cdot \boldsymbol{\nabla})v_{i;j} = -\nabla_j n_i T_i - \nabla_l \pi_{lj}^{(i)} + e n_i E_j - R_j. \tag{2.8}$$

Equations (2.7) and (2.8) are completed by the continuity equations

$$\partial_t n_{e,i} + \boldsymbol{\nabla} \cdot (n_{e,i}\mathbf{v}_{e,i}) = 0, \tag{2.9}$$

and the energy balance equations

$$\begin{aligned} &\frac{3}{2} n_{e,i}(\partial_t + \mathbf{v}_{e,i} \cdot \boldsymbol{\nabla})T_{e,i} + n_{e,i}T_{e,i}\boldsymbol{\nabla} \cdot \mathbf{v}_{e,i} \\ &\quad = -\boldsymbol{\nabla} \cdot \mathbf{q}^{(e,i)} - \pi_{lj}^{(e,i)} \nabla_j v_{e,i;l} + Q_{e,i}. \end{aligned} \tag{2.10}$$

In the above equations, the terms $\boldsymbol{\nabla} nT$ represent the pressure forces of the electron and ion gases, and the stress tensors $\pi_{lj}^{(e,i)}$ are given by (the coefficients here as well as below are obtained by integrating the corresponding kinetic equations [111])

$$\pi_{lj}^{(e)} = -0.73\frac{n_e T_e}{\nu_e} w_{lj}^{(e)}, \quad \pi_{lj}^{(i)} = -0.96\frac{n_i T_i}{\nu_i} w_{lj}^{(i)}, \tag{2.11}$$

with the rate of strain tensors $w_{lj}^{(e,i)}$ given by

$$w_{lj}^{(e,i)} = \nabla_j v_{e,i;l} + \nabla_l v_{e,i;j} - \frac{2}{3}\delta_{lj}\boldsymbol{\nabla}\cdot\mathbf{v}_{e,i}. \tag{2.12}$$

Furthermore, the friction force $\mathbf{R}$ between the electrons and ions is

$$\mathbf{R} = \mathbf{R_u} + \mathbf{R}_T, \tag{2.13}$$

where $\mathbf{R_u}$ is associated with the force of relative friction (for $\omega \ll \nu_e$),

$$\mathbf{R_u} = -0.51 n_e m_e \nu_e \mathbf{u}, \tag{2.14}$$

which depends only on the relative velocity $\mathbf{u} = \mathbf{v}_e - \mathbf{v}_i$ between the electrons and ions. Note that the effective collision frequency here is $\nu_{\text{eff}} \simeq \nu_e$. Let us also stress that in the opposite limit, namely $\omega \gg \nu_e$, one can obtain the mathematically similar relation $\mathbf{R_u} \simeq -n_e m_e \nu_e \mathbf{u}$. However, in this case serious questions on the validity of the hydrodynamic description arise [111].

The thermal-gradient frictional force $\mathbf{R}_T$ apppearing in (2.13) is given by

$$\mathbf{R}_T = -0.71 n_e \boldsymbol{\nabla} T_e. \tag{2.15}$$

Furthermore, the heat fluxes $\mathbf{q}^{(e,i)}$ are

$$\mathbf{q}^{(e)} = \mathbf{q}_{\mathbf{u}}^{(e)} + \mathbf{q}_T^{(e)} = 0.71 n_e T_e \mathbf{u} - 3.16\frac{n_e T_e}{m_e \nu_e}\boldsymbol{\nabla} T_e \tag{2.16}$$

and

$$\mathbf{q}^{(i)} = -3.9\frac{n_i T_i}{m_i \nu_i}\boldsymbol{\nabla} T_i. \tag{2.17}$$

Finally, the heating powers $Q_{e,i}$ are

$$Q_e = -\mathbf{R}\cdot\mathbf{u} - Q_i \quad \text{and} \quad Q_i = 3\frac{m_e}{m_i} n_e \nu_e (T_e - T_i). \tag{2.18}$$

The linear dispersion was found [99] to be

$$\omega \approx v_s k - \mathrm{i}5A\frac{v_{Te}^2 k^2}{\nu_e} - 5Bkv_s\frac{v_{Te}^2 k^2}{\nu_e^2}, \tag{2.19}$$

where $v_s = (10T_e/3m_i)^{1/2}$ is the velocity of the *collisional ion sound* (note the factor 10/3 is absent in the case of nearly *collisionless ion sound* waves propagating with the velocity $c_s = (T_e/m_i)^{1/2}$), $v_{Te} = (T_e/m_e)^{1/2}$ is the electron thermal velocity,

$$A = \frac{3.16}{3} + \left(\frac{3.9}{3} + \frac{9.6}{3}\right)\frac{\nu_e m_e}{\nu_i m_i},$$

and

$$B = \frac{3m_i}{10m_e}\left(\frac{3.16}{3} + \frac{3.9}{3}\frac{\nu_e m_e}{\nu_i m_i}\right)^2 .$$

Equation (2.19) gives the frequency of the *ion-acoustic waves* in a *collision-dominated plasma.* We see that the waves exhibit collision-driven damping and dispersion, and that the contributions of the stress tensor as well as the temperature perturbations, usually neglected, are significant.

Furthermore, it is convenient to define the electrostatic potential φ by $\mathbf{E} = -\boldsymbol{\nabla}\varphi$. Thus we obtain [99]

$$(\partial_t + \partial_x - \partial_x^2 + 5\partial_x^3)\varphi + \partial_x(\varphi)^2 = 0, \tag{2.20}$$

where t, x, and φ have been normalized by $3.16m_e/15m_i\nu_e$, $3.16m_e v_s/15m_i\nu_e$, and $(5+2\cdot 0.71)m_i v_s^2/2e$, respectively. We also have assumed for convenience that $\nu_e m_e \ll \nu_i m_i$, so that, for example, $A \approx 3.16/3$. Equation (2.20), which does not contain any dimensionless parameters, is the *KdVB equation.* Unlike the KdVB equation for a weakly-collisional plasma, Eq.(2.20) is not reducible to a simple KdV equation in any limit. This is because the scaling (i.e., the normalization parameters) is fixed in the present problem, as is evident from the absence of free parameters which can be rescaled in order to neglect the dissipation term. Thus, the corresponding solutions, namely, shock-like structures (*shock waves*) with oscillating downstream tails, usually attributed to weakly collisional plasmas [100,103,110], occur in a strongly collisional plasma as a rule. It also follows that an *ion-acoustic soliton* of KdV type cannot appear in such a plasma. Physically, this result is expected, since when collisional effects dominate, dissipation is inevitable. The fact that thermal forces and inter-particle heat transfer dominate the nonlinear mechanism is also expected in some sense, since dissipation, similar to dispersion, is particularly sensitive to the large gradients associated with the shock wave.

2.1.2 Higher Order Dispersion Corrections

Consider influence of the *higher order dispersion* corrections on the structure of solutions of the KdV equation; these corrections sometimes arise naturally in the equation when studying propagation of nonlinear waves in some particular physical media. They are especially important when the factor β at the third-order dispersion term is very small or even equal to zero, e.g., for waves on the surface of a *shallow fluid* that takes place when $H^2 \to 3\sigma/\rho g$; for *FMS waves* propagating in a magnetized plasma – when $\tan^{-2}\theta \to m_e/m_i$ (see Sect. 1.1). Note that $\beta \to 0$ does not necessarily mean that dispersion in a medium is completely absent; usually it only indicates that in the expansion of the full dispersion equation in k (just how expressions (1.17) and (2.1) have been obtained), it is necessary to take into account the next order dispersion term to maintain the balance between the nonlinearity and dispersion (determining the existence of a soliton solution).

In this case, the dispersion equation can be written as

$$\omega = c_0 k \left(1 - \frac{\beta k^2 - \gamma k^4}{c_0}\right), \tag{2.21}$$

so that such generalized KdV equation is given by

$$\partial_t u + u\partial_x u + \beta\partial_x^3 u + \gamma\partial_x^5 u = 0. \tag{2.22}$$

This equation is called the *Kawahara equation* [84]. In this case the inflection point appears on the dispersion curve, and the dispersion dependence $\omega = \omega(k)$ becomes more complicated (see Fig. 2.1). As we can see from Fig. 2.1, the opposite dispersion signs appear in the regions of short and long waves: indeed, for small wave numbers k when $\mathrm{sgn}(\beta) = \mathrm{sgn}(\gamma)$, the wave dispersion is negative, and for large wave numbers k it is positive for $\beta, \gamma > 0$, and negative for $\beta, \gamma < 0$.

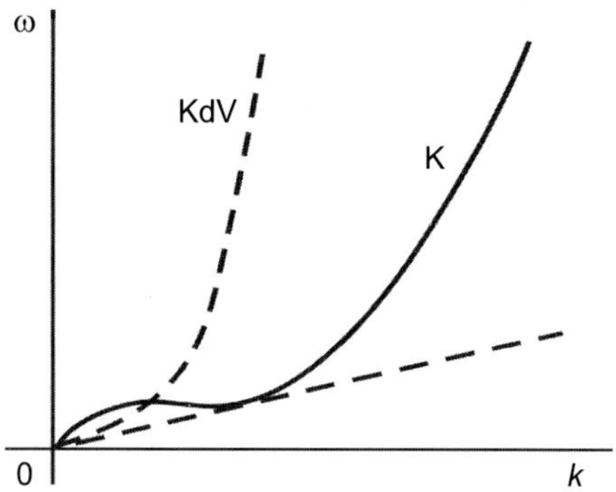

Fig. 2.1. Dispersion character for equation (2.22), the curve **K**

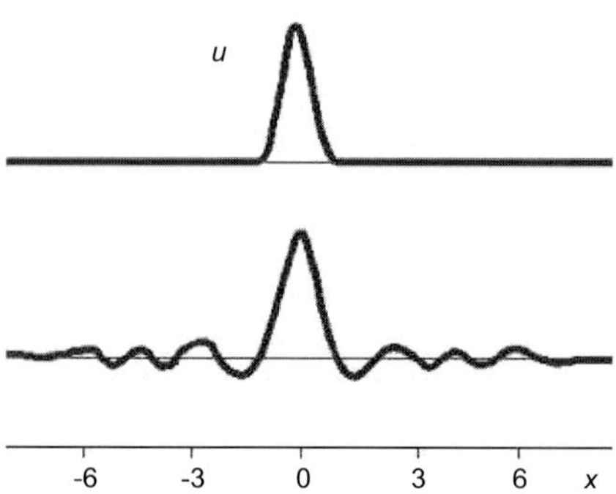

Fig. 2.2. Solutions of the Kawahara equation for different signs of the dispersion factors: 1 for $\gamma > 0$ and $\beta \leq 0$; 2 for $\gamma > 0$ and $\beta > 0$

Numerical simulations demonstrate that the Kawahara equation (2.22) can have two classes of solutions, see Fig. 2.2. In the case $\gamma > 0$ and $\beta \leq 0$[1], a soliton forms, with the monotonous exponential asymptotics which is similar to the asymptotics of the standard *KdV soliton*. However, in the case $\gamma > 0$ and $\beta > 0$, the soliton asymptotics acquires the oscillating character (see also Sect. 2.2). In this case the velocity V of such a soliton satisfies the conditions $V < 0$ and $V < V^{\rm ph}_{\rm min}$ where $V^{\rm ph}$ is the phase velocity of the small amplitude waves. From the dispersion relation (2.21), we obtain $V^{\rm ph} = -\beta k^2 + \gamma k^4$, and therefore the upper limit of the soliton velocity is $V < V^{\rm ph}_{\rm min} = -\beta^2/4\gamma$.

Kawahara investigated in detail the asymptotics of the oscillating solutions of a soliton type in 1972 [84].[2] Later in 1991, Karpman and Belashov [113,114] obtained analogous results for a two-dimensional analogue of the KdV equation generalized by the higher dispersion correction, namely, the *generalized KP equation* (*GKP equation*) (see Chap. 4 for details). In particular, numerical studies [113] showed that in the one-dimensional limit the amplitudes and wave lengths of the oscillations depend on $\beta|\gamma|^{-1/2}$, namely, the amplitudes of the oscillations decrease and the wave lengths increase when the ratio $(\beta|\gamma|^{-1/2})$ decreases. For $\beta|\gamma|^{-1/2} \to 0$, the asymptotics of the soliton approaches the monotonous one. We note that in the limit $\gamma \to 0$ for $\beta = \text{const}$, the *Kawahara equation* (2.22) converts to the usual KdV equation.

Finally, we note that more detailed analysis of all asymptotic cases including dependencies on the dispersion coefficients β and γ can be found in Sect. 2.2 and Sect. 4.4 (in particular, see upper part of Fig. 4.11). Results of numerical simulations of the evolution of solitons of the generalized KdV equation are considered in detail in Sect. 2.2.

2.1.3 Modified KdV Equations

Applying to the KdV equation $\partial_t u + 6u\partial_x u + \partial_x^3 u = 0$ (which is (1.19) with the change $u \to -u$) the *Miura transform* $u = v^2 + \partial_x v$ (this in fact means that we change the quadratic nonlinearity by the cubic one), we obtain the so-called *modified KdV equation* (*MKdV equation*):

$$\partial_t v + 6v^2\partial_x v + \partial_x^3 v = 0. \tag{2.23}$$

This equation, like KdV equation, possesses an infinite number of the polynomial conservation laws.[3] The prove of such connection of these equations

---

[1] Note that the case of the opposite signs of β and γ can be simply obtained from $x \to -x$, $u \to -u$, and therefore we consider here only the cases of either the same or the opposite sign of the dispersion factors.

[2] For a more complex generalized KdV equation the asymptotic analysis was done by Belashov and Tunina in 1997, see Ref. [112] and Sect. 2.2 for details.

[3] Note that modified equations of a more general class $\partial_t w + 6w^p\partial_x w + \partial_x^3 w = 0$, with a higher power of nonlinearity $p = 3, 4, \ldots$, have only three of the polynomial conservation laws [13].

(obtained using the direct comparison of the conservation laws) is based on the following Miura statement [115]:

Theorem. *If v is a solution of the modified KdV equation $Qv = 0$, then*

$$u \equiv v^2 + \partial_x v \tag{2.24}$$

is the solution of the KdV equation $Pu = 0$ where the differential operators are

$$Q \equiv \partial_t + 6v^2 \partial_x + \partial_x^3 \qquad \text{and} \qquad P \equiv \partial_t + 6u\partial_x + \partial_x^3.$$

Proof. Substituting (2.24) into the left-hand side of the KdV equation we obtain

$$Pu = (2v + \partial_x)Qv,$$

thus, if $Qv = 0$ then u satisfies the KdV equation.

It is interesting to note that unlike the KdV equation, the interchange $v \to -v$ in (2.23) does not change the MKdV equation, therefore the latter describes solutions of positive as well as negative polarity without its change. Similar property takes place for generalized MKdV equations for any even nonlinearity power (we discuss this problem in detail in Sect. 2.2.3 for arbitrary nonlinearity index p).

The transform (2.24) can be used to prove the existence of the infinite series of conservation laws for both KdV and MKdV equations. Such a proof, based on generalization of (2.24),

$$u \equiv w + \varepsilon \partial_x w + \varepsilon^2 w^2,$$

where w satisfies the *Gardner equation*

$$\partial_t w + 6(w + \varepsilon^2 w^2)\partial_x w + \partial_x^3 w = 0,$$

was given in [90]. The transform (2.24) is analogous to the *Cole–Hopf transform* which converts the *Burgers equation* into the heat conductivity (or diffusion) equation (see Sect. 1.1.3 and Ref. [2]). In our case, however, the transform (2.24) links two nonlinear equations, one of which (KdV) can be solved by the *IST method*, see Sect. 1.2.

Using the direct method of finding exact solutions of nonlinear evolution equations proposed by Hirota [94] and supposing that $v \to 0$ for $|x| \to \infty$ we can obtain the one-soliton solution of the MKdV equation (2.23) given by

$$v(t, x) = k \cosh^{-1} \xi, \tag{2.25}$$

where $\xi = k(x - x_0 - k^2 t)$. Now, we can construct the one-soliton solution of the KdV equation using the transform (2.24)

$$u(t,x) = \frac{k^2}{2}\cosh^{-2}\frac{\xi}{2} \tag{2.26}$$

(compare with the solution of (1.19) given in Sect. 1.1.2). We can see in (2.25) and (2.26) that the amplitudes of the MKdV and KdV solitons are k and k^2, their widths, $1/k$, are therefore inversely proportional to the amplitude and to the square root of the amplitude, respectively, and the velocities of the solitons are proportional to k^2 in both cases.

Note, that a result similar to (2.26) can be easily obtained using the direct *Hirota method.* This method also enables us to construct the N-soliton solution of the MKdV equation which is given by [26]

$$v = \mathrm{i}\partial_x\left(\ln\frac{f_N^*}{f_N}\right), \tag{2.27}$$

where

$$f_N = \sum_{\bar{\mu}}\exp\left[\sum_{i=1}^{N}\mu_i\left(\xi_i + \mathrm{i}\frac{\pi}{2}\right) + \sum_{1\le i<j}^{N}\mu_i\mu_j A_{ij}\right] \tag{2.28}$$

(compare with (1.21)). The factors A_{ij} are defined by

$$\mathrm{e}^{A_{ij}} = \left(\frac{k_i - k_j}{k_i + k_j}\right)^2. \tag{2.29}$$

For example, for the one-soliton solution we have $f_1 = 1 + \mathrm{e}^{\xi + \mathrm{i}\pi/2}$, and for two-soliton solution we obtain

$$f_2 = 1 + \mathrm{e}^{\xi_1 + \mathrm{i}\pi/2} + \mathrm{e}^{\xi_2 + \mathrm{i}\pi/2} + \mathrm{e}^{\xi_1 + \xi_2 + \mathrm{i}\pi + A_{12}}. \tag{2.30}$$

Now, using the *Miura transform* we can construct the N-soliton solution of the KdV equation obtaining, as a result of this procedure, expressions given in Sect. 1.1.2.

Regarding the problem of existence of other classes of solutions of the MKdV equation, consider the results also obtained by the same direct method. Suppose that the asymptotic boundary conditions are non-zero, i.e., $v \to v_0$ for $|x| \to \infty$, and following [116,117] instead of (2.27) we obtain

$$v = v_0 + \mathrm{i}\partial_x\left(\ln\frac{G_N}{F_N}\right), \tag{2.31}$$

where G_N and F_N are defined by the same expressions as (2.28), with the change $\mathrm{i}\pi/2 \to \varphi_i$ and $\mathrm{i}\pi/2 \to \psi_i$, respectively, where φ and ψ are defined by

$$\mathrm{e}^{\varphi_j} = 1 - \mathrm{i}k_j/2v_0, \qquad \mathrm{e}^{\varphi_j} = 1 + \mathrm{i}k_j/2v_0,$$

and $\xi_i = k_i\left[x - (6v_0^2 + k_i^2)t - x_{0i}\right]$. Thus, assuming that

$$F_1 = 1 + \mathrm{e}^{\eta_1 + \varphi_1},$$
$$G_1 = 1 + \mathrm{e}^{\eta_1 + \psi_1},$$

we obtain the one-soliton solution of the *MKdV equation* (2.23) in the implicit form [26]:

$$v = v_0 + \frac{k_1^2}{\sqrt{4v_0^2 + k_1^2} \cosh \xi_1 + 2v_0}. \tag{2.32}$$

Assuming that

$$F_2 = 1 + e^{\xi_1 + \varphi_1} + e^{\xi_2 + \varphi_2} + e^{\xi_1 + \xi_2 + \varphi_1 + \varphi_2 + A_{12}}$$

and

$$G_2 = 1 + e^{\xi_1 + \psi_1} + e^{\xi_2 + \psi_2} + e^{\xi_1 + \xi_2 + \psi_1 + \psi_2 + A_{12}},$$

where $e^{A_{12}}$ is defined by (2.29), we arrive at the two-soliton solution, etc.

Now set up the problem of finding the rational soliton solutions of the MKdV equation. Consider the limit $k_j \to 0$ with the proper choice of the phase constants. For $N = 1$ we choose $e^{k_1 x_{01}} = 1$. Then we obtain

$$F_1, G_1 \sim -k_1 \left(x - 6v_0^2 t \pm \frac{i}{2v_0} \right) \tag{2.33}$$

(the upper and the lower signs correspond to functions F_1 and G_1, respectively) which lead to the one-dimensional non-singular rational solution [26]

$$v = v_0 - \frac{4v_0}{4v_0^2 \left(x - 6v_0^2 t \right)^2 + 1}. \tag{2.34}$$

For $N = 2$, we choose

$$e^{k_1 x_{01}} = -e^{k_2 x_{02}} = \frac{k_1 + k_2}{k_1 - k_2} \left(1 + \frac{k_1 k_2}{8v_0^2} \right),$$

assume $k_i \to 0$, and analogously obtain

$$F_2, G_2 \sim -\frac{1}{6} k_1 k_2 (k_1 + k_2) \left[\eta^3 + 12t - \frac{3}{4v_0^2} \eta \pm \frac{3i}{2v_0} \left(\eta^2 + \frac{1}{4v_0^2} \right) \right], \tag{2.35}$$

where $\eta = x - 6v_0^2 t$. Substituting (2.35) into (2.31), we also obtain the solution in another form of the non-singular *rational soliton* [26]:

$$v = v_0 - \frac{12v_0 \left[\eta^4 + (3/2v_0^2)\eta^2 - 3/16v_0^4 - 24\eta t \right]}{4v_0^2 \left[\eta^3 + 12t - (3/4v_0^2)\eta \right]^2 + 3\left(\eta^2 + 1/4v_0^2 \right)^2}. \tag{2.36}$$

Finally, we note that the MKdV equation can have one more class of solutions of the form of a so-called *compacton* [118]. Unlike the solitons, which, although highly localized, still span infinitely, these solitary waves have compact support, and they vanish identically to zero outside a finite region. Moreover, their collisions are elastic, like those for solitons. The compacton-like solution of (2.23) presented for the first time in Ref. [119][4] is given by

[4] Note that in Ref. [119], the compacton-like solution is presented for the MKdV equation $\partial_t v + v^2 \partial_x v + \partial_x^3 v = 0$.

$$v(t,x) = \frac{8}{\sqrt{3}} \frac{k\cos^2 k(x-4k^2t)}{[1-2\cos^2 k(x-4k^2t)/3]} \tag{2.37}$$

within the region $\left|x-4k^2t\right| \leq \pi/2k$, and zero outside this region. We can see that for this solution both the width and the amplitude are proportional to the square root of the velocity. The evolution of solution (2.37) is similar to that of the MKdV solitons. It is necessary to stress the fact that the second derivative of (2.37) is discontinuous at the boundaries, but $v(t,x)$ is the strong solution of (2.23), the feature similar to that of the *compacton*. Thus the judicious truncation of the well-known periodic solution, by confining it within the fundamental strip, leads to the compact support for (2.37).

Stability of the compacton solutions of a more general class of equations, namely

$$\partial_t u + \partial_x(u^p) + \partial_x^3(u^q) = 0,$$

was considered in Ref. [120]. It was shown there by using the linear stability analysis, as well as the Lyapunov stability criterion, that these solutions are stable for arbitrary nonlinear parameters p and q. However, as was noted in Ref. [119], the analysis requires an additional check for the MKdV equation.

To conclude, we note that the MKdV equation can be further generalized analogously to the KdV equation, taking into account possible dissipation, instability, and the higher order dispersion effects. We consider this problem in detail below in Sect. 2.2 where the MKdV equation is studied as a particular example of the more general case of generalized KdV equations.

2.1.4 Higher Order Dispersive Nonlinearity

In this section, we demonstrate how a new KdV-type nonlinear evolution equation, which admits a solitary wave solution, appears for surface waves in a plasma [121–124].

Nonlinear Surface Plasma Waves. Consider a semi-infinite cold plasma $(0 < z < \infty)$ containing free electrons in a positively-charged background of heavy or lattice ions. The plasma is bounded at $z = 0$ by a dielectric of constant permittivity ε_d. In this case, p-polarized electromagnetic waves with the field components B_y, E_x, and E_z can propagate along the x-direction on the interface. Making use of the z-direction structure of the linear mode, one can express the electric fields associated with the nonlinear *surface waves* and apply the corresponding boundary conditions [124,125]. The thickness of the surface layer at $z = 0$ is assumed to be smaller than any other characteristic length.

From the Maxwell equations, the electric field is written $\mathbf{E}$ in terms of the current density $\mathbf{j}$. To describe the electron dynamics, we use the cold plasma fluid equations. To the second order, the quantities needed to obtain the Fourier components of the nonlinear current densities are

$$\mathbf{v}^{nl}_{\omega,k}(z) = \frac{\mathrm{i}e^2}{2m_e^2\omega}\int \boldsymbol{\nabla}\left(\frac{\mathbf{E}_1\cdot\mathbf{E}_2}{\omega_1\omega_2}\right)\mathrm{d}^{(2)}\omega\mathrm{d}^{(2)}k, \tag{2.38}$$

and

$$(n^l\mathbf{v}^l)_{\omega,k}(z) = \frac{\mathrm{i}e^2}{2m_e^2}\int\left[\mathbf{E}_1\boldsymbol{\nabla}_2\cdot\left(\frac{n_0\mathbf{E}_2}{\omega_1\omega_2^2}\right)\right.$$
$$\left.+\mathbf{E}_2\boldsymbol{\nabla}_1\cdot\left(\frac{n_0\mathbf{E}_1}{\omega_1^2\omega_2}\right)\right]\mathrm{d}^{(2)}\omega\mathrm{d}^{(2)}k, \tag{2.39}$$

where $\mathrm{d}^{(2)}\omega = \mathrm{d}\omega_1\mathrm{d}\omega_2\delta(\omega-\omega_1-\omega_2)$, $\mathrm{d}^{(2)}k = \mathrm{d}k_{x,1}\mathrm{d}k_{x,2}\delta(k_x-k_{x,1}-k_{x,2})$, and the right hand side of (2.38) contains contributions from the convective derivatives as well as the $\mathbf{v}\times\mathbf{B}$ force terms in the momentum equation. Here, $\mathbf{E}_j = \mathbf{E}(z;\omega_j,k_{x,j})$ and $\boldsymbol{\nabla}_j = \hat{\mathbf{z}}\partial_z + \mathrm{i}k_{x,j}\hat{\mathbf{x}}$. Thus, the Fourier component of the second order current density is given by

$$\mathbf{j}^{nl}_{\omega,k}(z) = -\frac{\mathrm{i}e^3}{2m_e^2}\int\left[n_0\boldsymbol{\nabla}\left(\frac{\mathbf{E}_1\cdot\mathbf{E}_2}{\omega\omega_1\omega_2}\right)+\mathbf{E}_1\boldsymbol{\nabla}_2\cdot\left(\frac{n_0\mathbf{E}_2}{\omega_1\omega_2^2}\right)\right.$$
$$\left.+\mathbf{E}_2\boldsymbol{\nabla}_1\cdot\left(\frac{n_0\mathbf{E}_1}{\omega_1^2\omega_2}\right)\right]\mathrm{d}^{(2)}\omega\mathrm{d}^{(2)}k. \tag{2.40}$$

Furthermore, we apply the inverse Fourier transform to the corresponding equations; defining a pseudo-potential φ by $E_x = -\partial_x\varphi$ (note that φ is not the usual electrostatic potential in this case), we obtain

$$(\partial_t+\partial_x)\varphi + \frac{\varepsilon_d}{2}\partial_x^3\left(\varphi+\varphi^2\right) = 0, \tag{2.41}$$

where we have normalized t, x, and φ by ω_{pe}^{-1}, $c/\omega_{pe}\varepsilon_d^{1/2}$, and $mc^2/3e\varepsilon_d$, respectively.

Equation (2.41) appears as a new type of evolution equation governing the nonlinear *surface waves* in a plasma [124]. Unlike the KdV equation considered above, the nonlinear term here incorporates dispersive characteristics. This and similar equations have been discussed in [121,122,124,126]. Depending on the initial conditions, (2.41) admits solutions describing the quasi-stationary propagation of finite amplitude surface waves. In particular, for $V>1$, it has the localized propagating solution

$$\varphi = -\frac{3}{4}\cosh^{-2}\left[\xi-\xi^*\mp\frac{3}{2}\left(\frac{4}{3}\varphi+1\right)^{1/2}\right], \tag{2.42}$$

where $\xi = (x-Vt)/L$, $L^2 = 2\varepsilon_d/(V-1)$, and ξ^* is an integration constant. Equation (2.42) is transcendental in φ, but it can be inverted to yield

$$\xi = \pm\frac{3}{2}\left(\frac{4}{3}\varphi+1\right)^{1/2} \pm \cosh\left(-\frac{4}{3}\varphi\right)^{1/2}, \tag{2.43}$$

where $\xi^* = 0$. We note that the latter, as well as the square root, can take on either the positive or the negative sign. Equation (2.43) can be easily evaluated numerically, with the upper and lower signs corresponding to $\xi > 0$ and $\xi < 0$, respectively. The solution is shown in Fig. 2.3. The soliton resembles the well-known $\cosh^{-2}(\xi)$ profile, but it is broader at the center because of the amplitude-dependent rescaling of the argument of the $\cosh^{-2}$ function by the additional term $3\left(1+4\varphi/3\right)^{1/2}/2$. The dashed curve in Fig. 2.3 represents the expression (2.42) without the rescaling term. It is not the KdV soliton profile because of the very different amplitude-width relation.

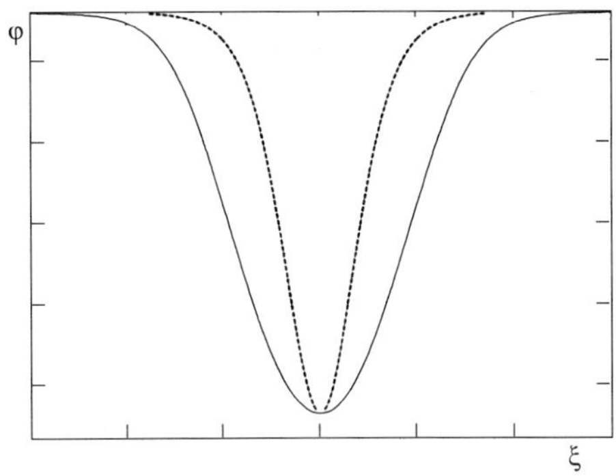

Fig. 2.3. The soliton represented by (2.43). The dashed curve is for the function $-(3/4)\cosh^{-2}(\xi)$ which is, however, not the KdV soliton profile because of the difference in the amplitude-width relations

Alternatively, one can construct functions such as

$$\xi = -2\xi_0 + \cosh\left(-\frac{4}{3}\varphi\right)^{1/2} - \frac{3}{2}\left(\frac{4}{3}\varphi+1\right)^{1/2}, \quad -\frac{1}{2} < \varphi < 0, \tag{2.44}$$

$$\xi = \mp\cosh\left(-\frac{4}{3}\varphi\right)^{1/2} \pm \frac{3}{2}\left(\frac{4}{3}\varphi+1\right)^{1/2}, \quad -\frac{3}{4} < \varphi < -\frac{1}{2}, \tag{2.45}$$

and

$$\xi = 2\xi_0 - \cosh\left(-\frac{4}{3}\varphi\right)^{1/2} + \frac{3}{2}\left(\frac{4}{3}\varphi+1\right)^{1/2}, \quad -\frac{1}{2} < \varphi < 0, \tag{2.46}$$

which also satisfies the required curvature change at $\varphi = -1/2$, but has isolated singular first derivatives at these points. Note again that ξ is double-valued since φ is symmetric in ξ. Figure 2.4 gives the profile of the soliton represented by (2.44)–(2.46). Again, the dashed curve represents the inverse hyperbolic-secant profile in the absence of the amplitude-dependent part of the argument.

In the original units, the amplitude of the soliton depends only on the dielectric properties of the bounding medium, while the width of the solution (2.42) is independent of the latter. In fact, unlike the KdV soliton, the amplitude, speed, and width of the solitons represented by (2.42) are not related

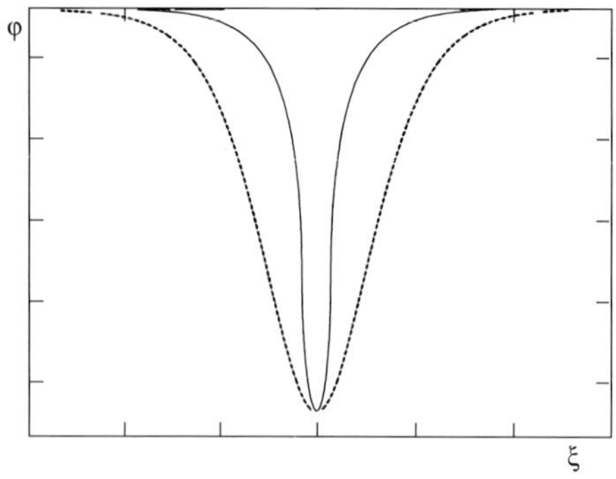

Fig. 2.4. The soliton represented by (2.44)–(2.46). The dashed curve is for the function $-(3/4)\cosh^{-2}(\xi)$

to each other. It is not known if other localized solutions also exist. Furthermore, although the velocity is larger than the speed of light ($= c/\varepsilon_d^{1/2}$) in the bounding medium, no radiation into the dielectric wall seems to occur, although within our ordering the formulation does not preclude coexisting electromagnetic waves with similar spatial scales.

The interaction properties and stability behaviors of the soliton found here are still unknown and remain to be investigated. The mathematical properties of the new nonlinear evolution equation (2.41), in particular with respect to its conservation properties and the inverse scattering method, should also be of interest.

Nonlinear Surface Plasma Waves on a Boundary of a Partially-Ionized Plasma. Many experiments are associated with *surface waves* on the boundary of ionizing plasmas [127–131]. The latter are low-temperature plasmas in which ionization and recombination processes are dominant. They appear near the wall regions of most plasmas since impurity atoms or particles are present there and the cooling effect of the wall prevents their full ionization. Due to the relatively large amplitudes observed for the ionizing surface waves, the nonlinear properties of the latter are of particular interest.

Here, a simple model [123] for studying nonlinear ionizing surface wave propagation in a partially-ionized *low-temperature plasma*, in which collisional effects such as ionization, recombination, and friction are dominant, is presented. We consider the lowest order (namely, the second order in the fields) nonlinear problem and investigate the evolution of finite amplitude electromagnetic surface waves. We show that the waves are governed by the above nonlinear equation with the "dispersive" nonlinearity in this case as well.

Consider a semi-infinite ($0 < z < \infty$) partially-ionized low-temperature plasma in a positively-charged background of ions and neutral particles. Again, the plasma is bounded at $z = 0$ by a dielectric of constant permittivity ε_d. The evolution of the electron density in the plasma is governed by the equation

$$\partial_t n_e + \boldsymbol{\nabla} \cdot (\mathbf{v}_e n_e) = n_e(N - n_e)\langle\sigma(\mathbf{v}_e)\mathbf{v}_e\rangle - \beta n_e^2 + \nu_i n_e, \tag{2.47}$$

where N is the density of the neutral particles, σ is the electron-neutral collision cross-section (which is, in general, dependent on the electron velocity $\mathbf{v}_e$), the angular brackets denote averaging over the directions of the latter, and β and ν_i are the recombination and (external) ionization rates, respectively. For simplicity, we assume that the coefficients σ, β, and ν_i are constants. For a *collision-dominated plasma*, the term with the divergence of the flux can be neglected.

The evolution of the electron fluid velocity $\mathbf{v}_e$ is governed by the usual cold plasma fluid momentum equation

$$\partial_t \mathbf{v}_e + (\mathbf{v}_e \cdot \boldsymbol{\nabla})\mathbf{v}_e = -\frac{e}{m_e}(\mathbf{E} + \frac{1}{c}\mathbf{v}_e \times \mathbf{B}) - \nu_{en}\mathbf{v}_e. \tag{2.48}$$

Noting that the right hand side of (2.47) is of lower order for a low-temperature collisional plasma, we obtain from (2.47) the steady-state, or unperturbed, electron density,

$$n_e^{(0)} = \frac{\nu_i}{\beta}, \tag{2.49}$$

which arises from an imbalance of recombination and ionization. From (2.48), we obtain the electron velocity

$$\mathbf{v}^{(1)}_{e;\omega,k}(z) = -\frac{\mathrm{i}e\mathbf{E}_{\omega,k}(z)}{m_e(\omega + \mathrm{i}\nu_{en})}. \tag{2.50}$$

Thus, the linear (electron) current density is

$$\mathbf{j}^{(1)}_{\omega,k}(z) = -en_e^{(0)}\mathbf{v}^{(1)}_{e;\omega,k}(z) = \frac{\mathrm{i}}{4\pi}\frac{\omega_{pe}^2\mathbf{E}_{\omega,k}(z)}{\omega + \mathrm{i}\nu_{en}}. \tag{2.51}$$

The nonlinear electron velocity is

$$\mathbf{v}^{(2)}_{e;\omega,k}(z) = \frac{\mathrm{i}e^2}{2m^2(\omega + \mathrm{i}\nu_{en})}\int \frac{\boldsymbol{\nabla}(\mathbf{E}_1 \cdot \mathbf{E}_2)}{(\omega_1 + \mathrm{i}\nu_{en})(\omega_2 + \mathrm{i}\nu_{en})}\mathrm{d}^{(2)}\omega \mathrm{d}^{(2)}k. \tag{2.52}$$

The right-hand side of (2.52) contains contributions from the convective derivative as well as the $\mathbf{v} \times \mathbf{B}$ and frictional force terms in the electron momentum equation. For the x-component of the nonlinear current density, we obtain (compare with (2.40))

$$\begin{aligned} j^{nl}_{x;\omega,k}(z) = \frac{e^3 n_e^{(0)}}{2m_e^2}\int &\frac{E_{x,1}E_{x,2}}{(\omega_1 + \mathrm{i}\nu_{en})(\omega_2 + \mathrm{i}\nu_{en})} \\ &\times \left(\frac{k}{\omega} + \frac{\mathrm{i}N\sigma_i/3}{\omega_1 + \mathrm{i}\nu_i} + \frac{\mathrm{i}N\sigma_i/3}{\omega_2 + \mathrm{i}\nu_i}\right)\mathrm{d}^{(2)}\omega \mathrm{d}^{(2)}k. \end{aligned} \tag{2.53}$$

Consider two limiting cases: (a) when the frequencies of the waves are much smaller than the (external) ionization frequency ν_i, $\omega_{1,2} \ll \nu_i$, and

(b) the opposite case, when $\omega_{1,2} \gg \nu_i$. Introducing a pseudo-potential φ_a defined by $E_x = -\partial_x \varphi_a$ (compare with the previous section), we obtain for the case (a) ($\omega_{1,2} \ll \nu_i$) again the equation (2.41), but here the normalization factors of t, x, and φ_a are ω_{pe}^{-1}, $c/\omega_{pe}\varepsilon_d$, and $mc^2/e\varepsilon_d(1 + 2cN\sigma_i/3\nu_i\sqrt{\varepsilon_d})$, respectively. For the case (b) ($\omega_{1,2} \gg \nu_i$), we first assume $N\sigma_i \gg k_x$, which implies that the width of the soliton is much larger than the electron mean free path for collisions with the neutrals. We now define, in contrast to the case (a), a non-potential function φ_b by $E_x = -\partial_x^2 \varphi_b$. It can easily be verified that the function φ_b, when normalized by $3mc^2/e\varepsilon_d N\sigma_i$, also satisfies (2.41). It is clear, however, that the result (due to the different orders of the derivatives in the definitions of φ_a and φ_b) of the case (b) differs considerably from that of the case (a). Thus, we have solutions similar to those considered above. It should be stressed, however, that because of the different underlying physics, these results are physically very different from each other.

Coupled Solitary Waves in Plasma Slabs. Of special importance to a *solid-state plasma* is the problem of wave propagation in certain dielectric or other electronic materials, in which the free electrons can support symmetric and antisymmetric *surface plasmons* at the boundaries. Here, we show that interaction of finite-amplitude symmetric and antisymmetric surface plasmons on a *plasma slab*, which models an electronic device, can result in the coupled propagation of *surface solitons* of the type represented by (2.41). Let us consider a slab, sharply bounded at $z = \pm a$ by a linear dielectric with constant permittivity ε_d, containing free electrons in a stationary (i.e., within the time scale of interest), positive background of heavy or bounded (lattice) ions. It is well known from the linear theory that in this case, either symmetrical or antisymmetrical (depending on the relative signs of E_x at $z = \pm a$) p-polarized electromagnetic surface waves can propagate on the boundaries. That is, $E_x(a) = \pm E_x(-a)$ for the symmetric (upper sign) and antisymmetric (lower sign) surface modes, respectively.

We are interested in the lowest order nonlinear problem. Making use of the z-direction mode structure from the linear theory [132,133], we can write the electric field of the nonlinear symmetrical surface modes in the form

$$E_{x,p}(z) = E_{x,p}^l \frac{\cosh(\alpha_p z)}{\cosh(\alpha_p a)} + E_x^{nl}(z) \tag{2.54}$$

in the plasma, and

$$E_{x,d}(z) = E_{x,d}^l \exp[\alpha_d(a \mp z)] \tag{2.55}$$

in the dielectric. Here, $\alpha_{p,d} = (k_x^2 - \omega^2 \varepsilon_{p,d}/c^2$, $\varepsilon_p = 1 - \omega_{pe}^2/\omega^2$, and the upper and lower signs denote the regions $z > a$ and $z < -a$, respectively, of the bounding-dielectrics. Similarly, for the antisymmetric mode we let

$$E_{x,p}(z) = E_{x,p}^l \frac{\sinh(\alpha_p z)}{\sinh(\alpha_p a)} + E_x^{nl}(z), \tag{2.56}$$

and

$$E_{x,d}(z) = \pm E^{l}_{x,d} \exp[\alpha_d (a \mp z)]. \tag{2.57}$$

As mentioned, the bounding dielectric is linear. We have also made use of the fact that the lowest order field structure in the plasma is that from the linear theory. In the following, we outline the derivation of the equation governing the nonlinear evolution of the electric field for the symmetric mode. The procedure for the antisymmetric mode is similar. We shall again take into consideration the singular surface currents [124] at the walls.

Consider the coupling to the antisymmetric waves. Since the nonlinear current density is quadratic in the fields, the simultaneous presence of an antisymmetric mode can affect the evolution of the symmetric mode. From the Maxwell equations, we obtain for the antisymmetric mode the relation

$$E^{a,l}_{z,p} = \frac{\mathrm{i}k_x}{\alpha_p} E^{a,l}_{x,p} \coth(\alpha_p a), \tag{2.58}$$

where the superscript a denote quantities associated with the antisymmetric waves. In the following, the symmetric mode shall be denoted by the superscript s. Thus, taking into consideration both the antisymmetric and symmetric components of the electric fields, and again defining the pseudo-potential φ by $E^{l}_{x,p} = -\partial_x \varphi$, we finally obtain

$$(\partial_t + \partial_x)\,\varphi_s + \frac{\varepsilon_d}{2}\coth^2\left(\frac{\omega_p a}{c}\right) \times \partial_x^3 \left[\varphi_s + \varphi_s^2 + \left(\frac{1}{3} + \frac{2}{3}\coth^2\left(\frac{\omega_p a}{c}\right)\right)\varphi_a^2\right] = 0 \tag{2.59}$$

and

$$(\partial_t + \partial_x)\,\varphi_a + \frac{\varepsilon_d}{2}\tanh^2\left(\frac{\omega_p a}{c}\right) \times \partial_x^3 \left[\varphi_a + 2\left(\frac{1}{3} + \frac{2}{3}\tanh^2\left(\frac{\omega_p a}{c}\right)\right)\varphi_a\varphi_s\right] = 0, \tag{2.60}$$

where we have normalized t, x, and $\varphi_{a,s}$ with ω_{pe}^{-1}, $c/\omega_{pe}\varepsilon_d$, and $mc^2/3e\varepsilon_d$, respectively. The linear parts of (2.59) and (2.60) describe the linear propagation of the symmetric and antisymmetric electromagnetic *surface waves* in an electronic *plasma slab* [132,133].

Equations (2.59) and (2.60) resemble *coupled KdV equations*, except that here, as in the evolution equations considered above, the nonlinear terms also contain higher spatial derivatives. That is, again, the nonlinear and dispersive characteristics of the waves are strongly coupled. The equations admit various nonlinear solutions, in particular the quasi-stationary *coupled solitons*

$$\varphi_s = \varphi_0 \cosh^{-2}\left[(x - Vt)/L + f(\varphi_s)\right], \tag{2.61}$$

and

$$\varphi_a = B\varphi_s, \tag{2.62}$$

where $V > 1$ is a constant, $\varphi_0 = -9/(8(1+2h))$, $h = \tanh^2(\omega_{pe}a/c)$, $L^2 = 2\varepsilon_d h/(V-1)$, $f(\varphi_s) = -2\varphi_0(1-\varphi_s/\varphi_0)^{1/2}$, and $B^2 = [2h^2(1+2h)-3]h/(2+h)$. Note that $B > 0$ is to be taken, so that in the limit $a \to \infty$, one recovers $\varphi_s + \varphi_a \to 2\varphi$, where φ is the solution for the semi-infinite plasma case.

Thus, the new type of solitary waves discussed here can propagate in a coupled manner on the surfaces of a dielectric slab. Again, their widths are dependent on the properties of the container material as well as the speed. This phenomenon can be useful in the diagnostics of electronic materials as well as for information transmission, and may also affect the properties of interfacial double layers [134], which are of recent interest in several branches of physics and chemistry. The stability and interaction characteristics of these solitons are still unknown, however, and have to be investigated before any practical application can be attempted.

2.2 Structure and Evolution of Solutions of Generalized KdV Equations

In this section, we consider evolution of solitons of the KdV equation generalized by the terms accounting for dissipation, higher order dispersion correction, and instability (Sect. 2.2.1), and also consider the soliton evolution in a medium with stochastic fluctuations of the wave field within the limits of the stochastic KdV equation, Sect. 2.2.2. Section 2.2.3 is devoted to the qualitative and asymptotic analysis of all possible classes of solutions of the generalized KdV equation.

2.2.1 Evolution of Solitons of Generalized KdV Equations

Consider now evolution of solitons of the *generalized KdV equation*, taking into account dissipation, instability, and *higher order dispersion* effects in a medium. The equation can be written as

$$\partial_t u + u\partial_x u - \nu\partial_x^2 u + \beta\partial_x^3 u + \delta\partial_x^4 u + \gamma\partial_x^5 u = 0. \tag{2.63}$$

Here, ν and δ are the constants characterizing dissipation and instability (self-excitation) in the medium, respectively, β and γ are the factors at the dispersion terms of the third and fifth order, respectively. They can have, in principle, any sign, but we suppose that $\beta > 0$ and $|\gamma| \geq 0$. It is easy to see that the case $\beta < 0$ and $\gamma > 0$, for example, can be obtained from the case $\beta > 0$ and $\gamma < 0$ by the simple transform $x \to -x$, $u \to -u$).

The generalized KdV equation (2.63) describes long-wavelength waves in various physical systems, such as *surface waves* in a viscous fluid flowing down an inclined plane [135] and unstable *drift waves* in a plasma (dissipative trapped-ion wave modes with dispersion due to the finite ion banana width) [136]. For these applications it is usually assumed that $\gamma = 0$ [137] and, therefore, the dispersion effects of the high order are not included. Equation (2.63) was also effectively used to study nonlinear soliton-like *internal gravity waves* in the *F-layer of the Earth's ionosphere* [81,138–140] including effects caused by powerful seismic processes and *volcanic eruptions* [141–143], as well as the motion of perturbation sources as the fronts of the *solar terminator* and the *solar eclipse* [81,140,144]. In the case $\beta = \gamma = 0$, (2.63) reduces to the equation describing chemical reactions with the turbulent-like behavior [145–147].

It is not difficult to obtain the dispersion relation for the linearized version of (2.63) by considering the Fourier mode: $u \sim \exp\left[-\mathrm{i}(kx - \omega t)\right]$

$$\omega = c_0 k \left[1 + \frac{\mathrm{i}k}{c_0} \left(\nu + \delta k^2\right) - \frac{k^2}{c_0} \left(\beta - \gamma k^2\right) \right], \tag{2.64}$$

where $k \equiv k_x$. It is clear that in the case $\beta = \gamma = 0$, the small-amplitude sinusoidal waves are linearly unstable (in the "soliton" sense) for both long-wavelength (growing) and short-wavelength (damped) waves.[5] The maximum growth rate is observed for the wave number $k_{\max} = (\nu/2\delta)^{1/2}$ [137]. However, the simultaneous presence of the instability and dispersion indicates the possibility of a steady state, because the energy influx due to the wave's self-excitation is transferred through the mode coupling to the short wavelength modes and is balanced by the damping due to the term with the fourth-order derivative. It is also interesting to investigate the role of the higher order dispersion term in this process because, as it was established in Ref. [148], this type of term starts to play a dominant role when the smaller-scale modes in the wave field appear.

Equation (2.63) with $\gamma = 0$ was first investigated numerically by Kawahara [137]. It was observed that for the strongly dispersive case, the temporal evolution of the initial disturbance in the form of both the Gaussian random noise and the periodic function $\cos \pi x$ is ended by the formation of a row of one-dimensional solitary pulses of equal equilibrium amplitudes. The width of each pulse is determined by the relative influence of the instability and the dissipation effects, and the equilibrium amplitudes increase as the effects of dispersion increase.

Having in mind the results of Ref. [137], we investigate numerically in more detail the role of all terms of the full generalized equation (2.63). Mainly, we use the *explicit scheme with $O(\tau^2, h^4)$ approximation* (1.82) with $N = 1001$ and $h = 0.01$ (see Sect. 1.3), taking into account the difference terms approximating the second, fourth and fifth derivatives, namely,

$$\partial_x^2 u \approx -\frac{1}{12h^2} \left(u_{i+2} - 16u_{i+1} + 30u_i - 16u_{i-1} + u_{i-2}\right),$$

$$\partial_x^4 u \approx \frac{1}{2h}(\Delta_{i+1}^3 u - \Delta_{i-1}^3 u), \text{ and } \partial_x^5 u \approx \frac{1}{2h^2}(\Delta_{i+1}^3 u - 2\Delta_i^3 u + \Delta_{i-1}^3 u),$$

[5] In the terminology of Ref. [137] the damped small amplitude waves are stable.

where Δ_i^3 is the difference approximation of the third derivative. Two types of boundary conditions are chosen: zero and periodical. Different initial conditions are assumed for different runs of numerical simulations (cf. Table 2.1). We see that the cases B1 and C1 correspond to those considered in Ref.

Table 2.1. The cases investigated numerically

Run	Initial condition	ν	β	δ	γ
A1		0		0	-10^{-4}
A2	$u(x,0)=\mathrm{e}^{-36x^2}$	0	2.3×10^{-3}	0	10^{-4}
A3		0.01		10^{-6}	-10^{-4}
A4		0.01		5×10^{-6}	10^{-4}
B1	Gaussian				0
B2	random	-0.01	0	5×10^{-6}	10^{-4}
B3	values				-10^{-4}
C1	Gaussian				0
C2	random	-0.01	4.8×10^{-4}	5×10^{-6}	10^{-4}
C3	values				-10^{-4}
D1		<0			≤ 0
D2	$u(x,0)=\mathrm{e}^{-x^2/l^2}$	<0	$\gg l^2/12$	>0	>0
D3		>0			<0
D4		>0			>0

[137], and the rest in the series B and C correspond to the problem of the influence of a small dispersion correction on the dynamics of the soliton-like solutions. We have run all series of the numerical experiments and obtained the following results.

Figures 2.5 and 2.6 show temporal evolution of the pulse u in the simulation runs A1,2 and A3,4 for the different and the equal signs of β and γ when $\nu=\delta=0$ and $\nu,\delta\neq0$, respectively. Figure 2.7 shows temporal evolution of an initially stochastic (Gaussian random) wave in runs B1 and B2. We can see from Fig. 2.7a that, similar to simulations [137], the energy is transferred over time from the high frequency wave components (with $k\sim k_{\max}$) to the low-frequency components which gradually acquire the roughly triangular shape and interact with each other without any further growth of the average amplitude. The wave field is not regular in this case and has no organized structure up to the time moment $t=6.0$ (compare with the results [137]). As was noted previously [137], this is a reproduction of the turbulent-like behavior in chemical reactions.

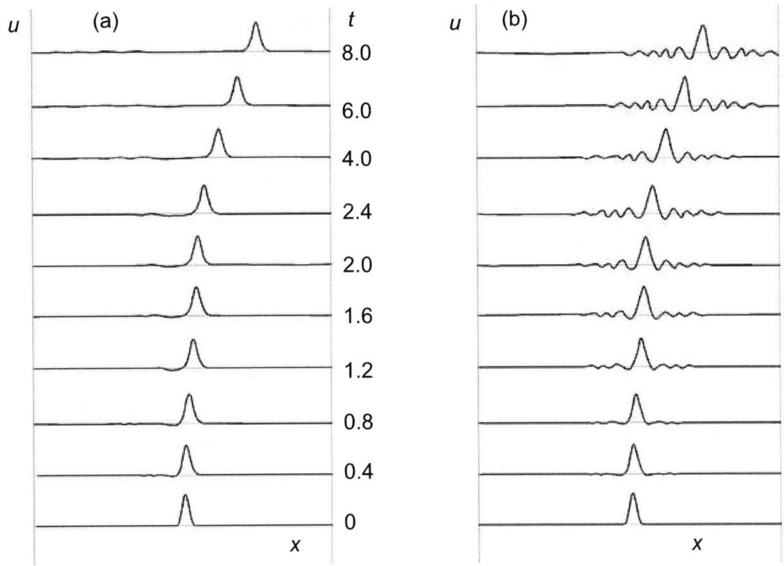

Fig. 2.5. Temporal evolution of initial pulse $u(x, 0) = \exp(-36x^2)$ for two cases. **a.** run A1. **b.** run A2

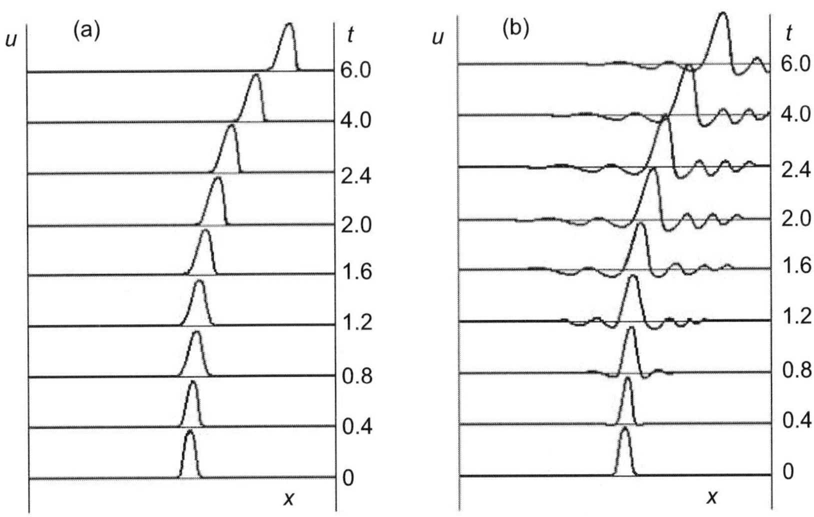

Fig. 2.6. Temporal evolution of initial pulse $u(x, 0) = \exp(-36x^2)$ for two cases. **a.** run A3. **b.** run A4

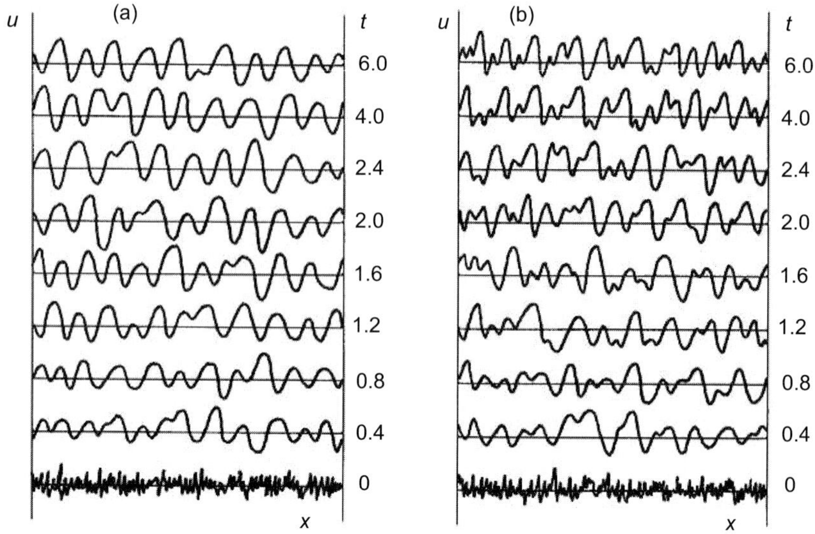

Fig. 2.7. Temporal evolution of an initially random wave for two cases. **a.** run B1. **b.** run B2

If we take into account the higher order dispersion processes, we obtain a different character of the wave evolution, Fig. 2.7b. In the case when the dispersion correction is positive, $\gamma > 0$, we observe that the role of the small correction is in selection with time of the small-scale wave structures, run B2 and case (b) in the figure. In the opposite case, where $\gamma < 0$, i.e., when the dispersion is negative for all wave numbers k (run B3), we observe the same qualitative character of the evolution as in Fig. 2.7a. Such sort of behavior generally agrees with the results presented in Figs. 2.5 and 2.6.

Figure 2.8 shows the results of runs C1 and C2 of numerical simulations, for the same initial condition. In this case, temporal evolution of the initially (Gaussian) random wave in the case (a) corresponds qualitatively to the results of Ref. [137]. We can see that, similar to simulations [137], the energy is also first transferred to the lower frequency region and the wave amplitudes grow while interacting with each other with time up to $t \approx 0.4$, and a row of pulses of equal amplitude is formed at $t \approx 4.0$. We note that when the amplitudes of the pulses become equal, the row (composed of the pulses) starts to travel as a whole (see also [137]). This is a manifestation of the role of dispersion effects in the self-organization process – formation of an organized structure from chaos in a nonlinear medium.

Taking into account finer dispersion effects, we arrive at the following results. Similar to the results of run B2, we observe that the influence of the higher order dispersion effects is manifested in the selection of small wave scales for $\beta, \gamma > 0$ when the sign of the dispersion for a large k is the opposite to that for a small k, run C2 and case (b) in Fig. 2.8. When β and γ have the

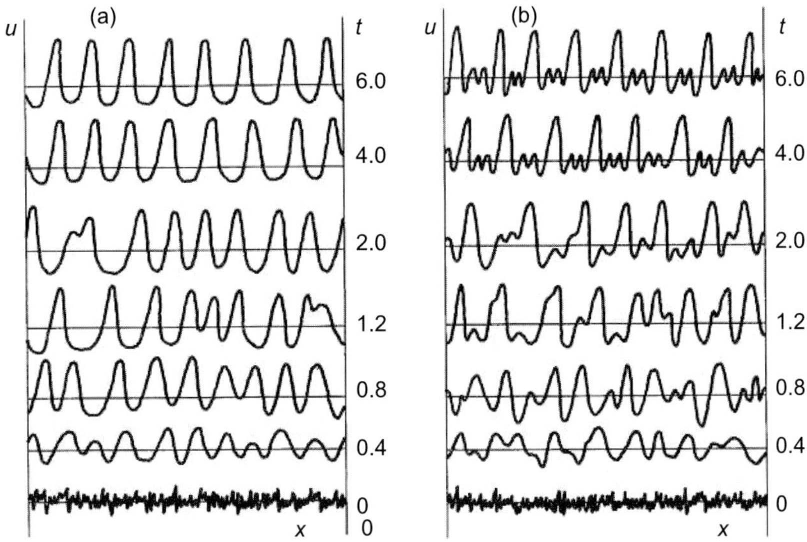

Fig. 2.8. Temporal evolution of an initially random wave for two cases. **a.** run C1. **b.** run C2

opposite sign, as in run C3, the dispersion is negative for all wave numbers k, and we have the case qualitatively similar to that presented in Fig. 2.8a.

Our studies also include the case of stronger dispersion for the initial condition in the form of one pulse $u(x,0) = \exp(-x^2/l^2)$ and various values of factors in (2.63), in simulation runs D (see Table 2.1); for $\nu < 0$, $\beta, \delta > 0$, and $\gamma = 0$, the series D1, that corresponds to the case considered [137] for the initial condition $u(x,0) = \cos \pi x$. For all situations considered, at the initial stage of evolution we observe formation of multi-soliton solutions from the initial pulse up to $t \sim 0.4$.[6] The dynamics of the evolution at this stage is similar to that at the "non-stationary" stage when the solitons of the standard KdV equation are formed (see Sects. 1.2.3 and 1.3.5). The difference between the cases $\gamma < 0$ and $\gamma > 0$ (when $\beta > 0$) is the same as in simulation runs A1 and A2 (see above): in the last case, the solitons acquire oscillating structure of their tails and fronts in the process of their formation. Nonlinear effects then start to dominate, and the non-stationary pulses interact with each other while growing up to $t \approx 4.0$. Then the formed pulses with smooth ($\gamma < 0$) and oscillating ($\gamma > 0$) asymptotics almost align into a row of the soliton-like pulses of equal amplitude and travel as a whole without changing their form. Note also that in the case $\nu > 0$ we observe that the profiles of the soliton-like pulses are steeper in the direction of their propagation, in contrast to the case $\nu < 0$.

---

[6] Recall that decrease of the dispersion parameter β in the KdV equation leads to formation of the multi-soliton solutions (see Sect. 1.3.5).

Concluding the results obtained in all these simulations, we state the following. The results of simulation runs A1 and A2 clearly demonstrate the formation and evolution of the solitons of the *generalized KdV equation* for $\nu = \delta = 0$, with monotonous and oscillating asymptotics considered in Sect. 2.1.1. When $\nu > 0$ and $\delta > 0$, the fronts of such solitons become steeper, as shown in runs A3 and A4. Runs B1, C1, and D1 of our numerical simulations as well as Ref. [137] show that rows of soliton-like pulses are formed in these cases. The saturated amplitudes of the pulses are constant for a fixed set of ν, β, δ, and γ irrespective of the form of the initial condition, but the number of pulses that are formed depends on both the initial condition and the value of the dispersion parameter β. For $\gamma = 0$, this result was first obtained in Ref. [137] which showed, for example, that for the initial sinusoidal condition, five pulses appear for $\beta = 2.0 \times 10^{-3}$, and nine pulses for the "stochastic" initial condition with other conditions unchanged. When $\gamma = 0$, the equilibrium amplitude increases and each soliton-like pulse begins to resemble the usual KdV soliton when the dispersion becomes strong. Vice versa, with the dispersion decreasing, the symmetry of every pulse is broken because of the steepening of its front. The higher order dispersion correction leads to formation of small scale oscillations on the tails and fronts of the soliton-like pulses in the case where $\gamma > 0$. In the opposite case we have the larger structures with smooth asymptotics of the pulses.

We note here the following result obtained by Kawahara: if we introduce the parameter $\varepsilon \equiv \beta/(\nu\delta)^{1/2}$ representing the relative contribution of the dispersion then, although the critical value of ε for the transition from a turbulent-like to an equilibrium state occurs was not yet fixed, the equilibrium soliton-like pulses exist at least for $\varepsilon \geq O(1)$ [137]. Our simulation runs B1 and C1, as well as in simulations reported in [137], yield $\varepsilon = 2.15$ and $\varepsilon = 8.89$, respectively.

There are also other possibilities to investigate possible solutions of the generalized KdV equation (2.63). For example, it is possible to consider the case when $\beta\partial_x^3 \sim u\partial_x u \gg \nu\partial_x^2 u \sim \delta\partial_x^4 u \sim \gamma\partial_x^5 u \sim O(\varsigma)$, which is close to the case of the usual KdV equation and becomes its equivalent in the limit when $\nu \to 0$, $\delta \to 0$, and $\gamma \to 0$. This case can be investigated, for example, by using the two-time asymptotic expansion [92,149]. In particular, for $\gamma = 0$ studies in [137] obtained

$$u(t, x; \varsigma) = u^{(0)}(\eta, t, T) + \varsigma u^{(1)}(\eta, t, T) + O(\varsigma^2), \tag{2.65}$$

where

$$\eta \equiv [N(T)/12\beta]^{1/2} \left[x - \int_0^t (N_0 + N(T)/3)\, \mathrm{d}t \right], \tag{2.66}$$

$T \equiv \varsigma t$ is the small time scale, η is the (normalized) new space coordinate in the moving reference frame, and N is assumed to vary slowly in time. Substituting Eqs. (2.65) and (2.66) to the soliton solution of the KdV equation [137],

$$u(t,x) = N_0 + N \cosh^{-2}\left[(N/12\beta)^{1/2}(x - (N_0 + N/3)t)\right], \qquad (2.67)$$

we can obtain the ordinary differential equation for $N(T)$, namely,

$$\mathrm{d}_T N = \frac{4\delta}{189\beta^2}\left(\frac{21\nu\beta}{5\delta} - N\right)N^2, \qquad (2.68)$$

and investigate temporal evolution of the parameters of the solution (2.67) of the perturbed KdV equation.[7] For $\gamma \neq 0$, similar analysis becomes more complicated, and another approach should be used (see Sect. 2.2.3).

To conclude, we note that all the cases considered above for the generalized KdV equation (2.63) do match well the scheme based on classification of possible solutions by their phase portraits and asymptotics. This methodology is considered below in Sect. 2.2.3.

2.2.2 Soliton Evolution in Media with Stochastic Fluctuations of the Wave Field

Introductory Notes. From the point of view of possible applications, one of important problems of generalization of equations describing wave processes in real physical media is that related to the proper account of possible random factors. In reality, indeed, there are often small-scale as well as large-scale fluctuations of the basic wave characteristics caused by numerous factors of various origin (like dynamic chaos, developed turbulence, etc.). Their combined influence on the medium is obviously of the chaotic character. With a good approximation we can consider this influence as that of stochastic fluctuations of the wave field. Naturally, such fluctuations directly influence the propagation of regular oscillatory and solitary waves induced in the medium by various sources (such as, in the ionosphere plasma, the *solar terminator*, *solar eclipse*, man-made explosions, magnetic substorms, and *volcanic eruptions*). This section is devoted to the theoretical study of such an influence.

To exclude from our consideration the influence of other factors such as dissipation and instabilities of various types (defined by the particular type of the medium), which can obscure the effects of the stochastic field fluctuations on the propagation of the nonlinear waves, we consider below the problem within the frames of the "classic" KdV equation, taking into account the term describing the *stochastic fluctuations of the wave field* [151]:

$$\partial_t u - 6u\partial_x u + \partial_x^3 u - \eta(t) = 0. \qquad (2.69)$$

As it is known from previous sections, (2.69) with $\eta(t) = 0$ describes evolution of nonlinear waves and solitons in various dispersive media. For clarity, we consider the influence of stochastic fluctuations of the wave field on a *KdV soliton* since for $\eta(t) = 0$ it is a stable structure propagating in the medium

[7] For details see Refs. [137,150].

without change of its shape and profile. The approach applied below is general enough (thus we use it for any type of wave described not only by the KdV equation but also the more general *KP equation*).

The stochastic term $\eta(t)$ in (2.69) describes the external noise when the characteristic length of a soliton, l_s, is much smaller than the coherent length of the noise, l_n. This is a particular case of a more general one when the external noise is described by the term $\eta(x,t)$. Being the simplest case for analytical investigation, it still allows us to obtain an exact result and provides us information which is useful for the study of more general cases when $l_s > l_n$ or $l_s < l_n$.

Equation (2.69) (the so-called *stochastic KdV equation*) was first investigated by Wadati [152], and here we generally follow his approach. Also, the problem of the structure and dynamics of multidimensional solitons of the KP equation class is discussed in detail later in Sect. 4.4. Thus we now consider the influence of the stochastic fluctuations of the wave field on the soliton dynamics in the one-dimensional case.

General Approach. First, we note that (2.69) is related to the KdV equation,

$$\partial_t U - 6U\partial_\xi U + \partial_\xi^3 U = 0, \tag{2.70}$$

via the *Galilean transform*:

$$u(t,x) = U(t,\xi) + W(t), \qquad W(t) = \int_0^t \eta(t)\mathrm{d}t, \tag{2.71}$$

where

$$\xi = x + m(t) \quad \text{and} \quad m(t) = 6\int_0^t W(t)\mathrm{d}t.$$

Therefore (2.69) represents an integrable system and can be integrated using the *IST method* (see Sect. 1.2 and Ref. [24], for exmaple). Following an analysis as shown in [152] we also assume that the external noise $\eta(t)$ is Gaussian,

$$\begin{aligned} \langle \eta(t_1)\eta(t_2)\ldots\eta(t_n)\rangle &= 0 && (n \text{ odd}), \\ &= \Sigma\Pi\,\langle \eta(t_i)\eta(t_j)\rangle && (n \text{ even}), \end{aligned} \tag{2.72}$$

as well as white, i.e., $\langle \eta(t)\eta(t')\rangle = 2\varepsilon\delta(t-t')$. The angle brackets here stand for statistical average and the symbol $\Sigma\Pi$, as in Ref. [152], means that we choose $n/2$ pairs (t_i, t_j), compose the averages of every pair $(n/2)\langle\eta(t_i)\eta(t_j)\rangle$, and sum over all different $(n-1)!!$ choices. In this case for $W(t)$ we have

$$\begin{aligned} &\langle W(t)\rangle = 0, \qquad \langle W(t_1)W(t_2)\rangle = 2\varepsilon\min(t_1,t_2), \\ &\langle \exp\left[cW(t)\right]\rangle = \exp\left[c^2\left\langle W^2(t)\right\rangle/2\right], \end{aligned} \tag{2.73}$$

where c =const. Now let us consider the problem in a more general setup. Let the functional of $U(t,\xi)$ be given by

$$F\left[U(t,\xi)\right] = F\left[U(t,\xi), \partial_\xi U(t,\xi), \ldots\right] = F(t,\xi). \tag{2.74}$$

Consider the Fourier transform

$$F(t,\xi) = \frac{1}{2\pi}\int_{-\infty}^{\infty} \mathrm{d}k \tilde{F}(t,k)\mathrm{e}^{\mathrm{i}kx} \quad \text{and} \quad \tilde{F}(t,k) = \int_{-\infty}^{\infty} \mathrm{d}x F(t,\xi)\mathrm{e}^{-\mathrm{i}kx},$$

and obtain (taking into account fluctuations of ξ) [151]

$$\tilde{F}(t,k) = \tilde{F}_0(t,k)\mathrm{e}^{\mathrm{i}km(t)}, \tag{2.75}$$

where

$$\tilde{F}_0(t,k) = \tilde{F}(t,k)\Big|_{m=0} = \int_{-\infty}^{\infty} \mathrm{d}x F(t,x)\mathrm{e}^{-\mathrm{i}kx}. \tag{2.76}$$

Taking the statistical average we have

$$\left\langle \tilde{F}(t,k)\right\rangle = \tilde{F}_0(t,k)\tilde{G}(k). \tag{2.77}$$

Here, for $\tilde{G}(k) = \langle \exp\left[\mathrm{i}km(t)\right]\rangle$ we can write the following using (2.72) and (2.73):

$$\tilde{G}(k) = \exp\left[-k^2\left\langle m^2(t)\right\rangle/2\right], \quad \left\langle m^2(t)\right\rangle = 24\varepsilon t^3, \quad t > 0. \tag{2.78}$$

Equation (2.77) shows that the averaged spectrum (2.75) of the functional $F[U(t,\xi)]$ is the product of $\tilde{F}(t,k)$ without noise (2.76) and the Gaussian distribution (2.78). Thus we have

$$\langle F\left[U(t,\xi)\right]\rangle = \langle F(t,\xi)\rangle = \frac{1}{2\pi}\int_{-\infty}^{\infty} \mathrm{d}k\tilde{F}_0(t,k)\tilde{G}(k)\mathrm{e}^{\mathrm{i}kx}. \tag{2.79}$$

Using the convolution theorem we can also obtain from solution (2.79) [152]

$$\langle F\left[U(t,\xi)\right]\rangle = \int_{-\infty}^{\infty} \mathrm{d}s F\left[U(t,s)\right] G(x-s), \tag{2.80}$$

where

$$G(s) = \frac{1}{2\pi}\int_{-\infty}^{\infty} \mathrm{d}k\tilde{G}(k)\mathrm{e}^{\mathrm{i}ks} = \left[2\pi\left\langle m^2(t)\right\rangle\right]^{-1/2}\exp\left[-s^2/2\left\langle m^2(t)\right\rangle\right].$$

The newly obtained expressions (2.79) and (2.80) can now be used for investigation of the dynamic behavior of the solitons of Eq. (2.69).

KdV Soliton Dynamics. As an example, consider the case where $F[U(t,\xi)]$ (2.74) is a functional of the one-soliton solution of the *stochastic KdV equation* (2.69). Note that calculating $\tilde{F}_0(t,k)$ and $\tilde{G}(k)$ by using expressions (2.76) and (2.78), one can easily find $\langle \tilde{F}(t,k)\rangle$ and then obtain $\langle u(t,x)\rangle$. Here we use, however, a more visible and simple technique proposed by Wadati [152]. Consider the solution

$$U(t,x) = -2\nu^2 \cosh^{-2}\left[\nu(x-x_0) - 4\nu^3 t\right], \tag{2.81}$$

where ν =const is an an eigenvalue corresponding to the soliton (see Sect. 1.2). Considering the change $\xi = x + m(t)$ as well as (2.71), write the solution as

$$u(t,x) = W(t) - 2\nu^2 \cosh^{-2}\left[\nu(x-x_0) - 4\nu^3 t + 6\nu \int_0^t W(t')\mathrm{d}t'\right].$$

Furthermore, taking the statistical average and using (2.72) and (2.73), we obtain

$$\begin{aligned}\langle u(t,x)\rangle &= -2\nu^2 \left\langle \cosh^{-2}\left[\nu(x-x_0) - 4\nu^3 t + 6\nu \int_0^t W(t')\mathrm{d}t'\right]\right\rangle \\ &= 8\nu^2 \sum_{n=1}^{\infty} (-1)^n n \left\langle \exp\left[2n\left(\nu(x-x_0) - 4\nu^3 t + 6\nu \int_0^t W(t')\mathrm{d}t'\right)\right]\right\rangle .\end{aligned}$$

The second and third relations of (2.73) give us [152]

$$\begin{aligned}&\left\langle \exp\left[12nk \int_0^t W(t')\mathrm{d}t'\right]\right\rangle \\ &\qquad = \exp\left[\frac{1}{2}(12nk)^2 \int_0^t \mathrm{d}t_1 \int_0^t \mathrm{d}t_2 \left\langle W(t_1)W(t_2)\right\rangle\right] \\ &\qquad = \exp\left(48n^2\nu^2\varepsilon t^3\right), \qquad t > 0.\end{aligned}$$

Thus we have

$$\langle u(t,x)\rangle = 8\nu^2 \sum_{n=1}^{\infty} (-1)^n n e^{na+n^2 b}, \tag{2.82}$$

where

$$a = 2\left[\nu(x-x_0) - 4\nu^3 t\right] \quad \text{and} \quad b = 48\nu^2\varepsilon t^3. \tag{2.83}$$

We note that expression (2.82) has been obtained under the assumption that the "noise" is Gaussian. For a non-white type of the "noise", the expression for b is more complex. Thus we obtain from (2.82) [152]

$$\partial_b \langle u(t,x)\rangle = \partial_a^2 \langle u(t,x)\rangle \quad \text{and} \quad \langle u(t,x)\rangle|_{b=0} = -2\nu^2 \cosh^{-2}(a/2). \tag{2.84}$$

It follows from the first equality of (2.84) that the dynamic behavior of the soliton of the stochastic KdV equation is described by the diffusion equation where b plays the role of time and a plays the role of the space coordinate.

We note that (2.82) can be written in the form of a Fourier transform, and then the solution of (2.84) is given by [152]

$$\langle u(t,x)\rangle = -8\nu^2 \int_{-\infty}^{\infty} \frac{\mathrm{d}k}{2\pi} \frac{\pi k}{\sinh \pi k} \mathrm{e}^{-bk^2+iak} . \tag{2.85}$$

Expression (2.85) provides a spectral representation of the solution of the stochastic KdV equation with the stochastic Gaussian fluctuations of the wave field; in this case, the Fourier transform of the statistical average $\langle u(t,x)\rangle$ is the product of a pure soliton part $\propto -8\nu^2\pi k/\sinh(\pi k)$ and the diffusion part $\propto \exp(-bk^2)$. Using the convolution theorem, we can rewrite the solution (2.85) as [152]

$$\langle u(t,x)\rangle = -\frac{\nu^2}{\sqrt{\pi b}} \int_{-\infty}^{\infty} \mathrm{d}s \cosh^{-2}(s/2) \exp\left[-(a-s)^2/4b\right] . \tag{2.86}$$

On the basis of the result (2.85), consider now the dynamic behavior of the soliton in the presence of the Gaussian "noise." We obtain [152]:
(a) for $b \equiv 48\nu^2\varepsilon t^3 < 1$,

$$\langle u(t,x)\rangle = -2\nu^2 \sum_{n=0}^{\infty} \frac{1}{n!} b^n \frac{\partial^{2n}}{\partial a^{2n}} \cosh^{-2}(a/2); \tag{2.87}$$

(b) for $b > 1$,

$$\langle u(t,x)\rangle = -\frac{4\nu^2}{\sqrt{\pi}} \left(1 + \sum_{n=1}^{\infty} \frac{(2^{2n}-2)B_n\pi^{2n}}{(2n)!} \frac{\partial^n}{\partial b^n}\right) \frac{1}{\sqrt{b}} \mathrm{e}^{-a^2/4b} , \tag{2.88}$$

where B_n is the Bernulli number. Expressions (2.87) and (2.88) show that when $t = 0$, $\langle u(t,x)\rangle$ is defined by the right-hand side of (2.81) with $t = 0$, and for $t \to \infty$ we have

$$\langle u(t,x)\rangle = -\frac{\nu}{\sqrt{3\pi\varepsilon}} t^{-3/2} \exp\left[-\frac{(x-x_0-4\nu^2 t)^2}{48\varepsilon t^3}\right] .$$

As we see from the last expression, the soliton is deformed in its evolution as a result of the influence of the external noise, and its characteristic scale along the direction of propagation as well as its amplitude are asymptotically changed, respectively, as $t^{3/2}$ and $t^{-3/2}$. Note that this is not the consequence of the diffusion effects because the area occupied by the soliton is invariant, i.e., the integral $\int_{-\infty}^{\infty}\langle u(t,x)\rangle \mathrm{d}x$ is conserved; this is easy to check by the direct calculation, namely,

$$\int_{-\infty}^{\infty} \langle u(t,x)\rangle \,\mathrm{d}x = -8\nu^2 \int_{-\infty}^{\infty} \mathrm{d}x \int_{-\infty}^{\infty} \frac{\mathrm{d}k}{2\pi}\frac{\pi k}{\sinh \pi k}\mathrm{e}^{-bk^2+\mathrm{i}ak}$$

$$= -8\nu^2 \int_{-\infty}^{\infty} \frac{\mathrm{d}k}{2\pi}\frac{\pi k}{\sinh \pi k}\mathrm{e}^{-bk^2} 2\pi\delta(2\nu k) = -4\nu .$$

To conclude, we note that in this section we considered the influence of *stochastic fluctuations of the wave field* (the external white "noise") $\eta(t)$ on the structure and evolution of the KdV soliton. In a more general case, the KdV equation can be written as [152]

$$\partial_t u - 6u\partial_x u + \partial_x^3 u + \gamma u - \eta(t,x) = 0, \tag{2.89}$$

but the above analysis remains valid when the soliton's characteristic time scale is small, $t_s \ll 1/\gamma$, as well as the soliton's characteristic size, $l_s \ll l_n$ (l_n is the coherence length of the noise). In the case when $l_s \sim l_n$, the *Galilean transform* (2.71) is no longer valid, and it is necessary to generalize the IST method as it was done for the KdV equation in Refs. [95,96], for example (see also Sect. 1.2). Note that it is not possible to obtain the exact (analytical) solutions of (2.89) in this case, and the only way to investigate the dynamics of solutions of (2.89) is a numerical integration which can be successfully done with the use of methods considered in Sect. 1.3. Finally, we note that for the two-dimensional analog of the KdV equation – the KP equation – the corresponding results of the study of the soliton dynamics in a medium with low-frequency stochastic fluctuations of the wave field are significantly different from those obtained above for the one-dimensional case (see Sect. 4.4.4 and Ref. [83] for details).

2.2.3 Qualitative Analysis and Asymptotics of Solutions of Generalized KdV-Class Equations

Since a numerical simulation cannot cover the wide range of all possible cases for the generalized KdV equation and, even more, to classify its solutions, consider here the classes of possible solutions of the *generalized KdV equation*,

$$\partial_t u + \alpha u\partial_x u - \mu\partial_x^2 u + \beta\partial_x^3 u + \delta\partial_x^4 u + \gamma\partial_x^5 u = 0, \tag{2.90}$$

as well as the asymptotics of the solutions along the direction of the wave propagation. We employ here the methods of qualitative analysis, usually used in the theory of dynamic systems, and the asymptotic analysis of the structure of solutions for $|x| \to \infty$.

Changing $u \to (6/\alpha)u$, we can rewrite (2.90) as

$$\partial_t u + 6u\partial_x u - \mu\partial_x^2 u + \beta\partial_x^3 u + \delta\partial_x^4 u + \gamma\partial_x^5 u = 0. \tag{2.91}$$

Taking into account the more general case, we extend the class of equations (2.91) by introducing the arbitrary positive exponent p of the nonlinear term

$$\partial_t u + 6u^p \partial_x u - \mu \partial_x^2 u + \beta \partial_x^3 u + \delta \partial_x^4 u + \gamma \partial_x^5 u = 0. \tag{2.92}$$

Equation (2.92) for $\mu = \delta = \gamma = 0$ is the usual KdV equation when $p = 1$, and it is the *modified KdV equation* when $p = 2$. The asymptotics of (2.92) with $\mu = \delta = 0$, and $p = 1$ was first investigated in Refs. [84,113]; it was shown that depending on the sign of β and γ, solitary solutions with monotonous or oscillating asymptotics can take place. The full equation (2.92), including, in addition to the *higher order dispersion* correction, also terms describing dissipation, $\mu \neq 0$, an instability, $\delta \neq 0$, and an arbitrary exponent of the nonlinearity, does not represent an integrable system (i.e., the known analytical methods such as the IST method, are not applicable to this equation). Thus, (2.92) was investigated by the asymptotic and qualitative analyses [112], and, as a result, the sufficiently full classification of its solutions was constructed. In this section we mainly follow the ideas and technique of Ref. [112].

We note that from the physical point of view, the cases $p = 1, 2$ in (2.92) are the most interesting, and applications for $p > 2$ are presently unknown. However, since equations of the family (2.92) with an arbitrary integer $p > 0$ demonstrate, to a considerable extent, similar mathematical properties, we use here a general approach elucidating, apart from other, the dependence of the characteristics of the solutions on the nonlinearity's exponent.

Basic Principles. Transforming to $\xi = x - Vt$ and integrating (2.92) in ξ, we obtain

$$-Vu + \frac{6}{p+1} u^{p+1} - \mu \partial_\xi u + \beta \partial_\xi^2 u. + \delta \partial_\xi^3 u + \gamma \partial_\xi^4 u = 0. \tag{2.93}$$

Taking into account that $\mu > 0$, $\delta > 0$ (according to the physical sense of the terms with the first and third space derivatives describing dissipation and instability), and assuming without loss of generality that $\gamma > 0$ and $\beta = \pm 1$, after the change $u = Vw$, $\xi \to |V|^{-1/4}\xi$, we convert (2.93) to

$$\begin{aligned} &\mathrm{sgn}(V)\gamma \partial_\xi^4 w + \mathrm{sgn}(V)\delta \left|V\right|^{-1/4} \partial_\xi^3 w + \mathrm{sgn}(V)\beta \left|V\right|^{-1/2} \partial_\xi^2 w \\ &\quad -\mathrm{sgn}(V)\mu \left|V\right|^{-3/4} \partial_\xi w - w + \frac{6s}{p+1} \left|V\right|^{p-1} w^{p+1} = 0, \end{aligned} \tag{2.94}$$

where

$$s = \begin{cases} \mathrm{sgn}(V) & \text{for even } p\,, \\ 1 & \text{for odd } p\,. \end{cases}$$

Depending on the signs of V and β in (2.94), the following four cases can be considered:

(a) $V > 0$, $\beta = 1$:

$$\gamma\partial_\xi^4 w + \delta |V|^{-1/4} \partial_\xi^3 w + |V|^{-1/2} \partial_\xi^2 w - \mu |V|^{-3/4} \partial_\xi w - w + \frac{6s}{p+1} |V|^{p-1} w^{p+1} = 0 \tag{2.95}$$

(b) $V > 0$, $\beta = -1$:

$$\gamma\partial_\xi^4 w + \delta |V|^{-1/4} \partial_\xi^3 w - |V|^{-1/2} \partial_\xi^2 w - \mu |V|^{-3/4} \partial_\xi w - w + \frac{6s}{p+1} |V|^{p-1} w^{p+1} = 0 \tag{2.96}$$

(c) $V < 0$, $\beta = 1$:

$$-\gamma\partial_\xi^4 w - \delta |V|^{-1/4} \partial_\xi^3 w - |V|^{-1/2} \partial_\xi^2 w + \mu |V|^{-3/4} \partial_\xi w - w + \frac{6s}{p+1} |V|^{p-1} w^{p+1} = 0 \tag{2.97}$$

(d) $V < 0$, $\beta = -1$:

$$-\gamma\partial_\xi^4 w - \delta |V|^{-1/4} \partial_\xi^3 w + |V|^{-1/2} \partial_\xi^2 w + \mu |V|^{-3/4} \partial_\xi w - w + \frac{6s}{p+1} |V|^{p-1} w^{p+1} = 0 \tag{2.98}$$

However, as we see from (2.92), the velocity of the wave, V, depends on the equation's coefficients and it is restricted by

$$V \begin{cases} < V_{\min}^{\mathrm{ph}} = -1/4\gamma - 2\mu\,(\mu/3\delta)^{1/2}/3, & \beta = 1, \\ > V_{\max}^{\mathrm{ph}} = -1/4\gamma - 2\mu\,(\mu/3\delta)^{1/2}/3, & \beta = -1. \end{cases} \tag{2.99}$$

The right-hand sides of inequalities (2.99) for $\mu = \delta = 0$ correspond to results obtained in Ref. [113], and comparing these relations with expressions (2.95) and (2.98) leads to contradictions in the cases (a) and (d) outside the region

$$[(a-1)/4\mu]^2 < 2\gamma < [(a+1)/4\mu]^2, \quad \text{where} \quad a = \sqrt{1+16\mu\delta}. \tag{2.100}$$

Besides, it is necessary to note that condition (2.100) makes sense only for $\delta \neq 0$, i.e., when V^{ph}, in principle, can be non-negative in the second inequality (2.99).[8] In these cases (2.95) and (2.98) do not have solutions of the soliton type even if $\mu = \delta = 0$. Therefore, we limit ourselves below by considering the cases (b) and (c), as well as (d) within the region defined by (2.100). Since it is extremely difficult to study the full equations of type (2.95)–(2.98), we investigate below the role of different terms and groups of terms of these equations using methods of the qualitative and asymptotic analyses separately.

[8] From the necessary condition of the existence of the extremum V^{ph} for $\beta = -1$ follows that $(2/3)\mu = \mu - \delta/2\gamma$.

Qualitative Analysis and Asymptotics of Solutions. First, we note that every equation of the set (2.96)–(2.98) is equivalent to the set of the first order ordinary differential equations

$$\left\{\begin{array}{l} x_1 = \dot{w}, \qquad\qquad x_2 = \dot{x}_1, \qquad\qquad x_3 = \dot{x}_2, \\ \begin{pmatrix} + \\ - \\ - \end{pmatrix} \gamma \dot{x}_3 = \begin{pmatrix} - \\ + \\ + \end{pmatrix} \delta C x_3 \begin{pmatrix} + \\ + \\ - \end{pmatrix} C^2 x_2 \begin{pmatrix} + \\ - \\ - \end{pmatrix} \mu C^3 x_1 \\ \qquad\qquad +w - 6sC^{4(1-p)} w^{p+1}/(p+1), \end{array}\right. \tag{2.101}$$

where the dots stand for the ξ-derivatives, and the sign in the brackets corresponds to the above cases (b), (c), and (d), respectively; furthermore, $C = |V|^{-1/4}$. Solutions of equations (2.101) are stable if there exist *singular trajectories* of the *imaging point* in the phase space (w, x_1, x_2, x_3) of the set (2.101). Each trajectory of its kind is related to the state of equilibrium near the maximum of the soliton-like solution and at the boundaries $|\xi| = \infty$. Assuming as the boundary conditions that

$$w = \partial_\xi^n w = 0 \quad \text{and} \quad n = 1, 2, 3 \qquad\qquad \text{for } |\xi| \to \infty, \tag{2.102}$$

we can find from (2.101) the number as well as the coordinates of the *singular points*

$$w_1 = 0, \qquad\qquad w_j = \sqrt[p]{\frac{p+1}{6sC^{4(1-p)}}}, \tag{2.103}$$

where the points $w_1 = 0$ and w_j correspond, respectively, to $|\xi| = \infty$ and the bending points of the function $u(\xi)$; $j = 2$ for the odd and $j = 2, 3$ for the even p, in the last case $w_2 = -w_3$. Considering only real roots of (2.103), we immediately conclude (using the *Sturm's theorem*) that for the odd p there are two singular points, and for any even p there are three singular points. The distance between the singular points defines the amplitude of the soliton-like solution of (2.93). The nonlinearity's exponent p defines the character of the dependence $V = f(u)$: for $p > 1$ this dependence becomes nonlinear (Fig. 2.9) unlike the known linear one for $p = 1$ (e.g., in the case of KdV equation). As we see in Fig. 2.9, for the even p the solutions of (2.93) can have the positive as well as the negative polarity for any sign of V.

To investigate the types of the *singular points*, it is necessary to linearize the set (2.101) in the neighborhood of every point. Using the Taylor's expansion, we obtain from (2.101) [112]:

1. For the singular point $w_1 = 0$ (this corresponds to $u_1 = 0$ in (2.93) taking into account the boundary conditions (2.102)),

$$\left\{\begin{array}{l} x_1 = \dot{w}, \qquad\qquad x_2 = \dot{x}_1, \qquad\qquad x_3 = \dot{x}_2, \\ \begin{pmatrix} + \\ - \\ - \end{pmatrix} \gamma \dot{x}_3 = \begin{pmatrix} - \\ + \\ + \end{pmatrix} \delta C x_3 \begin{pmatrix} + \\ + \\ - \end{pmatrix} C^2 x_2 \begin{pmatrix} + \\ - \\ - \end{pmatrix} \mu C^3 x_1 + w \end{array}\right. \tag{2.104}$$

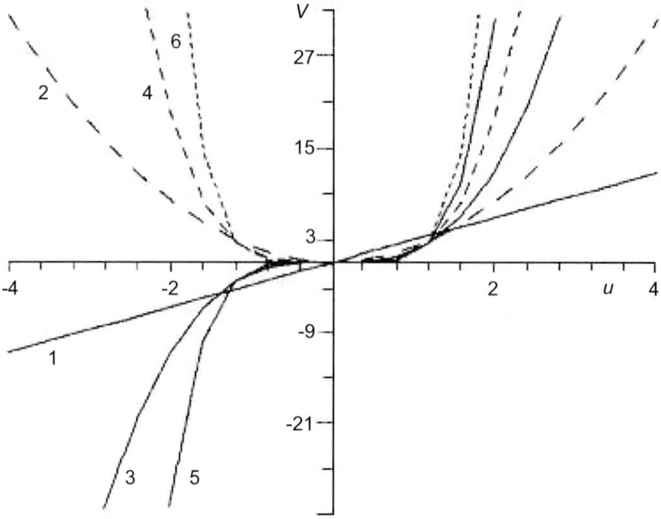

Fig. 2.9. Dependence $V = f(u)$ in Eq. (2.93) for different values of the nonlinearity exponent. Numbers on the curves correspond to $p = 1, 2, \ldots, 6$

2. For the singular point w_j (that corresponds to $u_j = \sqrt[p]{(p+1)V/6s}$ in (2.64))

$$\begin{cases} x_1 = \dot{w}, \qquad x_2 = \dot{x}_1, \qquad x_3 = \dot{x}_2, \\ \begin{pmatrix} + \\ - \\ - \end{pmatrix} \gamma \dot{x}_3 = \begin{pmatrix} - \\ + \\ + \end{pmatrix} \delta C x_3 \begin{pmatrix} + \\ + \\ - \end{pmatrix} C^2 x_2 \begin{pmatrix} + \\ - \\ - \end{pmatrix} \mu C^3 x_1 - pw \end{cases} \tag{2.105}$$

Since the sets (2.104)–(2.105) are essentially four-dimensional, we investigate them by expanding of the corresponding canonical systems into subsystems [153] (see Appendix 1). In this case, it is possible to consider the phase portraits of the linear sets (2.104)–(2.105) as projections of the singular points and trajectories onto two planes. For simplicity we consider separately the cases when $\mu = \delta = 0$ (the family of conservative equations) and $\beta = \gamma = 0$ (dissipative equations with instability).

Conservative Equations, $\mu = \delta = 0$. For the singular point $w_1 = 0$ we obtain that the eigenvalues of the matrices of subsets to the set (2.104) (see Appendix 1) corresponding to the phase planes $P1(w, x_1)$ and $P2(x_2, x_3)$ are defined by [112]

$$\lambda_{1,2}^{(P1,P2)} = \pm(2\gamma)^{-1/2} \left[\begin{pmatrix} + \\ - \end{pmatrix} C^2 \pm \sqrt{C^4 \begin{pmatrix} + \\ - \end{pmatrix} 4\gamma} \right]^{1/2} . \tag{2.106}$$

In the case (b), the characteristic roots λ_1 and λ_2 are real on the phase plane $P1$ and pure imaginary on the phase plane $P2$, besides, $\lambda_1 = -\lambda_2$ on both planes. In the case (c), taking into account conditions (2.99), the roots λ_1 and λ_2 are complex, with the positive and negative real parts on the planes $P1$ and $P2$, respectively, and $\lambda_1 = \lambda_2^*$. In the case (d), taking into account the second condition of (2.99) for $\mu = \delta = 0$, all four roots are real and $\lambda_1 = -\lambda_2$ on both planes. Therefore, the singular points $w_1 = 0$ of three types exist in the phase space, namely: the *saddle–center point*, the *stable focus–unstable focus point*, and the *saddle–saddle point* in the cases (b), (c), and (d), respectively.

Considering the matrix of subsets to the analogous set (2.105), we obtain eigenvalues for the singular points w_j defined by (2.103) in the cases (b), (c), and (d) for the subsets corresponding to the projections onto the phase spaces $P1(w, x_1)$ and $P2(x_2, x_3)$ [112]:

$$\lambda_{1,2}^{(P1,P2)} = \pm(2\gamma)^{-1/2}\left[\begin{pmatrix} + \\ - \end{pmatrix} C^2 \pm \sqrt{C^4 \begin{pmatrix} - \\ + \end{pmatrix} 4\gamma p}\right]^{1/2} . \tag{2.107}$$

We see from (2.107) that the character of the singular point depends on the nonlinearity exponent p defining the wave velocity $V = 6su^p/(p+1)$. Nevertheless, the conditions (2.99) and (2.100) remain valid in these cases as well. Analysis of (2.107) enables us to conclude the following. In the case (b), taking into account the second condition of (2.99), the eigenvalues λ_1 and λ_2 are complex (moreover, $\lambda_1 = \lambda_2^*$), with the positive real parts on the plane $P1$ and the negative real parts on the plane $P2$. In the case (c), λ_1 and λ_2 are real on the plane $P1$ and purely imaginary on the plane $P2$, and $\lambda_1 = -\lambda_2$ in both cases. In the case (d), the situation is analogous to that of the case (c). Therefore, only the singular points w_j of the type "stable focus–unstable focus" take place in the phase in the case (b), and of the type "saddle–center" takes place in the cases (c) and (d).

To investigate the global phase portraits including singular trajectories corresponding to the stable solutions of (2.101), studies included the use [112] of the *Bendixon–Dulac criteria* [154,155], as well as calculations of the first and the second *Lyapunov exponents* [156]. Omitting here cumbersome mathematical calculations, we note that in the case (b), as well as in the cases (c) and (d), the closed trajectories appear in the phase space. Expressions (2.106) and (2.107) enable us to obtain the parameters of these curves as well as their directions and, consequently, the angles with respect to the coordinate axes on both planes $P1$ and $P2$, and therefore to construct the global phase portraits. The examples of such phase portraits for the cases (b) and (c) for $p = 1, 2$ are shown in Figs. 2.10a,b and 2.11a,b.

Knowing the characteristic roots λ_1 and λ_2 (2.106) for the singular points $w_1 = 0$, accounting for (2.99) and the boundary conditions (2.102), we obtain the asymptotics of the solutions of (2.92) for the cases (b), (c), and (d) [113], namely:

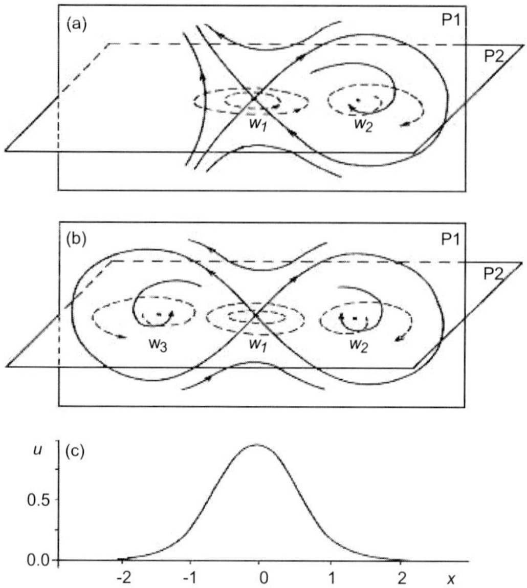

Fig. 2.10. The phase portrait of the solutions of Eq. (2.96) with $\mu = \delta = 0$ for $p = 1$ (**a**) and $p = 2$ (**b**), and the numerical solution of Eq. (2.92) with $\mu = \delta = 0$, $\gamma = 1$, $\beta = -1$, for $p = 1$ (**c**). The solid and dashed lines correspond to the phase trajectories in the planes $P1$ and $P2$, respectively

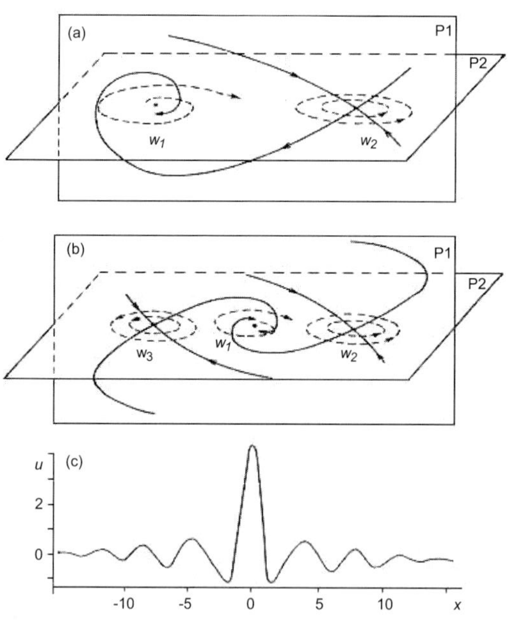

Fig. 2.11. The phase portrait of the solutions of (2.97) with $\mu = \delta = 0$ for $p = 1$ (**a**) and $p = 2$ (**b**), and the numerical solution of (2.92) with $\mu = \delta = 0$, $\gamma = 1$, $\beta = 3.16$, for $p = 1$ (**c**). Again, the solid and dashed lines correspond to the phase trajectories in the planes $P1$ and $P2$, respectively

1. For the cases (b) and (d),

$$w = A_1 \exp\left\{ (2\gamma)^{-1/2} \left[C^2 + \sqrt{C^4 \pm 4\gamma} \right]^{1/2} \xi \right\} \tag{2.108}$$

(the upper/lower sign corresponds to the case (b)/(d), respectively)

2. For the case (c),

$$w = A_2 \exp\left[\left(2C^{-1}\gamma^{1/2}\right)^{-1} \left(2C^{-2}\gamma^{1/2} - 1\right)^{1/2} \xi \right]$$
$$\times \cos\left[\left(2C^{-1}\gamma^{1/2}\right)^{-1} \left(2C^{-2}\gamma^{1/2} + 1\right)^{1/2} \xi + \Theta \right], \tag{2.109}$$

where A_1, A_2, and Θ are arbitrary constants. As we can see from expressions (2.108)–(2.109), solitons with both monotonic and oscillating asymptotics (depending on the sign of V and β) can serve as solutions of (2.92) with $\mu = \delta = 0$. (Note that for $\beta = 0$ and any $\gamma > 0$, the solution of (2.93) with $\mu = \delta = 0$ has the form $w = (A_1 + A_2\xi)\exp(\gamma^{-1/4}\xi)$ and it therefore also describes a soliton with the monotonous asymptotics [113].) The phase portraits shown in Fig. 2.10a,b correspond to the solitons with the monotonic asymptotics, and the phase portraits shown in Fig. 2.11a,b are for the solitons with the oscillating asymptotics. Figures 2.10c and 2.11c show the results of numerical integration of (2.92) with $\mu = \delta = 0$ for the initial condition $u = u_0 \exp(-x^2/l^2)$ that agrees with the results of the asymptotic analysis.

In conclusion, we consider a simple and illustrative case of the *linearized KdV equation* with the higher order dispersion correction, assuming in (2.92) that $p = 0$; this is possible far from the soliton humps. To simplify the analysis, we apply the scale transform $u \to u/6$ to (2.92) and take into account $\mu = \delta = 0$. Instead of the last equation of the set (2.104), we now have

$$\gamma \dot{x}_3 \begin{pmatrix} - \\ + \\ - \end{pmatrix} C^2 x_2 = 0.$$

Then, for the singular point $w_1 = 0$, the eigenvalues of the matrices of the subsets corresponding to the phase planes $P1$ and $P2$ are no longer defined by (2.106), but by the equalities

$$\lambda_{1,2}^{(P1)} = 0 \qquad \text{and} \qquad \lambda_{1,2}^{(P2)} = \pm C \left[\begin{pmatrix} + \\ - \\ - \end{pmatrix} \gamma \right]^{-1/2},$$

and, therefore, in the cases (b) and (d) the (singular) *saddle–center point* appears, and in the case (c) the *center–center point* appears.[9] The asymptotics of the solutions for $|\xi| \to \infty$ is given by:
1. In the cases (b) and (d),

$$w = A \exp\left(C\gamma^{-1/2}\xi\right) + B$$

2. In the case (c),

$$w = A \cos\left(C\gamma^{-1/2}\xi + \Theta\right) + B$$

Here, A, B, and Θ are arbitrary constants. We note that for the linearized KdV equation with $\beta = 0$ or $\gamma = 0$, the asymptotics at $|\xi| = \infty$ is a constant.

Dissipative Equations with Instability, $\beta = \gamma = 0$. Assume now that $\beta = \gamma = 0$ in the basic equations. Thus, instead of the sets (2.108)–(2.109) we obtain from (2.101) for the singular points $w_1 = 0$ and w_j the following equations:

$$\begin{cases} x_1 = \dot{w}, \qquad x_2 = \dot{x}_1, \\ \delta C \dot{x}_2 = \mu C^3 x_1 \begin{pmatrix} + \\ - \end{pmatrix} w, \end{cases} \tag{2.110}$$

and

$$\begin{cases} x_1 = \dot{w}, \qquad x_2 = \dot{x}_1, \\ \delta C \dot{x}_2 = \mu C^3 x_1 \begin{pmatrix} - \\ + \end{pmatrix} pw. \end{cases} \tag{2.111}$$

The sign in the brackets in these sets corresponds to $V > 0$ or $V < 0$, this is equivalent to the change $w_1 \leftrightarrow w_j$. Therefore we can investigate the sets only with one sign, e.g., the upper one corresponding to $V > 0$.

Since the sets (2.110)–(2.111) are three-dimensional, we can use in our study the above expansion (decomposition) method of the corresponding canonical systems. In both cases the expansion leads to a two-dimensional set and an equation (see Appendix 1). Such factorization enables us to obtain eigenvalues of the respective sets, taking into account correlation of the coefficients μ and δ and the velocity V [112]. We have:

1. When $\delta > (4/27)\mu^3 C^8$,

$$\lambda_1 = (2\delta C)^{-1/3} Q_1^+, \qquad \lambda_{2,3} = -(16\delta C)^{-1/3}\left[Q_1^+ \pm \mathrm{i}\sqrt{3}Q_1^-\right] \tag{2.112}$$

2. When $\delta = (4/27)\mu^3 C^8$,

$$\lambda_1 = (\delta C/4)^{-1/3}, \qquad \lambda_{2,3} = -(2\delta C)^{-1/3} \tag{2.113}$$

---

[9] Degeneration of the singular stable focus–unstable focus point corresponding to the eigenvalues (2.106) into the center–center point is the corollary of the transform $u \to u/6$ of the initial equation.

3. When $\delta < (4/27)\mu^3 C^8$,

$$\lambda_1 = (\delta C/4)^{-1/3}, \quad \lambda_{2,3} = -(2\delta C)^{-1/3}\left[\mathrm{Re}(Q^{\pm}) \mp \sqrt{3}\left|\mathrm{Im}(Q^{\pm})\right|\right], \tag{2.114}$$

where

$$Q_1^{\pm} = Q^+ \pm Q^-,$$

$$Q^{\pm} = \left[1 \pm \sqrt{1 - 4\mu^3 C^8/27\delta}\right]^{1/3}, \tag{2.115}$$

and $Q^{\pm}$ is real in the cases (1.) and (2.), and complex in the case (3.).

For the singular point w_j, these expressions are valid as well, if to change the sign of the eigenvalues $\lambda_{1,2,3}$ by the opposite one, and with the change $\delta \to \delta/p$ in (2.112)–(2.114) and the change $\delta \to \delta p^2$ in (2.115), .

It follows from (2.112)–(2.114) that the sets (2.110)–(2.111) have the saddle-type singular points with the coordinates w_1 and w_j for $\delta \leq (4/27)\mu^3 C^8$ and the singular *saddle–focus point* in the opposite case. Thus the considered state roughly corresponds to the equilibrium state of the three-dimensional system. In both cases, there is a two-dimensional *separatrix surface* with two isolated separatrices on both sides [112]. The node appears on the saddle-type separatrix surface, and the saddle appears on the saddle–focus separatrix surface. All other trajectories passing through a sufficiently small neighborhood of the saddle or the saddle–focus surfaces leave the neighborhood of the latter. However, this information about the local behavior of the solutions is not sufficient to construct the *global phase portrait.* Therefore, the fact should be invoked that for a nonlinear system, the directions of the separatrices of the corresponding linearizations give the directions of the nonlinear separatrices in the singular point. We can obtain these main directions from the linear transformation relating the linearized system with its canonical system (see Ref. [112] and Appendix 1).

Figures 2.12a,b and 2.13a,b show the general phase portraits for the cases $\delta \leq (4/27)\mu^3 C^8$ and $\delta > (4/27)\mu^3 C^8$, with $\beta = \gamma = 0$ and $p = 1, 2$, $V > 0$. It is clear that the phase portraits are the same for $p > 2$ as well, that follows from (2.112)–(2.115) with the above-mentioned changes. Using the eigenvalues λ_i ($i = 1, 2, 3$) we can obtain the asymptotics of the solutions of (2.92) with $\beta = \gamma = 0$ [112] from (2.112)–(2.114) for the singular points $w_1 = 0$ with the boundary conditions (2.102):

1. When $\delta > (4/27)\mu^3 C^8$,

$$\begin{aligned} w = A_1 \exp\left[(2\delta C)^{-1/3} Q_1^+ \xi\right] &+ \exp\left[-(16\delta C)^{-1/3} Q_1^+ \xi\right] \\ &\times \left\{A_2 \cos\left[\sqrt{3}(16\delta C)^{-1/3} Q_1^- \xi + \Theta_1\right]\right. \\ &\left.+ A_3 \sin\left[\sqrt{3}(16\delta C)^{-1/3} Q_1^- \xi + \Theta_2\right]\right\} \end{aligned} \tag{2.116}$$

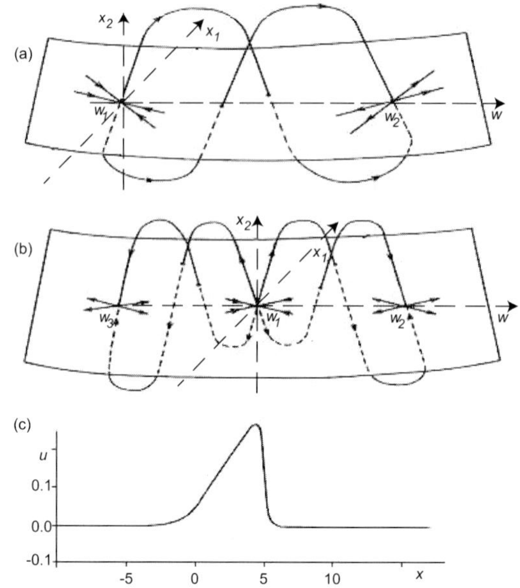

Fig. 2.12. The phase portrait of the solutions of (2.93) with $\beta = \gamma = 0$ and $V > 0$ for the case $\delta \leq (4/27)\mu^3 C^8$ for $p = 1$ (**a**) and $p = 2$ (**b**), and the numerical solution of Eq. (2.92) for $\mu = 0.1$, $\delta = 10^{-6}$, at $t = 3$ (**c**)

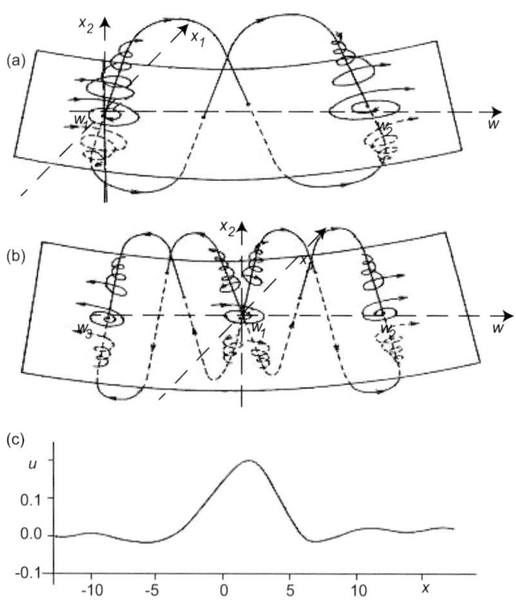

Fig. 2.13. The phase portrait of the solutions of (2.93) with $\beta = \gamma = 0$ and $V > 0$ for the case $\delta > (4/27)\mu^3 C^8$ for $p = 1$ (**a**) and $p = 2$ (**b**), and the numerical solution of (2.92) for $\mu = 0.01$, $\delta = 1$, at $t = 3$ (**c**)

2. When $\delta = (4/27)\mu^3 C^8$,

$$w = A_1 \exp\left[(\delta C/4)^{-1/3}\xi\right] + A_2(1 + A_3\xi)\exp\left[-(2\delta C)^{-1/3}\xi\right] \qquad (2.117)$$

3. When $\delta < (4/27)\mu^3 C^8$,

$$\begin{aligned} w = {} & A_1 \exp\left[(\delta C/4)^{-1/3}\mathrm{Re}(Q^\pm)\xi\right] \\ & + A_2 \exp\left\{-(2\delta C)^{-1/3}\xi\left[\mathrm{Re}(Q^\pm) - \sqrt{3}\left|\mathrm{Im}(Q^\pm)\right|\right]\right\} \\ & + A_3 \exp\left\{-(2\delta C)^{-1/3}\xi\left[\mathrm{Re}(Q^\pm) + \sqrt{3}\left|\mathrm{Im}(Q^\pm)\right|\right]\right\} \qquad (2.118) \end{aligned}$$

Here, A_1, A_2, A_3, Θ_1, and Θ_2 are arbitrary constants. It can be seen from these expressions that the solutions of (2.92) have oscillating asymptotics in case (1.) and the exponential one in cases (2.) and (3.). Numerical solutions of (2.92) for the initial condition $u = u_0 \exp(-x^2/l^2)$ are presented in Figs. 2.12c and 2.13c for cases (3.) and (1.), respectively.

Considering (analogously to the above case $\mu = \delta = 0$) the linearized equation with $\beta = \gamma = 0$, it is easy to obtain that the eigenvalues of the corresponding sets are given by

$$\lambda_1 = 0 \qquad \text{and} \qquad \lambda_{2,3} = \pm C(\mu/\delta)^{1/2}.$$

For $|\xi| = \infty$, they define the exponential asymptotics

$$w = A \exp\left[C(\mu/\delta)^{1/2}\xi\right].$$

In the special cases when $\mu = 0$ and $\delta = 0$, the asymptote is a constant.

Some Comments. To conclude, we note that here we have considered the special cases when $\mu = \delta = 0$ and $\beta = \gamma = 0$ in (2.92). For other values of the coefficients, one can observe more complicated wave structures being determined by the presence of all the considered factors in their totality. Numerical results obtained in Refs. [114,137] (see also the previous section) show that for $\beta, \mu, \delta \neq 0$, stable wave structures of the soliton type can also be formed with time in the presence of the Gaussian random fluctuations of the wave field for the harmonic initial conditions and for the initial conditions in the form of a solitary pulse. Moreover, the stable soliton structure can be formed also for $\gamma \neq 0$. An analytical study of such cases is quite complicated although for these cases one can also use the approach considered in this section. We note that the above results (see also [112]) can be very useful for solution studies and interpretation of multidimensional phase portraits of more complicated multidimensional model equations [83,113]. In particular, as we demonstrate below in Sect. 4.2, these results are useful for the study and classification of possible solutions of generalized KP-class equations.

2.3 Nonlinear Schrödinger and Zakharov Equations

In this section, we consider effects related to modulational processes [36], i.e., processes during which change of the wave envelope occurs as a result of the modulational instability, an instability of a wave with respect to its modulations. The final stage of this instability, in the one-dimensional case, leads to formation of envelope solitons. The canonic equation describing this type of processes is the nonlinear Schrödinger (NLS) equation. Its important generalization, including interactions via propagating lower-frequency perturbations, is called the Zakharov system of equations or Zakharov equations, with numerous applications in plasma physics and nonlinear optics. Thus we first, in Sect. 2.3.1, derive NLS equation for the simplest case of slow modulations of Langmuir wave in a plasma. The inverse scattering problem for NLS equation is briefly outlined in Sect. 2.3.2. Then in Sect. 2.3.3 we generalize NLS equation for faster modulations of Langmuir wave in a plasma, derive the Zakharov system of equations, and, in Sect. 2.3.4, demonstrate its solution in the form of Langmuir soliton. In the last segment, Sect. 2.3.5, we consider the interesting cases of near-sonic Langmuir solitons and study the influence of dissipative processes on their propagation.

2.3.1 Derivation of the NLS equation

To derive equations for the amplitude of the *wave envelope*, we assume that the density variations occur at very low frequencies in a medium. As an example, we consider the propagation of electrostatic Langmuir waves in a plasma. The characteristic time of the density inhomogeneity variations δn is supposed to be much larger than the period of the electron plasma oscillations. Furthermore, we consider only sufficiently small ($\delta n/n_0 \ll 1$) inhomogeneities.

Under these assumptions we introduce the slowly changing (compared to the wave period and wave length) quantity $\delta n(t, \mathbf{r})$ and call it as a density perturbation. Note that under assumption of a weak nonlinearity (when the ratio of collective wave energy to mean particle energy is significantly less than unity), this density perturbation is weak,

$$\frac{\delta n(t, \mathbf{r})}{n_0} \ll 1. \tag{2.119}$$

The starting equation is the dispersion equation for *Langmuir waves*,

$$\varepsilon_{\omega,\mathbf{k}}(n) E_{\omega,\mathbf{k}} = 0, \tag{2.120}$$

where $\varepsilon(n)$ is given by

$$\varepsilon_{\omega,\mathbf{k}}(n) = 1 - 3r_{De}^2 \mathbf{k}^2 - \frac{\omega_{pe}^2(n)}{\omega^2}. \tag{2.121}$$

Here, the electron plasma frequency $\omega_{pe}(n)$ is the function of the slowly-varying (weak) density inhomogeneity $\delta n \ll n_0$ ($n = n_0 + \delta n$ is the electron density):

$$\omega_{pe}^2(n) = \frac{4\pi n e^2}{m_e} = \frac{4\pi(n_0 + \delta n)e^2}{m_e} = \omega_{pe}^2 + \frac{\delta n}{n_0}\omega_{pe}^2 \qquad (2.122)$$

(we assume $\omega_{pe} = \sqrt{4\pi n_0 e^2/m_e}$ is the electron plasma frequency for the unperturbed plasma density). In accordance with (2.122), we rewrite (2.120) as

$$\varepsilon^{(0)}_{\omega,\mathbf{k}} E_{\omega,\mathbf{k}} = \frac{\delta n}{n_0}\frac{\omega_{pe}^2}{\omega^2} E_{\omega,\mathbf{k}}, \qquad (2.123)$$

where $\varepsilon^{(0)}$ is given by (2.121) with the plasma frequency ω_{pe} independent on δn. The linear part of (2.123) in the limit $r_{De}^2\mathbf{k}^2 \ll 1$ gives us the dispersion law for Langmuir waves

$$\omega_{\mathbf{k}} = \omega_{pe} + \frac{3}{2} r_{De}^2 \mathbf{k}^2. \qquad (2.124)$$

Furthermore, we transform (2.123) to the equation for the envelope amplitude given by

$$\mathbf{E} = \mathrm{Re}\mathbf{E}(t, \mathbf{r})\mathrm{e}^{-\mathrm{i}\omega_{pe} t}. \qquad (2.125)$$

Thus in the one-dimensional case we obtain

$$\left(\frac{\mathrm{i}}{\omega_{pe}}\partial_t + \frac{3r_{De}^2}{2}\partial_x^2\right) E(t,x) = \frac{1}{2}\frac{\delta n(t,x)}{n_0} E(t,x). \qquad (2.126)$$

The low-frequency motion is assumed to be quasi-neutral,[10] i.e.,

$$\delta n_e = \delta n_i \equiv \delta n. \qquad (2.127)$$

The equation for the low-frequency density variations can be obtained from the hydrodynamic continuity equation,

$$\partial_t \delta n = -n_0 \boldsymbol{\nabla} \cdot \mathbf{v}_i, \qquad (2.128)$$

and the momentum equation for plasma ions,

$$m_i \partial_t \mathbf{v}_i + \frac{T_e}{n_0}\boldsymbol{\nabla}\delta n = \mathbf{F}^{\mathrm{pond}} \equiv e\mathbf{E}^{\mathrm{pond}}. \qquad (2.129)$$

In the literature, the force on the right-hand side of equation (2.129) is referred to as the *ponderomotive force* or the striction force (see Refs. [36,157], for example). Its appearance is due to the ponderomotive (striction) field,

---

[10] We emphasize that assumption (2.127) is important since (2.123) contains the low-frequency variation of the electron density while (2.128) and (2.129) contain the variations of the ion density. We note also that we assume the density variations to be small, which allows us to neglect them on the right-hand side of (2.128).

$\mathbf{E}^{\text{pond}}$, which acts on the plasma electrons. The ponderomotive force produces charge separation when the electrons are pushed out by the high-frequency Langmuir field. The field $\mathbf{E}^{\text{pond}}$ can be determined from the electron momentum equation averaged over the high frequency ω_{pe}:

$$\langle \partial_t \mathbf{v}_e + (\mathbf{v}_e \cdot \boldsymbol{\nabla})\, \mathbf{v}_e \rangle = -\frac{e}{m_e} \mathbf{E}^{\text{pond}}. \tag{2.130}$$

In the first approximation we can use

$$\mathbf{v}_{e;\omega,\mathbf{k}} \approx \frac{\mathrm{i}e}{m_e \omega_{pe}} \mathbf{E}_{\omega,\mathbf{k}}, \tag{2.131}$$

where the high-frequency field $\mathbf{E}$ is given by expression (2.27). Substituting (2.131) into (2.130) and averaging over the high frequency we obtain

$$\mathbf{E}^{\text{pond}} = -\frac{e}{2m_e \omega_{pe}^2} \boldsymbol{\nabla} \left|\mathbf{E}(t, \mathbf{r})\right|^2. \tag{2.132}$$

Taking into account (2.132) we find,

$$\frac{\delta n}{n_0} = -\frac{|\mathbf{E}(t, \mathbf{r})|^2}{8\pi n_0 T_e} \tag{2.133}$$

for the slow ($|\partial_t|^2 \ll c_s^2 |\partial_x|^2$) density perturbations.

Thus we obtain the *nonlinear Schrödinger equation* (*NLS equation*) which in the one-dimensional case is given by

$$\left(\frac{\mathrm{i}}{\omega_{pe}} \partial_t + \frac{3r_{De}^2}{2} \partial_x^2 \right) E(t, x) = -\frac{|E(t, x)|^2}{16\pi n_0 T_e} E(t, x). \tag{2.134}$$

In the dimensionless units, e.g., $t \to \omega_{pe} t$, $x \to (2/3)^{1/2} r_{De} x$, and $u \to E/(16\pi n_0 T_e)^{1/2}$, the NLS equation can be written as

$$\mathrm{i}\partial_t u + \partial_x^2 u + |u|^2 u = 0. \tag{2.135}$$

Note that the above elementary derivation of the equations for the wave envelope and the density perturbations has both advantages and shortcomings. The particular advantage is in the clear physical meaning of the quantity δn. One has to take into account that this simple physical meaning can be lost in the next approximation (or when taking into account the *nonlinear Landau damping*) [36]. This is especially important for numerical simulations and computer experiments where one usually notes plasma density variations, and compares them with the solutions of the equations reasoning from the correspondence between δn and the density inhomogeneity variations. In fact, such a correspondence takes place only under many assumptions and simplifications performed in the exact equations (of course, this comment concerns only the case of weak nonlinearities; the computer simulations can be, in principle, performed for stronger nonlinearities as well).

2.3.2 IST for the NLS Equation. NLS Solitons

Taking into account both the modulationally-stable and -unstable cases, when the signs of the dispersive and nonlinear terms are the opposite and the same, respectively,[11] we perform in (2.135) the change $u \to s^{1/2}u$, and rewrite the NLS equation as

$$\mathrm{i}\partial_t u + \partial_x^2 u + s|u|^2 u = 0. \tag{2.136}$$

Thus for $s < 0$ we have the *modulationally-stable case* and for $s > 0$ the *modulationally-unstable case*. Note that the soliton solutions of (2.136) are possible only in the case $s > 0$, when as a result of the effect of the *modulational instability* [36], the wave packet transforms into a series of nonlinear solitary waves, each in the form of the *NLS soliton*. This conclusion was proved by the analysis of Zakharov and Shabat [158]. Following Lax [159], they demonstrated the application of the *IST method*, by analogy with the KdV equation (see above Sect. 1.2), to the NLS equation (2.136).

Note that this method is applicable for any equation $\partial_t u = \hat{\mathrm{S}}u$ which can be written as the *Lax representation*

$$\partial_t \hat{\mathrm{L}} = \mathrm{i}\left[\hat{\mathrm{L}}, \hat{\mathrm{A}}\right] = \mathrm{i}(\hat{\mathrm{L}}\hat{\mathrm{A}} - \hat{\mathrm{A}}\hat{\mathrm{L}}) \tag{2.137}$$

(see Sect. 1.2), where $\hat{\mathrm{L}}$ and $\hat{\mathrm{A}}$ are the linear differential (generally speaking, matrix) operators with the coefficients containing the function $u(t, x)$. Thus the main problem is to find such a representation.

The generalized *Zakharov–Shabat eigenvalue problem* is formulated as

$$\begin{aligned} \partial_x \psi_1 + \mathrm{i}\lambda\psi_1 &= u\psi_2, \\ \partial_x \psi_2 - \mathrm{i}\lambda\psi_2 &= v\psi_1, \end{aligned} \tag{2.138}$$

where ψ_1 and ψ_2 are the eigenfunctions of some $\hat{\mathrm{L}}$ operator. In the case of the NLS equation, (2.136) is the integrability condition of the set (2.138), and the sign of s determines the character of the relation between the functions u and v, namely, $v = -su^*$. We rewrite the eigenvalue problem (2.138) via the $\hat{\mathrm{L}}$ operator as

$$\hat{\mathrm{L}}\psi = \lambda\psi \tag{2.139}$$

and take the time derivative of (2.139)

$$\begin{aligned} \mathrm{i}\psi d_t\lambda + \mathrm{i}\lambda\partial_t\psi &= \mathrm{i}\hat{\mathrm{L}}\partial_t\psi + \mathrm{i}(\partial_t\hat{\mathrm{L}})\psi \\ &= \mathrm{i}\hat{\mathrm{L}}\partial_t\psi - (\hat{\mathrm{L}}\hat{\mathrm{A}} - \hat{\mathrm{A}}\hat{\mathrm{L}})\psi \\ &= \hat{\mathrm{L}}(\mathrm{i}\partial_t\psi - \hat{\mathrm{A}}\psi) + \lambda\hat{\mathrm{A}}\psi. \end{aligned}$$

It then follows that

[11] For the opposite signs of the dispersion and nonlinear terms, the system is hyperbolic and the periodic wave packets are unstable with respect to their modulations; for the same signs, the system is parabolic and the wave packets are modulationally stable [2] (see also Sect. 2.4).

$$\mathrm{i}\psi \mathrm{d}_t\lambda = (\hat{\mathrm{L}} - \lambda)(\mathrm{i}\partial_t\psi - \hat{\mathrm{A}}\psi).$$

If the function ψ at the initial moment satisfies (2.139) and its time evolution is described by the equation

$$\mathrm{i}\partial_t\psi = \hat{\mathrm{A}}\psi, \tag{2.140}$$

then it also satisfies (2.139) for $t > 0$ with the same value of λ. Equation (2.139) relates the function $u(t,x)$ to the scattering problem, and (2.140) gives the dependence of the scattering data on time. Note that (2.137) is the condition of joint integrability of the equations (2.139) and (2.140).

The main problem now is to construct matrix operators $\hat{\mathrm{L}}$ and $\hat{\mathrm{A}}$ so that they satisfy (2.137). Zakharov and Shabat have demonstrated [158] that these operators can be written as

$$\hat{\mathrm{L}} = \mathrm{i}\begin{bmatrix} 1+p & 0 \\ 0 & 1-p \end{bmatrix}\partial_x + \begin{bmatrix} 0 & u^* \\ u & 0 \end{bmatrix} \tag{2.141}$$

and

$$\hat{\mathrm{A}} = -p\begin{bmatrix} 1 & 0 \\ 0 & 1 \end{bmatrix}\partial_x^2 + \begin{bmatrix} |u|^2/(1+p) & \mathrm{i}\partial_x u^* \\ -\mathrm{i}\partial_x u & -|u|^2/(1-p) \end{bmatrix}, \tag{2.142}$$

in the case

$$s = \frac{2}{1-p^2} > 0$$

when stable soliton solutions are possible. The solution of the NLS equation is further constructed the same way as that for the KdV equation (see Sect. 1.2), though there is a difference here related to the necessity to solve the inverse problem for (2.139) because the operator $\hat{\mathrm{L}}$ is not the self-adjoint one in the case $s > 0$.[12] Thus the special technique for the singular integral equations, to some extent distinguished from the standard *GLM equation* approach considered above in Sect. 1.2, has been developed [158]. It uses the evolution of ψ described by (2.140) to obtain the information on the time evolution of the scattering matrix. Then, taking into account this information, we can construct the solution $u(t,x)$. Omitting details of the very complicated procedure of solving the corresponding GLM equation, we just present here the one-soliton solution for $s = 1$[13] in its explicit form as the *NLS soliton* [27]

$$\begin{aligned} u(t,x) = 2\eta \exp\left[-\mathrm{i}\left(2\xi x + 4(\xi^2 - \eta^2)t + \tau\right)\right] \\ \times \cosh^{-1}\left[2\eta(x - x_0) + 8\eta\xi t\right]. \end{aligned} \tag{2.143}$$

[12] In the opposite case where $s < 0$ we have $v = |s|u$, the eigenvalue problem (2.139) is the self-adjoint one, and the soliton solutions do not appear.

[13] In principle, without loss of generality, we can suppose that $s = \pm 1$; this can be obtained by the simple change of the variables $t \to |s|^{-1}t$ and $x \to |s|^{-1/2}x$.

Here, the envelope's amplitude is $A = 2\eta \exp(-\mathrm{i}\eta)$, $x_0 = \gamma/2\eta$ is the coordinate of the envelope's maximum at $t = 0$, and the soliton velocity is $v = -4\xi$. Furthermore,

$$\gamma = \ln\left(|D|/2\eta\right) \quad \text{and} \quad D = -\frac{\mathrm{i}b(k,t)}{a(k,t)},$$

where $b(k,t)a^{-1}(k,t)$ defines the amplitude of the backward scattering $r(k,t)$ (see Sect. 1.2 for details) and $D = -\mathrm{i}r(k,t)$ plays here the role of the normalization factor for the bound states, $k = \xi + \mathrm{i}\eta$, $\eta > 0$.

The solutions describing the interacting solitary waves can be also obtained in the explicit form; they correspond to the case when the total contribution to the solution is defined by the discrete spectrum only. The expression for $|u|^2$, similar to the KdV equation, is given by

$$|u|^2 \sim \mathrm{d}_x^2 \ln |P|, \tag{2.144}$$

where $|P|$ is the determinant consisting of the exponential functions and in the case of the NLS equation related to the operator $\mathrm{i}\partial_t + \partial_x^2$. Taking into account the theory developed in Chap. 1 for the KdV equation, we can conclude that the solitary waves conserve their structure as a result of their interaction (with possible time delays caused by the interaction).

The solution of the initial (Cauchy) problem is obtained using the same method as that for the KdV equation, and it is clear that for large t the contribution of the discrete spectrum dominates. This means that initial disturbances tend to evolve in a series of *solitary-type waves*. Note that the above analysis is limited by those solutions when $|u| \to 0$ for $|x| \to \infty$, but the conclusion that the series of the solitary waves are the end result of the evolution of modulationally-unstable wave packets, remains valid.

To conclude, we note that in the cases when $v \neq \pm u^*$, the generalized *Zakharov–Shabat eigenvalue problem* enables us to obtain solutions of other evolution equations. The examples are:

1. If $v = +u$ and u is complex, then the solution becomes singular for a finite time and cannot form the "secant-type" solitary wave. In this case, for the special choice of the scattering data (the frequency in the time dependence of the amplitude of the backward scattering $\Omega(k) = -8\mathrm{i}k$, see Sect. 1.2) we can obtain the integrable equation, which is the *complex modified KdV equation*

$$\partial_t u \pm 6u^2 \partial_x u - \partial_x^3 u = 0.$$

Its "soliton" solution is given by [27]

$$u(t,x) = \mathrm{e}^{\mathrm{i}\lambda(t,x)} f(t,x) + \mathrm{e}^{-\mathrm{i}\lambda(t,x)} g(t,x),$$

where

$$f = \xi(\cosh\omega - \sinh\omega)(\cosh^2\omega - \sin^2\lambda)^{-1},$$
$$g = \xi(\cosh\omega + \sinh\omega)(\cosh^2\omega - \sin^2\lambda)^{-1},$$

and

$$\omega(t,x) = 2\eta x + 8\eta\left(3\xi^2 - \eta^2\right)t,$$

$$\lambda = 2\xi x - 8\xi^2\left(\xi - 3\eta\right)t + \tau,$$

$$\gamma = \frac{\ln|D|}{4\xi} + \mathrm{i}\tau.$$

We can see that this solution becomes singular within a finite time period. Usually, models corresponding to such situations are non-physical.

2. If $v = \alpha u = \alpha u^*$ and α is real, then $k = \mathrm{i}\eta = -k$, $\eta > 0$, and

$$\bar{D} \equiv \frac{\mathrm{i}\bar{b}(\bar{k},t)}{\bar{a}(\bar{k},t)} = -D/\alpha = D^*/\alpha,$$

so that the amplitude $A = \pm 2\eta\alpha^{-1/2}$. In this case we have either the *modified KdV equation* ($\Omega = 8\mathrm{i}k^3$, $v = \pm u$),

$$\partial_t u \pm 6u^2\partial_x u + \partial_x^3 u = 0,$$

with the soliton solution

$$u(t,x) = 2\eta\alpha_{\pm}^{-1/2}\cosh^{-1}\left[2\eta(x - 4\eta^2 t) - \gamma\right],$$

$\alpha_\pm = \pm 1$, or the *sine–Gordon equation* (*SG equation*),

$$\partial_{tx} u = \pm\sin u,$$

with the solutions

$$u(t,x) = 2\eta\cosh^{-1}\left[(2\eta)^{-1}(4\eta^2 \pm t) - \gamma\right]$$

and

$$w(t,x) = 4\arctan\left\{\exp\left[\gamma - (2\eta)^{-1}(4\eta^2 \pm t)\right]\right\}.$$

In the last case the solution for w is called the "kink" [27]. The velocity of the MKdV soliton is $4\eta^2$, and the velocity of the *kink soliton* of SG equation, corresponding to the negative sign of the argument, is given by $1/4\eta^2$.

2.3.3 Zakharov System of Equations

Zakharov Equations. In the case of faster (low-frequency perturbations), we obtain from (2.126) and (2.129)–(2.132)

$$\left(\mathrm{i}\partial_t + \frac{3v_{Te}^2}{2\omega_{pe}}\partial_x^2\right) E(x,t) = \frac{\omega_{pe}}{2}\frac{\delta n(x,t)}{n_0}E(x,t), \tag{2.145}$$

and

$$\left(\partial_t^2 - v_s^2\partial_x^2\right)\frac{\delta n(x,t)}{n_0} = \partial_x^2\frac{|E(x,t)|^2}{8\pi m_i n_0}. \tag{2.146}$$

In the literature, (2.145) and (2.146) are usually cited as the one-dimensional *Zakharov equations* for the amplitude of the field envelope and the low-frequency density perturbation [160]. They are the basis of numerous investigations in plasma physics and nonlinear optics. Their derivation based on kinetic plasma theory [36] demonstrates that they can serve as an approximation to more general nonlinear equations. Historically, the Zakharov equations were first [160] obtained from the simple approach similar to that presented above and based on the hydrodynamic plasma equations where the averaging over the high frequency (the electron plasma frequency ω_{pe}) was used. Note that it is difficult to establish their applicability limits in this case. Also, the interpretation of the slow variable δn in this approach is evidently valid only within the approximation for which this approach is correct. In the generalized equations for the amplitude of the field envelope, δn can be referred to as the effective density variation which in general is not the actual low-frequency perturbation of the plasma density [36].

If we introduce the dimensionless variables

$$x = \frac{2}{3}\frac{m_e}{m_i}\frac{\omega_{pe}}{v_s}x, \quad t = \frac{2}{3}\frac{m_e}{m_i}\omega_{pe}t, \quad E = \frac{E}{\sqrt{32\pi n_0 T_e m_e/3m_i}},$$
$$\text{and} \quad \delta n = \frac{3}{4}\frac{m_i}{m_e}\frac{\delta n}{n_0}, \tag{2.147}$$

we can write the Zakharov equations (2.145) and (2.146) in the dimensionless form as

$$\begin{aligned}\left(\mathrm{i}\partial_t + \partial_x^2\right)E &= \delta n E,\\ \left(\partial_t^2 - \partial_x^2\right)\delta n &= \partial_x^2|E|^2.\end{aligned} \tag{2.148}$$

Conservation Laws. It is easy to check that the conservation laws for the energy $\mathcal{H}$ and the momentum $\mathcal{P}$ follow from the set of equations (2.148). Here,

$$\mathcal{H} = \int \mathrm{d}x \left[|\partial_x E|^2 + \delta n|E|^2 + \frac{\delta n^2 + v^2}{2}\right] \tag{2.149}$$

and

$$\mathcal{P} = \frac{1}{2}\int \mathrm{d}x \left[\mathrm{i}E\partial_x E^* - \mathrm{i}E^*\partial_x E + 2v\delta n\right], \tag{2.150}$$

where v is a dimensionless ("slow") hydrodynamic velocity related to δn by the continuity equation

$$\partial_t \delta n = -\partial_x v. \tag{2.151}$$

Restoring for the moment the dimensionality, we can write

$$v = \frac{4}{3}\frac{m_e}{m_i}\frac{v}{v_s}. \tag{2.152}$$

The conservation laws of the energy and the momentum are the consequence of homogeneity of the system in space and time. They play an important role in the development of *modulational interactions*. There also exists a third conservation law following from (2.148). This conservation law is the consequence of the gauge invariance for the transform of the type

$$E \to E\exp(\mathrm{i}\phi), \tag{2.153}$$

where the phase ϕ is real. This third conservation law describes conservation of the number of quanta of waves $\mathcal{N}$ (the number of "plasmons" in the case of Langmuir waves),

$$\mathcal{N} = \int \mathrm{d}x |E|^2. \tag{2.154}$$

The set of the integrals of motion (*conservation laws*) is useful, similar to the case involving KdV and MKdV equations, for investigation of behavior of particular solutions of equations (2.148) because it is difficult to find general solutions of these equations. For example, of interest are some types of special solutions, e.g., *solitary-type waves*, quasi-stationary solutions where density perturbations move with a constant velocity, a *self-similar solution*, etc. We can find variations of the functional $\mathcal{H}$ for these special solutions which allow us to obtain information about their stability. If the functional $\mathcal{H}$ reaches its minimum on one of these special solutions then this can serve as an evidence for the stability of such a solution even in the presence of some perturbations. These problems have been widely studied, see examples in [36,161–165].

Using the conservation law (2.154) we can easily visualize development of the modulational process which is completed (in the one-dimensional case) by establishing a balance between the ponderomotive force and the kinetic pressure of a plasma. Let us assume that a local density depletion, δn, appears in a plasma as a result of some fluctuation or by any other reason. This depletion results, according to (2.122), in a local decrease in the plasma frequency $\omega_{pe}(n)$, and some Langmuir waves are trapped in this local density perturbation. The local growth of the intensity of *Langmuir waves* leads to the increase of the *ponderomotive force* (2.132) which acts on plasma electrons. If this force varies sufficiently slowly as compared to ω_{pi}^{-1}, then the ions follow the electrons due to the ambipolar polarization force. This results in further increase of the plasma depletion which, in turn, leads to further accumulation of plasma oscillations. The above process explains the physical mechanism of the *modulational instability* [36].

In the one-dimensional case we can trace further development of the modulational process including its final stage. Let x_0 be the characteristic size of the density perturbation. Then the conservation law (2.154) allows us to conclude that $|E|^2 x_0 = \text{const}$. This means that in the density depletion the electric field squared varies inversely proportional to x_0 and, in turn, the ponderomotive force is $F^{\text{pond}} \propto \partial_x |E|^2 \propto x_0^{-2}$. The condition for the waves to be trapped is $\delta n \propto k^2 \propto x_0^{-2}$, where k^2 corresponds to the dispersion correction to the frequency of Langmuir waves. The force F_{kin} due to the kinetic plasma pressure has the dependence on x_0, and is given by $F_{\text{kin}} \propto \partial_x (T\delta n) \propto x_0^{-3}$. Thus the size of the density depletion decreases until the moment when the kinetic pressure force balances the ponderomotive force. This balance is stable. Indeed, the increase in x_0 leads to an associated increase in the ponderomotive force, which in turn tends to decrease x_0. The decrease in x_0 in comparison with its "balanced value," in turn, results in an increase in the pressure force which tends to enhance x_0.

2.3.4 Langmuir Solitons

The solution corresponding to the above balance (and minimizing the functional $\mathcal{H}$) is the *Langmuir soliton.* The distinctive feature of this soliton-type solution in comparison with other possible equilibrium solutions of Zakharov equations is that the density perturbation associated with it moves with a constant velocity and all perturbations vanish for $x \to \pm\infty$, i.e., the soliton appears as a localized wave packet moving with a constant speed. This, however, does not necessarily mean that the wave amplitudes has a constant profile moving with a constant velocity. Moreover, the phase of the waves inside the soliton can be varied in a rather complicated manner. Only the envelope field squared, $|E|^2$, has a property of the profile moving without a change in its shape. This is the direct consequence of the equations for the amplitude of the field envelope.

We can find the explicit soliton solution under the conditions

$$v_{Ti} \ll v_0 \ll v_{Te} \quad \text{and} \quad v_0 \neq v_s, \tag{2.155}$$

where v_0 is the velocity of the soliton. From the second equation of the set (2.148), we obtain

$$\delta n = \frac{|E|^2}{v_0^2 - 1}. \tag{2.156}$$

Here, the velocity v_0 is given in units of the ion-sound speed $v_s = \sqrt{T_e/m_i}$. Thus the density variations are negative (that corresponds to the density depletion) only if the soliton velocity is less than the ion-sound speed.

Substituting relation (2.156) in the first equation of set (2.148), we find

$$\left(\mathrm{i}\partial_t + \partial_x^2 + \frac{|E|^2}{1 - v_0^2} \right) E = 0. \tag{2.157}$$

This is the *NLS equation* that can be solved exactly by the IST method [158] (see Sect. 2.3.2). But this is not the case for the set (2.148) because (2.157) has been derived from (2.148) for the particular velocity v_0 of the soliton. Moreover, the speed v_0 is not contained in (2.157) only for $v_0 \ll 1$ (in the first approximation). Thus one can apply the *IST method* to find general solutions of (2.157) only in this limit, corresponding to the extreme subsonic motions. However, a particular solution moving with the constant speed v_0 can be easily obtained from (2.157) by separating the real and imaginary parts of this equation.

We note that the solutions vanishing for $x \to \pm\infty$ exist only for $v_0 < 1$. This is the so-called *bright soliton.* For $v_0 > 1$ there also exists a localized solution with a local decrease in the field intensity which approaches a constant amplitude for $x \to \pm\infty$. This solution is usually referred to as a *dark soliton.*

For the case $v_0 < 1$, the soliton solution of (2.157) is given by

$$E = E_0 \frac{\exp\left[-\mathrm{i}\Omega t + \mathrm{i}v_0(x - v_0 t)/2 + \mathrm{i}\psi_0\right]}{\cosh\left[E_0(x - v_0 t)(2(1 - v_0^2))^{-1/2}\right]}, \tag{2.158}$$

where

$$\Omega = -\frac{v_0^2}{4} - \frac{E_0^2}{2(1 - v_0^2)} \tag{2.159}$$

and ψ_0 is the initial soliton phase. This solution depends on two important parameters, namely E_0 and v_0. As we have already noticed, the density depletion corresponding to the soliton does not spread out (due to dispersion effects) because the linear wave dispersion is balanced by the modulational self-contraction.

Solution (2.158) can be also written as

$$E = E_0 \frac{\exp\left[\mathrm{i}k_0 x - \mathrm{i}\Omega^N t\right]}{\cosh\left[k_N(x - 2k_0 t)\right]}, \tag{2.160}$$

where

$$k_0 = v_0/2, \tag{2.161}$$

$$\Omega^N = k_0^2 - k_N^2, \tag{2.162}$$

and

$$k_N = \frac{E_0}{\sqrt{2(1 - v_0^2)}}. \tag{2.163}$$

The values k_0 and Ω^N have the meanings of the "central" wave vector and the nonlinear frequency shift, respectively. We note that for $k_N > k_0$, the nonlinear dispersion dominates the linear one. The condition $E_0 \gg 1$, which in the dimensional units can be presented as

$$1 \gg \frac{|E_0|^2}{16\pi n_0 T_e} \gg \frac{m_e}{m_i}, \tag{2.164}$$

corresponds to $k_N \gg k_0$. The left inequality in (2.164) implies that the expansion parameter used in our consideration is less than unity.

The soliton is a particular solution of (2.157). For given $\mathcal{N}$ and $\mathcal{P}$, however, the soliton solution corresponds to a minimum of the energy $\mathcal{H}$. The general expressions for the number of waves $\mathcal{N}$, the momentum $\mathcal{P}$, and the energy $\mathcal{H}$ of a single soliton with $E_0 \gg 1$ are given by

$$\mathcal{N} = 2E_0\sqrt{2(1-v_0^2)}, \tag{2.165}$$

$$\mathcal{P} = \frac{4}{3}v_0 E_0^3 \frac{1}{1-v_0^2}\sqrt{\frac{2}{1-v_0^2}}, \tag{2.166}$$

and

$$\mathcal{H} = -\frac{1}{3}E_0^3 \frac{1}{1-v_0^2}\sqrt{\frac{2}{1-v_0^2}}(1-5v_0^2). \tag{2.167}$$

For an arbitrary E_0 (not for $E_0 \gg 1$ only as in (2.165)–(2.167)) the energy of the soliton is

$$\mathcal{H} = -\frac{\sqrt{2}E_0^3(1-5v_0^2)}{3(1-v_0^2)^{3/2}} + \frac{E_0}{\sqrt{2}}v_0^2\sqrt{1-v_0^2}. \tag{2.168}$$

If the energy is negative then the soliton formation is favorable. One can see from (2.167) that $\mathcal{H} < 0$ at least for

$$v_0 < 1/\sqrt{5}. \tag{2.169}$$

When the soliton velocity is close to unity, some of the above expressions can be formally singular. In this case we have to take into account the nonlinear terms caused by the presence of the low-frequency potential. Consideration of such terms enables us to obviate the appearance of singularities in the description of the solitons. We will discuss that below in the section devoted to near-sonic solitons.

The exact condition when the general set of envelope equations can be reduced to the NLS equation is given by

$$\sqrt{\frac{T_i}{T_e}} \ll \frac{|\Delta\omega|}{|\Delta\mathbf{k}|} \ll 1, \tag{2.170}$$

where $\Delta\omega$ and $\Delta\mathbf{k}$ are the frequency and the wave vector of low-frequency density perturbations, respectively.

It is possible to establish some general relations describing the properties of the solutions. We mention one of them in which an initial distribution of the field will (asymptotically as $t \to \infty$) disintegrate into a set of solitons. Using this theorem and the conservation laws it is possible to find, for given values of the integrals $\mathcal{N}$ and $\mathcal{H}$, the number of solitons in which the initial disturbance evolves. If the final solitons have $v_0 \ll 1/\sqrt{5}$, the energy of each of them is

$$\mathcal{H}_0 = -\frac{\sqrt{2}}{3} E_0^3. \tag{2.171}$$

The total energy of them is $N\mathcal{H}_0$ where N is the number of the solitons, while the total number of the waves is given by

$$\mathcal{N} = N\mathcal{N}_0 = 2\sqrt{2} N E_0. \tag{2.172}$$

Thus for the given values of the integrals $\mathcal{N}$ and $\mathcal{H}$, the total number of the solitons created as a result of development of the modulational interaction is

$$N = \sqrt{-\frac{\mathcal{N}^3}{48\mathcal{H}}}. \tag{2.173}$$

The absolute minimum of the energy $\mathcal{H}$ is reached for a single soliton at rest, and is given by

$$\mathcal{H}_0 = -\frac{\mathcal{N}^3}{48}. \tag{2.174}$$

2.3.5 Near-Sonic Solitons

When the soliton velocity approaches the speed of ion sound, effects of nonlinearity and dispersion of the ion-acoustic (low-frequency) field modify the *Langmuir soliton.* Indeed, only in this case do the nonlinearity and the sound dispersion become comparable with the small linear term in the evolution equations (which determines the linear dispersion of the ion sound $\omega_{\mathbf{k}} = |\mathbf{k}|v_s$). The corresponding restriction on the deviation of the soliton velocity from the sound velocity is presented below. The one-dimensional equations in this case can be written in terms of the dimensionless electric field E and the potential ϕ of the low-frequency field. These equations are [166]

$$\mathrm{i}\partial_t E + \partial_x^2 E + E\Delta = \frac{1}{\mu}\phi E - |E|^2 E + \hat{\mathrm{N}}_h(\phi, E), \tag{2.175}$$

and

$$\partial_x^2\phi - \partial_t^2\phi + \mu\partial_t^2|E|^2 + \frac{\mu}{3}\partial_x^4\phi + \partial_x^2\phi^2 + \frac{6T_i}{T_e}\partial_x^2\phi$$
$$+\mu\partial_x^2(\phi|E|^2) + 2\mu\partial_x\left(|E|^2\partial_x\phi\right) + \partial_x\hat{\mathrm{D}}_l(\phi) + \mu\hat{\mathrm{N}}_l\partial_x E = 0. \tag{2.176}$$

In (2.176) we have taken into account the actual dispersion of the low-frequency field, namely the deviation of its linear dispersion from $\omega_{\mathbf{k}} = |\mathbf{k}|v_s$. This is important for investigation of the solitons or other nonlinear structures propagating with the velocities close to the sound speed.

The operator $\hat{\mathrm{N}}_h$ in the equation for the high-frequency field and operators $\hat{\mathrm{D}}_l$, $\hat{\mathrm{N}}_l$ in the equation for the low-frequency potential are introduced to describe the dissipative processes. The explicit expressions for these operators are given below. Furthermore, $\mu = m_i/m_e$ in (2.175) and (2.176) and Δ for the high-frequency field in (2.175) is given by

$$\Delta = 2\frac{\omega_0 - \omega_{pe}}{\mu\omega_p e}, \tag{2.177}$$

where ω_0 is the frequency of the high-frequency wave or structure. The dimensionless field of the low-frequency field, ϕ, corresponds to $-\phi e/T_e$.

Solitons Without Dissipation. First, we start with equations (2.175) and (2.176) where we neglect the dissipative terms. Moreover, we neglect the third-order nonlinear terms in (2.176). We have in this case

$$\mathrm{i}\partial_t E + \partial_x^2 E + E\Delta = \frac{1}{\mu}\phi E - E|E|^2 \tag{2.178}$$

and

$$\partial_x^2\phi - \partial_t^2\phi + \mu\partial_t^2|E|^2 + \frac{\mu}{3}\partial_x^4\phi + \partial_x^2\phi^2 + \frac{6T_i}{T_e}\partial_x^2\phi = 0. \tag{2.179}$$

This set of equations was originally derived in [167]. In the near-sonic limit we can use for the fields propagating in the positive x-direction

$$(\partial_x + \partial_t)(\partial_x - \partial_t) \approx 2\partial_x(\partial_x + \partial_t). \tag{2.180}$$

Taking into account (2.180) we obtain the simplified set of equations

$$\mathrm{i}\partial_t E + \partial_x^2 E + E\Delta = \frac{1}{\mu}\phi E - |E|^2 E \tag{2.181}$$

and

$$\partial_t\phi + \partial_x\phi + \frac{\mu}{2}\partial_x|E|^2 + \frac{1}{2}\partial_x\phi^2 + \frac{\mu}{6}\partial_x^3\phi + \frac{3T_i}{T_e}\partial_x\phi = 0. \tag{2.182}$$

Equations (2.181) and (2.182) were obtained for the first time by Nishikawa [168].

If we neglect the sound dispersion terms and the quadratic (in the low-frequency field) nonlinearities in the above set of equations, then we restore the *Zakharov equations* written in terms of E and ϕ. Indeed, in the dimensionless variables we obtain

$$\delta n \approx \phi - \mu|E|^2. \tag{2.183}$$

Using this expression in (2.178) and (2.179) we return to the usual Zakharov equations.

Substituting the soliton solution of the Zakharov equations (2.158) to (2.178) and (2.179), one can show that it is possible to neglect the low-frequency nonlinearity and sound dispersion terms for

$$\left(1 - \frac{v_0^2}{v_s^2}\right)^2 \gg \frac{|E_0|^2}{16\pi n_0 T_e}. \tag{2.184}$$

Thus a usual *Langmuir soliton* discussed in the previous sections can exist only if its velocity is not very close to the sound speed. As we will see later, the dissipative terms are small and can be considered as perturbations only for

$$\left(1 - \frac{v_0^2}{v_s^2}\right)^2 \gg \frac{m_e}{m_i}, \tag{2.185}$$

and

$$\frac{|E_0|^2}{16\pi n_0 T_e} \gg \frac{m_e}{m_i}. \tag{2.186}$$

Thus the soliton solutions obtained from the Zakharov equations are valid if

$$\left(1 - \frac{v_0^2}{v_s^2}\right)^2 \gg \frac{|E_0|^2}{16\pi n_0 T_e} \gg \frac{m_e}{m_i}. \tag{2.187}$$

In the opposite limit where

$$\frac{|E_0|^2}{16\pi n_0 T_e} \geq \left(1 - \frac{v_0^2}{v_s^2}\right)^2 \gg \frac{m_e}{m_i}, \tag{2.188}$$

it is impossible to neglect the dispersion and nonlinear terms in the equation for the low-frequency potential. The soliton can still exist in this limit and can be called a *near-sonic soliton* because

$$\frac{|E|^2}{16\pi n_0 T_e} \ll 1.$$

The set of equations (2.181) and (2.182) has the following soliton solution:

$$\phi = -\frac{6\mu|\Delta|}{\cosh^2\left[\sqrt{|\Delta|}(x - x_0 - v_0 t)\right]} \tag{2.189}$$

and

$$E = |E| \exp\left(\frac{\mathrm{i}}{2}v_0 x - \frac{\mathrm{i}}{4}v_0^2 t + \mathrm{i}\chi_0\right), \tag{2.190}$$

where

$$\mu|E|^2 = 8\mu\Delta\phi - \frac{4}{3}\phi^2 \tag{2.191}$$

and

$$v_0 + \frac{3T_i}{T_e} = 1 - \frac{10}{3}\mu|\Delta|. \tag{2.192}$$

Here, χ_0 is the initial phase and x_0 is the initial position of the soliton. Expression (2.192) gives the relation between the frequency shift Δ and the soliton velocity. The solitons can exist for $\Delta < 0$ which corresponds to the density depletion. Substituting this solution into (2.182), we can confirm the validity of the inequality (2.187) and also that all terms of (2.182) are of the same order of magnitude.

Dissipative Processes for Near-Sonic Solitons. Here, we take into account the linear *Landau damping* for the the low-frequency potential and the nonlinear damping for both the low-frequency and the high-frequency components of the field. The explicit expressions for the dissipative operators in (2.175) and (2.176) are given by

$$\hat{\mathrm{N}}_h(\phi, E) = E\delta\phi/\mu, \tag{2.193}$$

$$\hat{\mathrm{D}}_l(\phi) = \frac{1}{2}\int\limits_{-\infty}^{\infty} \frac{\mathrm{d}\omega \mathrm{d}k}{(2\pi)^2}\sqrt{\frac{\pi}{2}}\frac{\omega^2 v_s}{|k|v_{Te}}\phi_{\omega,k}\mathrm{e}^{-\mathrm{i}\omega t+\mathrm{i}kx}, \tag{2.194}$$

and

$$\hat{\mathrm{N}}_l(E) = -\int\limits_{-\infty}^{\infty} \frac{\mathrm{d}\omega \mathrm{d}k}{(2\pi)^2}\sqrt{\frac{\pi}{8}}\frac{\omega^2 v_s}{|k|v_{Te}}|E|^2_{\omega,k}\mathrm{e}^{-\mathrm{i}\omega t+\mathrm{i}kx}, \tag{2.195}$$

where

$$\delta\phi = \int\limits_{-\infty}^{\infty} \frac{\mathrm{d}\omega \mathrm{d}k}{(2\pi)^2}\sqrt{\frac{\pi}{2}}\frac{\mathrm{i}\omega v_s}{|k|v_{Te}}\phi_{\omega,k}\mathrm{e}^{-\mathrm{i}\omega t+\mathrm{i}kx}. \tag{2.196}$$

To find how the soliton parameters are changed due to the *dissipative effects*, we use the procedure similar to that already used for the usual Langmuir solitons. Namely we derive the conservation laws in the absence of dissipative processes. Next, we find the changes in the equations expressing these laws due to dissipative processes, assume that the dissipative processes are weak and, finally, substitute the soliton solution in these equations to determine the changes in the soliton parameters.

Expressions for the number of waves $\mathcal{N}$, the momentum $\mathcal{P}$, and the energy $\mathcal{H}$ for the set of equations (2.178) and (2.179) are given by

$$\mathcal{N} = \int |E|^2 \mathrm{d}x, \tag{2.197}$$

$$\mathcal{P} = \int \frac{1}{2}\left[\mathrm{i}\left(E\partial_x E^* - E^*\partial_x E\right) + \frac{\phi^2}{\mu^2}\right]\mathrm{d}x, \tag{2.198}$$

and

$$\begin{aligned}\mathcal{H} = &\int \left[|\partial_x E|^2 - |E|^2\Delta + \frac{1}{\mu}|E|^2\phi\right]\mathrm{d}x \\ &+ \int \left[\frac{\phi^2}{\mu^2} - \frac{1}{6\mu}\left(\partial_x\phi\right)^2 + \frac{1}{3\mu^2}\phi^2\right]\mathrm{d}x.\end{aligned} \tag{2.199}$$

One can check that due to the dissipative processes, the *near-sonic soliton* perceptibly decreases in amplitude and increases in velocity [166,169]. In other words, it is accelerated while destroying. Its velocity becomes closer to the sound speed so long as the necessary condition for the soliton's existence (2.188) is not violated. Such a behavior is the opposite to that of the usual solitons which do not change their amplitudes but significantly decrease their velocities.

All these examples aim to demonstrate the possibility of creation of different coherent structures under the conditions when the *modulational interactions* dominate. We have discussed mainly the one-dimensional structures which can be the final result of the development of the modulational interactions. In the three-dimensional situation the quasi-stationary structures are mainly dissipative, i.e., the modulational interactions develop until the dissipative effects are comparable to the nonlinear effects, and the modulational development stops when the dissipative effects are of the order of the nonlinear ones.

2.4 Derivative Nonlinear Schrödinger Equation

In this section, we introduce the derivative nonlinear Schrödinger (DNLS) equation, demonstrate its origin from the plasma magnetohydrodynamic (MHD) equations (Sect. 2.4.1), consider DNLS equation as an integrability condition for two linear differential equations (Sect. 2.4.2), and discuss stability of DNLS-solutions such as DNLS solitons (Sect. 2.4.3). Furthermore, we consider some numerical approaches to study the dynamics of Alfvén solitons (Sect. 2.4.4), and consider results of numerical simulations of the solitons' evolution (Sect. 2.4.5).

2.4.1 Origin of the DNLS Equation

For a *magnetized plasma*, consider the case when the magnetic pressure is much stronger than the plasma kinetic pressure, i.e., $(B^2/8\pi nT) \gg 1$. We are interested in the waves with velocities much less than speed of light, c. Then the basic plasma equations are given by

$$m_i \mathrm{d}_t \mathbf{v}_i = e\mathbf{E} + \frac{e}{c}\mathbf{v}_i \times \mathbf{B}, \tag{2.200}$$

$$m_e \mathrm{d}_t \mathbf{v}_e = -e\mathbf{E} - \frac{e}{c}\mathbf{v}_e \times \mathbf{B}, \tag{2.201}$$

$$\mathrm{d}_t n_{i,e} + \nabla \cdot (n_{i,e}\mathbf{v}_{i,e}) = 0, \tag{2.202}$$

$$\nabla \times \mathbf{B} = \frac{4\pi e}{c}\left(n_i \mathbf{v}_i - n_e \mathbf{v}_e\right), \tag{2.203}$$

and

$$\mathrm{d}_t \mathbf{B} = -c\nabla \times \mathbf{E}. \tag{2.204}$$

Assume that the plasma is quasi-neutral, i.e.,

$$n_i \approx n_e \equiv n. \tag{2.205}$$

This takes place if the characteristic frequencies are small compared to the ion plasma frequency ω_{pi}. Introducing the "mass velocity"

$$\mathbf{u} = \frac{m_i \mathbf{v}_i + m_e \mathbf{v}_e}{m}, \tag{2.206}$$

where $m = m_e + m_i \approx m_i$, adding (2.200) and (2.201), and using (2.203) and (2.205), we obtain

$$\mathrm{d}_t \mathbf{u} = -\frac{1}{4\pi n m} \mathbf{B} \times \boldsymbol{\nabla} \times \mathbf{B}. \tag{2.207}$$

From (2.200), (2.201), and (2.206) we obtain

$$\mathbf{E} = -\frac{1}{2} \mathbf{u} \times \mathbf{B} + \frac{m_i}{c} \mathrm{d}_t \mathbf{u} - \frac{m_e}{c} \mathrm{d}_t \mathbf{v}_e. \tag{2.208}$$

Substituting (2.208), (2.205), and (2.206) into the Maxwell equations (2.203) and (2.204), we find

$$\boldsymbol{\nabla} \times \mathbf{B} = \frac{4\pi n e}{c} \frac{m}{m_i} (\mathbf{u} - \mathbf{v}_e) \tag{2.209}$$

and

$$\mathrm{d}_t \mathbf{B} = \boldsymbol{\nabla} \times \mathbf{u} \times \mathbf{B} - \frac{m_i c}{e} \boldsymbol{\nabla} \times \mathrm{d}_t \mathbf{u} + \frac{m_e c}{e} \boldsymbol{\nabla} \times \mathrm{d}_t \mathbf{v}_e. \tag{2.210}$$

Equations (2.207), (2.209), and (2.210), together with the *continuity equation*,

$$\partial_t n + \boldsymbol{\nabla} \cdot (n\mathbf{u}) = 0, \tag{2.211}$$

constitute the full set of equations of a cold magnetized plasma for the low-frequency oscillations when $\omega \ll \omega_{pi}$. Two last terms in (2.210) are negligible when the frequencies are well below the ion cyclotron frequency,

$$\omega \ll \omega_{Bi} \equiv eB/m_i c. \tag{2.212}$$

If we omit them, the basic set of equations turns into well-known *magnetohydrodynamic equations* (*MHD equations*) for a cold plasma where the pressure term $\boldsymbol{\nabla} p$ is ignored.

Assuming that $\mathbf{B} = \mathbf{B}_0 + \mathbf{h}$ and $n = n_0 + n'$, and neglecting the second-order perturbations of $\mathbf{h}$, $\mathbf{u}$, and $\mathbf{u}'$, we obtain the equations that are linear with regard to these functions, leading to the dispersion law of the linear oscillations (see (0.19)):

$$\omega_{1,2} = \frac{v_A k/2}{\sqrt{1 + c^2k^2/\omega_{pe}^2}} \left[\left((1+\cos\theta)^2 + \frac{c^2k^2}{\omega_{pi}^2} \frac{\cos^2\theta}{1 + c^2k^2/\omega_{pe}^2} \right)^{1/2} \pm \left((1-\cos\theta)^2 + \frac{c^2k^2}{\omega_{pi}^2} \frac{\cos^2\theta}{1 + c^2k^2/\omega_{pe}^2} \right)^{1/2} \right], \tag{2.213}$$

where Θ is the angle between the wave vector and the magnetic field, and $v_A = B_0/\sqrt{4\pi nm}$ is the Alfvén velocity. As we noted in the Introduction, the signs in (2.213) correspond to two types of waves. In the long wavelength limit when $c^2k^2 \ll \omega_{pe}^2$ and $c^2k^2\cos^2\Theta \ll \omega_{pi}^2(1-\cos\Theta)^2$, the dispersion equation with the lower sign is given by $\omega = v_A k\cos\Theta$, i.e., turns into the dispersion equation for the Alfvén wave. Accordingly, the branch of oscillations corresponding to the "minus" sign in (2.213) is sometimes called the Alfvén branch.

Expanding (2.213) in a Taylor series in powers of the wave number k, applying the inverse Fourier transform to the obtained equation and retaining the nonlinear terms, we obtain an equation describing (in the one-dimensional approximation) the propagation of nonlinear Alfvén waves in a plasma – the *derivative nonlinear Schrödinger equation* (*DNLS equation*) –

$$\partial_t h + s\partial_x(|h|^2 h) - \mathrm{i}\partial_x^2 h = 0, \tag{2.214}$$

where $h = h(t,x)$ corresponds to the right-circularly polarized wave,

$$h = \frac{B_y + \mathrm{i}B_z}{2\sqrt{|1-\beta|B_0}},$$

and the magnitude of the magnetic field B_0 is defined by $\mathbf{B}_0 = B_0\hat{\mathbf{x}}$ (the magnetic field is in the x-direction). The variables x and t in (2.214) are the normalized space and time in the reference frame moving with the Alfvén velocity.

The DNLS equation plays an important role in space plasma physics. There are many publications devoted to DNLS solitons in, for example, solar wind as well as in a laboratory plasma. The DNLS equation can also be used for the study of nonlinear processes in the Earth's magnetosphere. Indeed, the first experimental confirmation of the DNLS solitary waves was demonstrated in the experiment MASSA [49]. The obtained data were interpreted in terms of the one-dimensional Alfvén solitons excited in the F-region of the Earth's ionosphere because of the formation of strong magnetic field-aligned currents caused by the motion of neutral particles against the charged ones in the dynamo region. The self-consistent (one-dimensional) model of this phenomenon was proposed in 1994 [170]. Propagation of nonlinear structures in the *ionospheric plasma* in this case is accompanied by plasma turbulence excited by the strong transverse electric currents at the front of the solitary wave leading to the "turbulent track" of the soliton. That proved to be correct in further experiments on board the satellite DE2 [171] when the satellite passed through the magnetic field tube over the Nevada site just after a strong underground nuclear explosion.[14]

[14] Noting relative simplicity and clarity of such one-dimensional approach, we nevertheless would like to remark, that this explanation was only qualitative and a full description of the observed phenomenon requires more complex two-dimensional and three-dimensional theory to be developed (see Chap. 4).

Thus, applications of the DNLS model in physics are numerous, but in this book we mostly consider only those related to the physics of the ionospheric plasma and the *magnetospheric plasma*. In next sections, we study mathematical properties of the DNLS equation and its possible solutions.

2.4.2 DNLS Equation as an Integrability Condition for Two Linear Differential Equations

Equation (2.214) is completely integrable. It has infinite sequence of the integrals of motion and can be solved using the IST method (see Sect. 1.2). This equation appears also as an integrability condition for a set of two linear differential equations; for the latter the eigenvalue problem can be formulated by

$$\begin{aligned} \partial_x v_1 + \mathrm{i}\xi^2 v_1 &= h\xi v_2, \\ \partial_x v_2 - \mathrm{i}\xi^2 v_2 &= g\xi v_1, \end{aligned} \tag{2.215}$$

where $g = -sh^*$ is the left-circularly polarized wave satisfying the DNLS equation with the positive dispersion term

$$\partial_t g + s\partial_x \left(|g|^2 g\right) + \mathrm{i}\partial_x^2 g = 0. \tag{2.216}$$

Note that this eigenvalue problem (2.215) differs from the generalized *Zakharov–Shabat eigenvalue problem*:

$$\begin{aligned} \partial_x \tilde{v}_1 + \mathrm{i}\lambda \tilde{v}_1 &= \tilde{h}\tilde{v}_2, \\ \partial_x \tilde{v}_2 - \mathrm{i}\lambda \tilde{v}_2 &= \tilde{g}\tilde{v}_1, \end{aligned} \tag{2.217}$$

which is applicable to the KdV equation (Sect. 1.2), the modified KdV equation (Sect. 2.1), and the NLS and sine-Gordon equations (Sect. 2.3), where $\tilde{h}$ always corresponds to a function whose evolution is described by (2.217).

Following [33], we can show that the eigenfunctions v_1 and v_1 in (2.215) can be transformed to the form they acquire in (2.217). Let us introduce [35]

$$H = h\mathrm{e}^{-2\mathrm{i}\mu} \quad \text{and} \quad G = -\frac{\mathrm{i}}{2}\left(\partial_x g + \frac{\mathrm{i}}{2}hg^2\right)\mathrm{e}^{2\mathrm{i}\mu}, \tag{2.218}$$

as well as

$$\mu = \frac{1}{2}\int_{-\infty}^{\infty} gh\mathrm{d}x \quad \text{and} \quad \lambda = \xi^2.$$

In addition, we also let

$$\begin{aligned} w_1 &= v_1 \mathrm{e}^{-\mathrm{i}\mu}, \\ w_2 &= \xi v_2 \mathrm{e}^{\mathrm{i}\mu} - (\mathrm{i}/2) v_2 g \mathrm{e}^{\mathrm{i}\mu}. \end{aligned} \tag{2.219}$$

Then the set of equations (2.215) transforms into

$$
\begin{aligned}
\partial_x w_1 + \mathrm{i}\xi^2 w_1 &= H w_2, \\
\partial_x w_2 - \mathrm{i}\xi^2 w_2 &= G w_1,
\end{aligned}
\tag{2.220}
$$

which is similar to (2.217). The introduced functions H and G satisfy the conditions

$$
\begin{aligned}
\mathrm{i}\partial_t H + \partial_x^2 H - 2H^2 G = 0, \\
\mathrm{i}\partial_t G - \partial_x^2 G + 2G^2 H = 0,
\end{aligned}
\tag{2.221}
$$

which are equations of the NLS type, if the relation between G and H is given by [33]

$$
G = -\frac{1}{4}\left|H\right|^2 H^* + \frac{\mathrm{i}}{2} s \partial_x H^* . \tag{2.222}
$$

Equations (2.214)–(2.216) and (2.220)–(2.222) are equivalent. Thus the considered transformation enables us to use the results obtained above for the *Zakharov–Shabat eigenvalue problem* (2.217). This is useful for integration of the DNLS equation by the *IST method.*

Here, we would like to note the following. One of the most important properties of equations that can be solved by the IST method is that any localized initial condition $|h| \to 0$ for $|x| \to \infty$ results in the N-soliton solution and the oscillating "tail" (radiation) spreading with time. The number of arising solitons is determined for every particular case by the initial condition and is related to the number of the discrete eigenvalues of the associated scattering problem. Their number is always finite and can be $0, 1, 2, \ldots$. It is very difficult to obtain this number analytically, but when there is only one dispersion relation for the linearized evolution equations with the functions $\tilde{h}$ and $\tilde{g}$, and the associated scattering problem is of standard form (2.217), the number of solitons is equal to the number of the corresponding discrete eigenvalues λ_i such that $\mathrm{Im}\lambda_i > 0$ [24]. Some conditions of the existence of these eigenvalues can be estimated by the largest value, λ_0, determined by the integral

$$
C_1 = 2\mathrm{i}\lambda_0 = \int_{-\infty}^{\infty} \tilde{g}\tilde{h}\mathrm{d}x. \tag{2.223}
$$

It is necessary to note that C_1 has the sense of the first *integral of motion* for those evolution equations which are the integration conditions of (2.217). Equality (2.223) suggests that there are no proper discrete eigenvalues if $\mathrm{Re}C_1 > 0$ [172], therefore any initial pulse will merely spread (without forming a soliton). This is in agreement with two facts, namely:

1. The standard *NLS equation* with the negative nonlinearity,

$$
\mathrm{i}\partial_t \tilde{h} - 2\tilde{h}^2\tilde{h}^* + \partial_x^2 \tilde{h} = 0, \tag{2.224}
$$

has no conventional soliton solutions [3,24]. In this case $\tilde{g} = \tilde{h}^*$ [172] and, as it is clear from (2.223), $C_1 > 0$.

2. For the *KdV equation*,

$$\partial_t \tilde{h} + 6\tilde{h}\partial_x \tilde{h} + \partial_x^3 \tilde{h} = 0, \tag{2.225}$$

where $\tilde{h} \in \mathbb{R}$, if the initial pulse $\tilde{h}(0,x) < 0$ for any x then no solitons form. In this case $\tilde{g} = -1$ [172], and (2.223) also gives $C_1 > 0$.

The *modulational instability* [36] of the finite length plane waves was investigated in detail in Ref. [173]. Four different regimes of the soliton formation for the initial condition in the form of a solitary pulse were established. The weakly and strongly stable cases considered in [173] correspond to the case $C_1 > 0$, and different signs of the first integral of motion correspond to different physical situations: $C_1 > 0$ to the *modulationally-stable case* and $C_1 < 0$ to the *modulationally-unstable case.*

Numerical simulations [174] demonstrated that an initial pulse corresponding to the modulated circularly polarized Gaussian wave packet,

$$h(0,x) = A_0 \exp\left(\frac{2\pi \mathrm{i} x}{\lambda}\right) \exp\left(-x^2/l^2\right), \tag{2.226}$$

collapses or spreads with time, depending on the relative sign of the nonlinear and dispersive terms in (2.214). This can be due to the fact that the inverse scattering problem has no corresponding discrete eigenvalues. In this case, the collapse-type instability can reflect the fact that the inverse scattering problem has no solution, and this leads to the infinite (for a finite time) solutions of (2.214). Here we note that the detailed analysis of such situations is very complicated.

2.4.3 Stability of DNLS Solitons

The evolution equations (2.214) and (2.216) can be written in the Hamiltonian form

$$\begin{pmatrix} \partial_t h \\ \partial_t g \end{pmatrix} = \begin{pmatrix} 0 & 1 \\ 1 & 0 \end{pmatrix} \partial_x \nabla \mathcal{H}, \tag{2.227}$$

where the *Hamiltonian* is given by

$$\mathcal{H} = 2C_1 = \int_{-\infty}^{\infty} \left[\frac{1}{2}|h|^4 + \varepsilon s g h \partial_x \varphi\right] \mathrm{d}x, \qquad \varphi = \arg h, \tag{2.228}$$

and the signs of $\varepsilon = \pm 1$ correspond to the DNLS equation with the negative (2.214) or the positive (2.216) dispersion term, respectively.

Here we employ the method [50,175] to analyze the stability of the DNLS solutions and investigate the lower boundedness of the Hamiltonian $\mathcal{H}$ under the deformations conserving the momentum $\mathcal{P} = (1/2)\int |h|^2 \mathrm{d}x$ when the *variational equation* is given by

$$\delta\left(\mathcal{H} + v\mathcal{P}\right) = 0. \tag{2.229}$$

Here, v is the *Lagrange factor* and (2.229) reflects the fact that all finite solutions of the DNLS equation are the stationary points of the Hamiltonian for a fixed $\mathcal{P}$. Apply the transform

$$h(x) \to \zeta^{-1/2} h(x/\zeta), \qquad \zeta \in \mathbb{R}, \tag{2.230}$$

conserving the momentum as follows:

$$\mathcal{P} = \frac{1}{2}\int |h|^2\,\mathrm{d}x \to \frac{1}{2}\int \zeta^{-1}\,|h|^2\,\zeta\mathrm{d}x = \frac{1}{2}\int |h|^2\,\mathrm{d}x.$$

Substituting (2.230) into (2.228), we obtain

$$\mathcal{H} \to \int\limits_{-\infty}^{\infty}\left[\frac{1}{2}\zeta^{-2}\,|h|^4 + \varepsilon s\zeta^{-1}\,|h|^2\,\zeta^{-1}\partial_x\varphi\right]\zeta\mathrm{d}x = \zeta^{-1}(a+b), \tag{2.231}$$

where $a = (1/2)\int |h|^4\mathrm{d}x$ and $b = \varepsilon s \int |h|^2\partial_x\varphi\mathrm{d}x$.

Furthermore, we solve the set

$$\begin{cases} \partial_\zeta H = 0, \\ \partial_\zeta^2 H > 0, \end{cases}$$

where the equality stands for the necessary condition for the existence of an extremum, and the inequality is the sufficient condition for the existence of the minimum of $\mathcal{H}$, and obtain

$$\frac{a+b}{\zeta^2} = 0 \quad \text{and} \quad \frac{a+b}{\zeta^2}\cdot\frac{2}{\zeta} > 0. \tag{2.232}$$

Obviously, the necessary condition to satisfy the inequality is

$$a+b > 0, \qquad \zeta > 0, \tag{2.233}$$

and, in this case, for $\zeta \to +\infty$ we have $\partial_\zeta\mathcal{H} \to 0$ from the right. Therefore, for $\zeta \to +\infty$ we obtain $\partial_\zeta^2\mathcal{H} > 0$, i.e., we have the boundary minimum at $\zeta = +\infty$. In the opposite case, when

$$a+b < 0, \qquad \zeta < 0, \tag{2.234}$$

we have the boundary minimum at $\zeta = -\infty$.

Note, that the change $\zeta \to -\zeta$ is equivalent to the change of sign of the coefficient s in (2.214) and (2.217) determining the wave "polarity." Therefore, condition (2.233) is valid for the "positive" wave or soliton, and (2.234) is valid for the "negative" wave. Thus the necessary conditions (2.233) and (2.234) are the sufficient conditions for the Hamiltonian to be limited from below. Therefore, in the cases (2.233) and (2.234), the solitons of the DNLS equation are stable.

2.4.4 Numerical Approaches to Study Dynamics of Alfvén Solitons

First we note that solutions of the DNLS equation cannot always be obtained analytically in a closed form because the IST method imposes sufficiently strong limitations on the initial and boundary conditions (first of all, the locality of the potential $|h(x,t)| \to 0$ for $|x| \to \infty$). Thus development of numerical methods and numerical codes for integration of the DNLS equation is a very actual and important problem. In this section, we consider a few methods used for numerical modeling of the dynamics of one-dimensional Alfvén solitons described by the DNLS equation (2.214).

Dawson–Fontán Method. Consider now briefly a rather effective method for the numerical integration of the DNLS equation based on the *Ablowitz–Ladik scheme* [176] used for simulation of the evolution of the one-dimensional Alfvén soliton in Ref. [33]. Assume in (2.215) that

$$\partial_t v_1 = Av_1 + Bv_2 \quad \text{and} \quad \partial_t v_2 = Cv_1 + Dv_2, \tag{2.235}$$

where $D = -A$, and A, B, and C are some functions of h and g. Introducing the change of the original functions by the standard functions of the Ablowitz–Ladik method (2.219), we obtain evolution equations for the new functions H and G (2.221). They represent the integrability conditions of (2.220) together with the corresponding evolution equations for w_1 and w_2 of type (2.235) where

$$\begin{aligned} A &= -\mathrm{i}HG - 2\mathrm{i}\lambda^2, \\ B &= \mathrm{i}\partial_x H + 2H\lambda, \\ C &= -\mathrm{i}\partial_x G + 2G\lambda. \end{aligned} \tag{2.236}$$

It is clear from (2.221) that the evolution equations for H and G are similar to the NLS system (2.235) if conditions (2.236) hold. The difference between (2.221) and the usual NLS equation is in the different relation between H and G given for the DNLS equation by (2.222), unlike the relation $G = \mp H^*$ for the NLS equation.

Similarity between equations (2.221) for H and G and the Schrödinger equation gives a simple way to guess their discrete versions using an extension of the NLS results to the transformed problem (2.221). Following [176] (and similar to the case of NLS equation [177]), choose for A, B, C, and D the expansion in powers of the eigenvalues z and $1/z$:

$$\begin{aligned} A_n^m &= A_n^{(-2)} z^{-2} + A_n^{(0)} + A_n^{(2)} z^2, \\ B_n^m &= B_n^{(-1)} z^{-1} + B_n^{(1)} z^1, \\ C_n^m &= C_n^{(-1)} z^{-1} + C_n^{(1)} z^1, \\ D_n^m &= D_n^{(-2)} z^{-2} + D_n^{(0)} + D_n^{(2)} z^2. \end{aligned} \tag{2.237}$$

Then we can write the first equation of (2.221) in the implicit discrete form on the grid (m, n), where m refers to the time grid point and n refers to the space grid point,

$$
\begin{aligned}
\mathrm{i}\Delta^m H_n^m = &-\frac{\Delta t}{2(\Delta x)^2}\left(H_{n+1}^m - 2H_n^m + P_{n-1}H_{n-1}^m + P_n H_{n+1}^{m+1}\right.\\
&\left. -2H_{m+1}^n + H_{n-1}^{m+1}\right)\\
&+\frac{\Delta t}{2}\left[P_n G_n^m H_n^m H_{n+1}^{m+1} + P_{n-1}G_n^{m+1} H_{n-1}^m H_n^{m+1}\right.\\
&+\frac{1}{2}\left(H_n^m H_{n+1}^m G_n^m + H_{n-1}^{m+1} H_n^{m+1} G_n^{m+1} + H_{n-1}^m H_n^{m+1} G_n^m\right.\\
&\left.\left. +H_n^m \sum_{k=-N}^{n} \Delta^m \tilde{S}_k^m - H_n^{m+1} \sum_{k=-N}^{n-1} \Delta^m \tilde{T}_k^m\right)\right], \qquad (2.238)
\end{aligned}
$$

where

$$
\begin{aligned}
\Delta^m H_n^m &= H_n^{m+1} - H_n^m,\\
\tilde{S}_k^m &= H_{k+1}^m G_k^m + H_k^m G_{k-1}^m,\\
\tilde{T}_k^m &= H_{k-1}^m G_k^m + H_k^m G_{k+1}^m,
\end{aligned}
$$

and

$$
P_n = \prod_{k=-\infty}^{n} \left[\left(1 - H_k^{m+1} G_k^{m+1} (\Delta x)^2\right) / \left(1 - H_k^m G_k^m (\Delta x)^2\right)\right].
$$

The main problem for the scheme (2.238) is that the relation between G and H (2.222) is nonlinear, and the method provides no answer to the question how one can write it in the finite difference form. However, since in the discrete version of the *Ablowitz–Kaup–Newell–Segur eigenvalue problem* [172] the term $h\xi v$ is converted into $H_n^m w_n^m$, following [176] we can assume that

$$
G_n^m = \frac{\mathrm{i}s}{2}\left(\frac{H_{n+1}^{m^*} - H_{n-1}^{m^*}}{2\Delta x}\right) - \frac{1}{4} H_n^m \left(H_{n+1}^{m^*}\right)^2.
$$

The algorithm [176] realizing the simulation according to the scheme (2.238) consists of two nested iterations. The outer iteration accounts for the fact that expression (2.238) is a highly implicit one, and it contains nonlinear terms with the values H_n at the next time layer $(m+1)$. Therefore, as usual, it is necessary to assume $H_n^{m+1} = H_n^m$ in these terms and solve the linear system at the new time layer, that the internal iteration is intended for. Since the corresponding matrix of factors at unknown values is not inverted, the calculations are conducted via the iterative procedure at this iteration. In both iterations, the well known Crank–Nicholson method [178] is used. The procedure is finished when the condition

$$
\left|H_n^{m+1,j} - H_n^{m+1,j+1}\right| / \left|H_n^{m+1,j}\right| < \varepsilon
$$

is satisfied. Here, ε is the prescribed accuracy and j is the number of the outer iteration. The value of the obtained $H_n^{m+1,j+1}$ is considered to be the approximate solution at the time moment $t = (m+1)\Delta t$ at the point $x = n\Delta x$.

The main advantage of the considered algorithm is that, since the limitation of the time step $(\Delta x)^2/\Delta t$ is quite weak, it allows one (unlike earlier used methods) to proceed with a sufficiently large step on t in the calculations. The latter significantly minimizes the time expenditures for the calculations. Unfortunately, the considered procedure is rather cumbersome overall, and the advantage noted above does not secure a significant effect for multidimensional problems related to the 3-DNLS-class equations with the integral diffraction term (see Chap. 4). Nevertheless for problems connected with the study of evolution of one-dimensional Alfvén waves [33,176] this method in practice has well proven itself (see Sect. 2.4.5).

Dynamic Spectral Methods. Consider an approach to the integration of DNLS equation that can be called the *dynamic spectral method*, first proposed and used for the multidimensional KdV-class equations known as the Kadomtsev–Petviashvili (KP) class of equations [81,98][15] We assume DNLS equation in the form (2.214) and execute the Fourier transform on the x-coordinate:

$$\begin{aligned} H(t,\xi) &= (2\pi)^{-1} \textstyle\int h(t,x)\exp(-\mathrm{i}x\xi)\mathrm{d}x, \\ h(t,x) &= \textstyle\int H(t,\xi)\exp(\mathrm{i}x\xi)\mathrm{d}\xi. \end{aligned} \tag{2.239}$$

Thus we obtain

$$\partial_t H + fG + gH = 0, \tag{2.240}$$

where $f = \mathrm{i}s\xi$, $g = \mathrm{i}\xi^2$, $G = W * H$, and $W = \mathcal{F}[|h|^2]$. We therefore have obtained a simpler equation on the complex functions with the pure imaginary coefficients. The H values at $t = 0$ are defined by the Fourier transform of the initial condition, $h(0,x) = \psi(x)$, of the Cauchy problem for the DNLS equation (2.214), and $G|_{t=0} = W|_{t=0} * \mathcal{F}[\psi]$, $W|_{t=0} = \mathcal{F}[|\psi|^2]$. The arising problem of finding the convolution values on the next time layers is solved by executing the Fourier transform of the squared modulus of the inverse Fourier transform of H,

$$W = \mathcal{F}\left[\left|\mathcal{F}^{-1}[H]\right|^2\right].$$

Now consider another approach. Assume that the function describing the magnetic field is represented as $h = p\bar{h}$, where $p = 1 + \mathrm{i}e$, and e is the "eccentricity" of the polarization ellipse of the Alfvén wave. The *DNLS equation* in this case is given by

$$\partial_t \bar{h} + 3s|p|^2\bar{h}^2\partial_x\bar{h} - \mathrm{i}\varepsilon\partial_x^2\bar{h} = 0, \tag{2.241}$$

[15] See for details Sect. 4.3.3.

where the signs of $\varepsilon = \pm 1$ correspond to (2.214) or (2.216), respectively. Applying the Fourier transform (2.239), we obtain the equation in the form (2.240)

$$\partial_t \bar{H} + q\bar{G} + r\bar{H} = 0, \tag{2.242}$$

where $q = 3\mathrm{i}s|p|^2\xi$, $r = \mathrm{i}s\xi^2$, $\bar{G} = \bar{W} * \bar{H}$, and $\bar{W} = \bar{H} * \bar{H}$. In this case the convolutions on the time layers $n = t/\tau = 1, 2, 3, \ldots$ are calculated using the convolution theorem [179] according to the scheme

$$\{H\} \to \{\mathcal{F}[H]\} \to \{V\} = \{\mathcal{F}[H]\mathcal{F}[H]\} \to \{W\} = \{\mathcal{F}^{-1}[V]\}.$$

Thus in both considered cases we should solve an equation of type (2.240) on the complex functions with the purely imaginary coefficients. That can be done by using the *Runge–Kutta method* [180]. For example, one of the possible often-used schemes is

$$H^{n+1} = H^n + (k_1 + 2k_2 + 2k_3 + k_4)/6, \tag{2.243}$$

where

$$k_1 = \tau\varphi(H^n), \qquad k_2 = \tau\varphi(H^n + k_1/2),$$
$$k_3 = \tau\varphi(H^n + k_2/2), \qquad k_4 = \tau\varphi(H^n + k_3),$$

and

$$\varphi(H^n) = -fG^n - gH^n.$$

This scheme approximates (2.240) with the accuracy $O(\tau^4)$. Here, H^n and G^n are the Fourier-image and the convolution at the time layer n at the point ξ of the spectral space, while $\tau = t_{n+1} - t_n$ is the time step.

With regard to (2.240) and (2.242), we note that the finiteness of the real region of the numerical integration leads to percolation of the spectral components in the spectrum of the function h (*Gibbs oscillations*). This is connected with the presence of the ruptures of the periodic extension of h on the boundaries of the region. Therefore, to suppress the rupture order and to coordinate a possibly larger number of derivatives at the boundaries in the numerical realization of the Fourier transform when we approximate an integral by a finite sum, it is necessary to introduce a multiplicative weight function into (2.239). For that, the direct transform (2.239) can be presented as

$$H_\sigma(t, \xi_n) = \frac{1}{N} \sum_{n=0}^{N-1} \sigma(n\Delta x) h(t, n\Delta x) \exp(-\mathrm{i}\xi_n n\Delta x), \tag{2.244}$$

where $\sigma(n\Delta x) = \sigma[(N-n)\Delta x]$ and $\xi_n = 2\pi n/N\Delta x$. If, as a result of the weight function, it is possible to achieve a smoothly varying tendency of the function to zero on the boundaries for a minimum distortion of the spectrum in the center of the integration region, the periodic extension of h becomes continuous up to the higher order derivatives.

We can choose various sorts of windows widely used in the spectral analysis as the weight function σ. For example, we can effectively apply the *Blackman–Harris windows* [181],

$$\sigma(j) = c_0 - c_1 \cos(2\pi j/N) + c_2 \cos(4\pi j/N) - c_3 \cos(6\pi j/N), \quad j = 0, 1, 2, \ldots, N-1,$$

which gave a good result in our numerical simulations for the multidimensional equations of the KP- and DNLS-classes (see Chap. 4).

The numerical simulations of the DNLS equation[16] enable us to conclude that in the one-dimensional case, both approaches described by (2.240) and (2.242) are approximately equivalent regarding their temporal characteristics, though the second approach has nevertheless some advantages which can appear to be essential to multidimensional problems.

Families of Explicit and Implicit Difference Schemes. The approach considered above, when applied to an elliptically polarized wave with the function h in (2.241) as $h = (1 + \mathrm{i}e)\bar{h}$, easily enables us to construct groups of rather simple explicit and implicit difference schemes. Consider it on the example of schemes with $O(\tau^2, \Delta^2)$ and $O(\tau^2, \Delta^4)$ approximation.

A simple way to realize the three-layered *explicit scheme with* $O(\tau^2, \Delta^2)$ *approximation* is given by

$$\frac{\bar{h}_i^{n+1} - \bar{h}_i^{n-1}}{2\tau} = \frac{\alpha}{2\Delta}\left(\bar{h}_i^n\right)^2\left(\bar{h}_{i+1}^n - \bar{h}_{i-1}^n\right) + \frac{\beta}{\Delta^2}\left(\bar{h}_{i+1}^n - 2\bar{h}_i^n + \bar{h}_{i-1}^n\right). \quad (2.245)$$

The *explicit scheme with* $O(\tau^2, \Delta^4)$ *approximation* can be written as

$$\frac{\bar{h}_i^{n+1} - \bar{h}_i^{n-1}}{2\tau} = \frac{\alpha}{12\Delta}\left(\bar{h}_i^n\right)^2\left(\bar{h}_{i+2}^n - 8\bar{h}_{i+1}^n + 8\bar{h}_{i-1}^n - \bar{h}_{i-2}^n\right)$$
$$-\frac{\beta}{12\Delta^2}\left(\bar{h}_{i+2}^n - 16\bar{h}_{i+1}^n + 30\bar{h}_i^n - 16\bar{h}_{i-1}^n + \bar{h}_{i-2}^n\right). \quad (2.246)$$

Here, $\alpha = -3s|p|^2$, $\beta = \mathrm{i}\varepsilon$, and $p = 1 + \mathrm{i}e$. The conditions of stability for (2.245) and (2.246) can be easily found using the Fourier method (see, for example, Sect. 1.3.1) [178], and for sufficiently small steps we obtain, respectively,

$$\tau \le \Delta^2 \left|2\varepsilon - \alpha v^2 \Delta\right|^{-1} \approx \frac{\Delta^2}{2|\varepsilon|} \quad (2.247)$$

and

$$\tau \le 3\Delta^2 \left|7\varepsilon - 4\alpha v^2 \Delta\right|^{-1} \approx \frac{3\Delta^2}{7|\varepsilon|}, \quad (2.248)$$

where $v = \max_{i,n} \left|\bar{h}_i^n\right|$.

---

[16] For some results see Sect. 2.4.5.

Consider now two relatively simple implicit schemes for the DNLS equation written in the form of (2.241). The *implicit scheme with $O(\tau^2, \Delta^2)$ approximation* can be written as

$$\frac{\bar{h}_i^{n+1} - \bar{h}_i^n}{\tau} = \frac{\alpha}{4\Delta}\left[\left(\bar{h}_i^{n+1}\right)^2\left(\bar{h}_{i+1}^n - \bar{h}_{i-1}^n\right) + \left(\bar{h}_i^n\right)^2\left(\bar{h}_{i+1}^{n+1} - \bar{h}_{i-1}^{n+1}\right)\right]$$
$$+\frac{\beta}{2\Delta^2}\left(\bar{h}_{i+1}^{n+1} - 2\bar{h}_i^{n+1} + \bar{h}_{i-1}^{n+1} + \bar{h}_{i+1}^n - 2\bar{h}_i^n + \bar{h}_{i-1}^n\right). \quad (2.249)$$

This scheme can be solved by use of various versions of the *sweep method*, e.g., according to the algorithms of the monotonous or non-monotonous sweep method [97]. Thus we obtain from (2.249) the set of algebraic equations

$$-a_i^1 \bar{h}_{i-1}^{n+1} + a_i^2 h_i^{n+1} - a_i^3 h_{i+1}^{n+1} = f_i, \qquad i = 1, 2, \ldots, N, \qquad (2.250)$$

where

$$a_i^1 = \frac{\alpha}{4\Delta}\left(\bar{h}_i^n\right)^2 - \frac{\beta}{2\Delta^2},$$
$$a_i^2 = \frac{1}{\tau} - \frac{\alpha}{4\Delta}\left[\bar{h}_i^n\left(\bar{h}_{i+1}^n - \bar{h}_{i-1}^n\right)\right] + \frac{\beta}{\Delta^2},$$
$$a_i^3 = -\frac{\alpha}{4\Delta}\left(\bar{h}_i^n\right)^2 - \frac{\beta}{2\Delta^2},$$

and

$$f_i = \frac{1}{\tau} + \frac{\beta}{2\Delta^2}\left(\bar{h}_{i+1}^n - 2\bar{h}_i^n + \bar{h}_{i-1}^n\right).$$

For $i = 1$ and $i = N$ we can choose the zeroth boundary conditions (in the case of sufficiently rapid decrease of the solution $|\bar{h}| \to 0$ for $|x| \to \infty$) or, taking into account the oscillating character of the function $\bar{h}$ in the majority of cases, the periodic boundary conditions. In either case, we have a three-diagonal matrix, and we can write (2.250) in the matrix form $A \cdot \mathbf{h} = \mathbf{f}$, where

$$A = \left\| \begin{array}{ccccccc} a_1^1 & a_1^2 & 0 & 0 & \ldots & 0 & 0 \\ a_2^1 & a_2^2 & a_2^3 & 0 & \ldots & 0 & 0 \\ 0 & a_3^2 & a_3^3 & a_3^4 & \ldots & 0 & 0 \\ \ldots & \ldots & \ldots & \ldots & \ldots & \ldots & \ldots \\ 0 & 0 & 0 & 0 & \ldots & a_n^{n-1} & a_n^n \end{array} \right\|, \qquad (2.251)$$

$$\mathbf{h} = \left\| \begin{array}{c} h_1 \\ h_2 \\ \ldots \\ h_n \end{array} \right\|, \quad \text{and} \quad \mathbf{f} = \left\| \begin{array}{c} f_1 \\ f_2 \\ \ldots \\ f_n \end{array} \right\|.$$

The values $\bar{h}_i$ can be found, for example, with the help of the *sweep equations* [97]:

$$\delta_1 = \kappa_1, \quad \gamma_1 = \mu_1, \quad \delta_{i+1} = \frac{a_i^3}{a_i^2 - a_i^1 \delta_i}, \quad \gamma_{i+1} = \frac{f_i + a_i^1 \gamma_i}{a_i^2 - a_i^1 \delta_i},$$
$$i = 1, 2, \ldots, N-1, \quad (2.252)$$
$$\bar{h}_N = \frac{\mu_2 + \kappa_2 \gamma_n}{1 - \delta_n \kappa_2}, \quad \bar{h}_i = \delta_{i+1} \bar{h}_{i+1} + \gamma_{i+1}, \quad i = N-1, N-2, \ldots, 1,$$

where $\kappa_{1,2}$ and $\mu_{1,2}$ are determined from equations $\bar{h}_1 = \kappa_1 \bar{h}_2 + \mu_1$ and $\bar{h}_N = \kappa_2 \bar{h}_{N-1} + \mu_2$, respectively, taking also into account the boundary conditions. For example, for the zeroth boundary conditions we have $\kappa_1 = a_1^3/a_1^2$, $\mu_1 = f_1/a_1^2$, $\kappa_2 = a_N^1/a_N^2$, and $\mu_2 = f_N/a_N^2$.

Note, that (2.249) is correct when the matrix A is not degenerate, i.e., $\det A \neq 0$. Note also, that expressions (2.252) can be used only if the conditions

$$\left|a_i^2\right| \geq \left|a_i^1\right| + \left|a_i^3\right|, \qquad i = 1, 2, \ldots, N-1,$$
$$\left|\kappa_{1,2}\right| \leq 1, \quad \text{and} \quad \left|\kappa_1\right| + \left|\kappa_2\right| < 2,$$

are satisfied. The latter (almost) always takes place for a sufficiently small step Δ, and gives the condition

$$\tau \leq 4\Delta \left|\sqrt{3}\alpha v^2\right|^{-1}.$$

As demonstrated by subsequent numerical experiments, this condition is sufficiently exact.

For a more accurate, *implicit scheme with $O(\tau^2, \Delta^4)$ approximation*, we can use the difference scheme given by

$$\begin{aligned}\frac{\bar{h}_i^{n+1} - \bar{h}_i^n}{\tau} = \frac{\alpha}{24\Delta} \Big[&\left(\bar{h}_i^{n+1}\right)^2 \left(\bar{h}_{i+2}^n - 8\bar{h}_{i+1}^n + 8\bar{h}_{i-1}^n - \bar{h}_{i-2}^n\right) \\ &+ \left(\bar{h}_i^n\right)^2 \left(\bar{h}_{i+2}^{n+1} - 8\bar{h}_{i+1}^{n+1} + 8\bar{h}_{i-1}^{n+1} - \bar{h}_{i-2}^{n+1}\right)\Big] \\ - \frac{\beta}{24\Delta^2} &\left(\bar{h}_{i+2}^{n+1} - 16\bar{h}_{i+1}^{n+1} + 30\bar{h}_i^{n+1} - 16\bar{h}_{i-1}^{n+1} + \bar{h}_{i-2}^{n+1}\right. \\ &\left. + \bar{h}_{i+2}^n - 16\bar{h}_{i+1}^n + 30\bar{h}_i^n - 16\bar{h}_{i-1}^n + \bar{h}_{i-2}^n\right). \qquad (2.253)\end{aligned}$$

This scheme can also be solved by using various versions of the sweep method, such as those considered in Sect. 1.3.2.

To conclude this subsection, we note, that numerical simulations using the explicit and implicit difference schemes with the representation $h = p\bar{h}$ need considerably fewer computer resources than simulations using directly the difference schemes approximating the DNLS equation in standard forms (2.214) and (2.216). This is related, first of all, to the real character of the function $\bar{h}$, unlike the function h in these equations. This advantage is especially important for multidimensional cases considered in Chap. 4.

2.4.5 Results of Numerical Simulations

To test the accuracy of the above schemes as well as their time expenses, we used the exact analytical solution of the DNLS equation [37]

$$h(t,x) = (A/2)^{1/2}\left(\mathrm{e}^{-Ax} + \mathrm{i}\mathrm{e}^{Ax}\right)\mathrm{e}^{-\mathrm{i}A^2 t}\cosh^{-2}(2Ax). \tag{2.254}$$

as the initial condition. We calculated the relative deviation of the mean of the numerical solution from the exact one at every time step:

$$\delta = \left|\left|h_\tau^{\mathrm{num}}\right|^2 - \left|h_\tau^{\mathrm{exact}}\right|^2\right| \Big/ \left|h_\tau^{\mathrm{exact}}\right|^2 .$$

The test was done for the equation's coefficients $s = -1$, $\varepsilon = 1$, and $s = 1$, $\varepsilon = -1$. Some results of the numerical test are presented in Figs. 2.14, 2.15 and Table 2.2. We can see that in both cases, the relative deviation

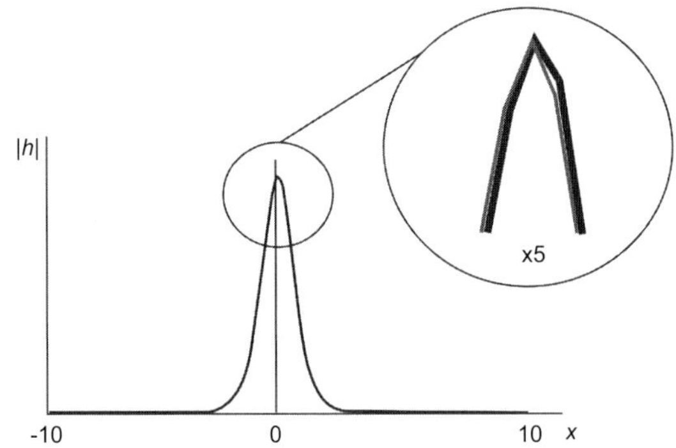

Fig. 2.14. Comparison of the numerical solution of the DNLS equation obtained using scheme (2.249) (black bold line) with the exact analytical solution (grey line) for $s = -1$ and $\varepsilon = 1$ at $t \approx 54$

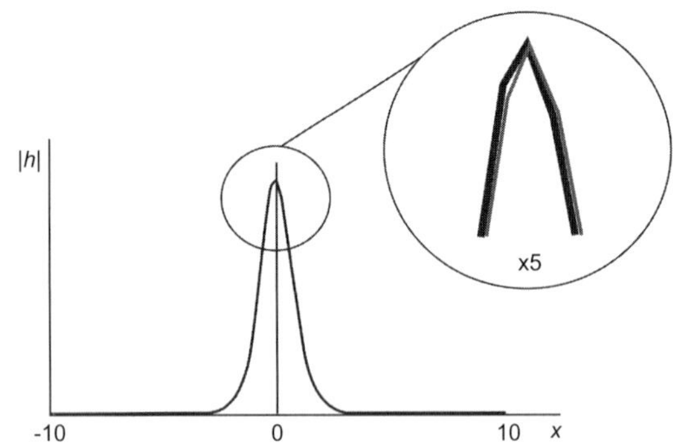

Fig. 2.15. Comparison of the numerical solution of the DNLS equation obtained using scheme (2.249) (black bold line) with the exact analytical solution (grey line) for $s = 1$ and $\varepsilon = -1$ at $t \approx 54$

at quite late time for $\Delta = 0.1$ does not exceed $\delta \sim 0.01$ for the schemes with the approximation $O(\tau^2, \Delta^2)$, and $\delta \sim 0.0005$ for the schemes with

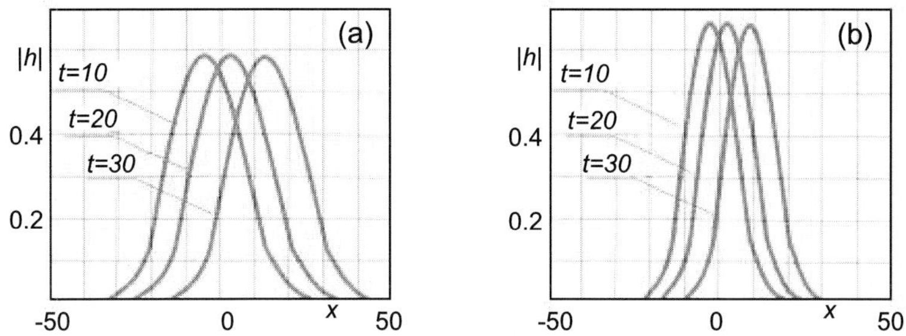

Fig. 2.16. Evolution of the solitary pulse (2.255) with $s = -1$ and $v = 0.5$. **a** $\gamma = \pi/4$, $\chi^2 = 0.1768$. **b** $\gamma = \pi/8$, $\chi^2 = 0.1353$

the approximation $O(\tau^2, \Delta^4)$. For schemes (2.240) and (2.242) realizing the spectral approach, the deviations are approximately equal and consist of $\delta \sim 0.01$. The schemes (2.246) and (2.253) with the approximation $O(\tau^2, \Delta^4)$ provide the best accuracy characteristics, while the accuracy of the solution using the spectral method (2.242) is a little bit lower, this is stipulated by an error arising because of the *Gibbs oscillations* due to the limited integration region in the numerical simulation.

Table 2.2. Some testing results at $t \approx 54$ for the schemes with $\delta = 0.1$

Scheme	(2.242)	(2.245)	(2.246)	(2.249)	(2.253)
τ	0.01	0.005	0.00125	0.01	0.01
δ	9.8×10^{-3}	9.7×10^{-3}	4.9×10^{-4}	9.1×10^{-3}	3.8×10^{-4}

To study the evolution of the *Alfvén soliton*, we used the difference schemes (2.242) and (2.253). As in Ref. [176], we employed the initial conditions of two types:

1. The *solitary pulse* looking like the one-soliton solution of the DNLS equation
$$h(0, x) = u \exp(\mathrm{i}\varphi), \tag{2.255}$$
$$u^2 = \frac{8\chi^2 \sin^2 \gamma}{\cosh(4\chi^2 \sin \gamma x) + \cos \gamma}, \quad \varphi = -2s\chi^2 \cos \gamma x - \frac{3s}{4} \int_{-\infty}^{x} u^2 \mathrm{d}x,$$
where $0 < \gamma < \pi$, corresponding to the soliton with the velocity $v = -4s\chi^2 \cos \gamma$;

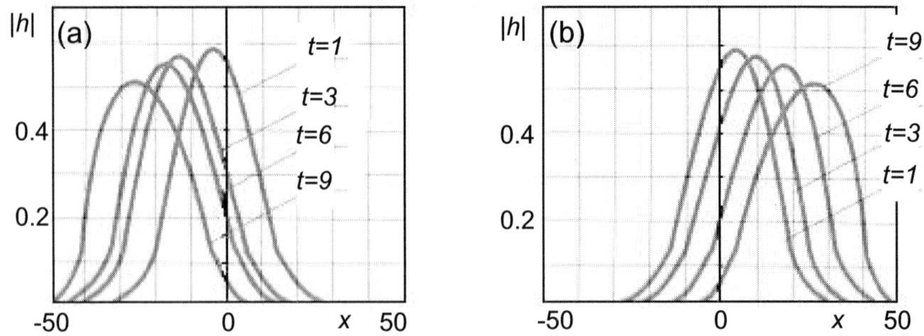

Fig. 2.17. Evolution of the solitary pulse (2.256) with $A_0 = 0.6$ and $l = 14$. **a** $s = 1$, $\lambda = 5$. **b** $s = -1$, $\lambda = -5$

2. The modulated plane wave

$$h(0, x) = A_0 \exp\left(\frac{2\pi i x}{\lambda}\right) \exp\left(-\frac{x^2}{l^2}\right), \qquad (2.256)$$

where λ is the wavelength, A_0 is the amplitude, and l is the modulation length of the Gaussian wave packet.

In the basic series of numerical simulations, the periodic boundary conditions were chosen, and in a number of cases we used the zeroth boundary conditions. Figures 2.16 and 2.17 present some examples of the results of our numerical simulation for various[17] signs of the coefficients of the DNLS equation (2.241) with $\varepsilon = 1$. We can see that in the case of a solitary wave looking like the one-soliton solution of the DNLS equation, the Alfvén soliton conserves its shape and momentum for different values of the amplitude. In the case of the initial plane modulated wave, we observe the broadening of the wave packet and increasing of the steepness of its profile together with the decreasing amplitude in the time evolution of the wave. The results obtained in our numerical experiments are similar to the results obtained in Ref. [176]; they allow us to study the structure and dynamics of the one-dimensional Alfvén waves with various polarization, and the developed numerical methods can form the basis for next generations of numerical codes for simulation of more complex three-dimensional systems taking into account higher order effects taking place in a magnetized plasma (see Chap. 4).

[17] The same as in Ref. [176].

3. Classic Two- and Three-Dimensional KP Models and Their Applications

3.1 (1+2)- and (1+3)-Dimensional KP Equation

In this section we discuss generalization of the KdV equation on a weakly non-one-dimensional case when the Kadomtsev–Petviashvili (KP) equation appears (Sect. 3.1.1). Furthermore, we consider various classes of solutions of the KP equation and investigate the problem of their stability in more detail in Sects. 3.1.2 and 3.1.3. Finally, we present some effective methods of numerical integration of the KP equation (Sect. 3.1.4) used in particular for study of the wave self-focusing phenomenon and the wave collapse in the classic KP model.

3.1.1 Generalization of the KdV Equation on Weakly Non-One-Dimensional Case

As we already know from Chap. 1, a wide class of one-dimensional nonlinear waves in weakly dispersive media (e.g., *surface waves* in shallow water, *ion-acoustic waves* and *magnetosonic waves* in a plasma) is described by the *KdV equation*

$$\partial_t u + 6u\partial_x u + \partial_x^3 u = 0. \tag{3.1}$$

This equation can be used with equal success for media with the negative dispersion (when the phase velocity of linear waves decrease with the increase of the wave number) as well as for media with the positive dispersion; the only difference is in the direction of the x-coordinate.

It is known that an important role in the evolution of an arbitrary initial disturbance is played by special solutions of (3.1) of the soliton type, namely

$$u(t, x) = af\left[\sqrt{a}\,(x - x_0)\right], \tag{3.2}$$

where a is the amplitude of the wave, $x_0 = at$ is its phase, and $f(\xi) = 3(\cosh \xi/2)^{-2}$. The soliton solution (3.2) of (3.1) is stable in the one-dimensional case. However, there is a problem: whether the stability of the soliton is conserved with respect to a feeble bending, when its amplitude a and phase x_0 become slowly varying functions of the y-coordinate, perpendicular to the direction of the soliton's propagation. This problem was first solved by

Kadomtsev and Petviashvili in 1970 [16], and the obtained equation was called the *Kadomtsev–Petviashvili equation* (*KP equation*).

When the strict one-dimensionality of the problem is broken, the KdV equation is distorted, and if the dependence of all functions on the y-coordinate is slow, this can be taken into account in the derivation of the basic equation from the full set of hydrodynamics equations. Let us demonstrate how to do that using the approach of Sect. 1.1.1. First, return to the three-dimensional nonlinear evolution *Boussinesq equations* (1.7):[1]

$$\partial_t \mathbf{v} + (\mathbf{v}\boldsymbol{\nabla})\,\mathbf{v} + \frac{c^2}{\rho}\boldsymbol{\nabla}\rho + \frac{2c_0\beta}{\rho_0}\boldsymbol{\nabla}\Delta\rho = 0,$$
$$\partial_t \rho + \boldsymbol{\nabla}(\rho\mathbf{v}) = 0. \tag{3.3}$$

As in Sect. 1.1.1, assume that v/c_0, $(\rho-\rho_0)/\rho_0$, and δ/λ are small values of the first order and suppose (as in the standard hydrodynamics) that

$$c^2(\rho) = c_0^2\,(\rho/\rho_0)^{\gamma-1}, \qquad \gamma = c_\mathrm{p}/c_\mathrm{v}$$

(see Sect. 1.1.1). Now, we look for the solution of (3.3) in the form of [83],

$$\rho(t,x,\psi') = \rho(v) + \varphi(t,x) - \psi'(t,y), \tag{3.4}$$

where

$$\begin{aligned}\rho(v) &= \pm c(v)d_v\rho,\\ c(v) &= c_0 + (\gamma-1)v/2,\end{aligned} \tag{3.5}$$

φ is a small value of the second order, $\psi'(t,y)$ is a small perturbation on the transverse coordinate, and the dependence of v on y is slow. Considering a wave propagating in the positive x-direction and, as in Sect. 1.1.1, ignoring the terms smaller than the second order, we can see that $\varphi(t,x)$ satisfies the equation $\partial_t\varphi + c_0\partial_x\varphi = 0$. In this case, substituting (3.4) into (3.3) and using (3.5), and excluding the potential φ from the obtained expressions, we obtain instead of (1.10) the equation

$$\partial_t v + \left(c_0 + \frac{\gamma+1}{2}v\right)\partial_x v + \beta\partial_x^3 v = \frac{c_0^2}{\rho_0}\partial_y\psi'. \tag{3.6}$$

Introducing now the new variables

$$\xi = x - c_0 t, \qquad u = \frac{\gamma+1}{2}v, \quad \text{and} \quad \psi = \frac{c_0^2}{\rho_0}\psi',$$

we can write (3.6) as

$$\partial_t u + u\partial_\xi u + \beta\partial_\xi^3 u = \partial_y\psi, \tag{3.7}$$

---

[1] Recall that these equations are valid for waves in shallow water as well as for waves in a plasma and other weakly dispersive nonlinear media.

where the term $\partial_y \psi$ describes the small (transverse) perturbation of the KdV equation. Its form, as was shown in [16], is determined by the equality

$$\partial_x \psi = \mp (c_0/2)\, \partial_y u,$$

where the upper sign corresponds to the negative dispersion and the lower sign corresponds to the positive dispersion. Finally, we obtain that generalization of the KdV equation on the weakly non-one-dimensional (two-dimensional) case is given by

$$\partial_t u + u\partial_\xi u + \beta \partial_\xi^3 u = \mp \frac{c_0}{2} \int\limits_{-\infty}^{\xi} \partial_y^2 u \mathrm{d}\xi. \tag{3.8}$$

For further convenience we proceed with the scale transform and rewrite (3.8) in the standard form as

$$\partial_t u + 6u\partial_x u + \beta \partial_x^3 u = 3\sigma^2 \int\limits_{-\infty}^{x} \partial_y^2 u \mathrm{d}x, \tag{3.9}$$

where $\sigma^2 = 1$ corresponds to the positive, and $\sigma^2 = -1$ corresponds to the negative dispersion (i.e., respectively, to the increase and the decrease of the phase velocity of the linear waves with the increase of the wave number). Equation (3.9) can be also written in the differential form as

$$\partial_x \left(\partial_t u + 6u\partial_x u + \beta \partial_x^3 u\right) = 3\sigma^2 \partial_y^2 u, \tag{3.10}$$

which is sometimes more convenient for a particular consideration.

Equations (3.9) and (3.10) represent the two-dimensional (strictly speaking, as it is often used, (1+2)-dimensional, meaning that there are one time and two space coordinates) *Kadomtsev–Petviashvili equation* (*KP equation*) in the integro-differential and differential forms, respectively. Note that the KP equations (3.9), (3.10) are written in the reference frame moving along the x-axis with the velocity c_0. Similar to the one-dimensional case described by the KdV equation, for the waves in shallow water, in a plasma and in a neutral gas, the function u stands for the disturbance of the velocity or the pressure (note that in a simple wave the latter are uniquely related to each other). The physical sense of the parameters ρ, β, and c_0 depends on the particular class of the studied phenomena (for example, on the type of the medium), and will be further explained in the corresponding sections.

The two-dimensional KP equations (3.9) and (3.10) can be easily generalized to the three-dimensional case. For this purpose it is necessary to consider the function $\psi'(t, y, z)$ defining the slow dependence of v on the variables y and z in (3.4). Thus the KP equation (3.9) in this case is given by

$$\partial_t u + 6u\partial_x u + \beta \partial_x^3 u = 3\sigma^2 \int\limits_{-\infty}^{x} \Delta_\perp u \mathrm{d}x, \tag{3.11}$$

and (3.10) can be written as

$$\partial_x\left(\partial_t u + 6u\partial_x u + \beta\partial_x^3 u\right) = 3\sigma^2 \Delta_\perp u, \tag{3.12}$$

where $\Delta_\perp = \partial_y^2 + \partial_z^2$. Below we mostly consider the KP equation in the form of (3.9) unless another form is specifically stated.

3.1.2 The KP Equation and its Solutions

The KP equation represents a completely integrable Hamiltonian system [20,24]. The latter is proved, for example, by the possibility to write it in the *Lax representation* [29,30] as

$$\partial_t \hat{\mathrm{L}} = \left[\hat{\mathrm{L}}, \hat{\mathrm{A}}\right],$$

where the differential operators are given by [24]

$$\hat{\mathrm{L}} = \sigma\partial_y - \hat{\mathrm{M}} \quad \text{and} \quad \hat{\mathrm{A}} = 4\partial_x^3 + 6u\partial_x + 3\partial_x u - 3\sigma w, \tag{3.13}$$

where $\hat{\mathrm{M}} = \partial_x^2 + u$ and $\partial_x w = \partial_y u$ (note that for $\sigma = 1$ the $\hat{\mathrm{L}}$ operator is the thermal conductivity operator while for $\sigma = \mathrm{i}$ it is the non-stationary *Schrödinger operator*).

The universality of the KP equation can be seen from the following derivation. Consider the dispersion law in a medium with the weak dependence of its "sound" velocity[2] on the wave number

$$\omega^2 = k^2 + \varepsilon k^4, \qquad \varepsilon k^2 \ll 1. \tag{3.14}$$

In the two-dimensional case $\mathbf{k} = (k_x, k_y)$; if, furthermore, $k_y \ll k_x$, we can write [24]

$$\omega^2 - k_x^2 = (\omega - k_x)(\omega + k_x) = \varepsilon k_x^4 + k_y^2.$$

In the moving reference frame ($\omega = k_x + \Omega$, where $\Omega \ll k_x$) we obtain $2k_x\Omega \cong \varepsilon k_x^4 + k_y^2$, i.e., the relation coinciding (with the accuracy up to a simple change of variables) with the Fourier transform of the linear part of (3.9) with $\varepsilon = -1/(3\sigma^2)$.

Due to the full integrability of the KP equation (as in the case of the KdV equation, having similar universality) we can to construct an infinite series of the *conservation laws.* The first three integrals obtained with the help of the Lax representation [24] from the solution of the spectral problem for the Schrödinger operator associated with KP equation (see Sect. 3.2) are given by

$$\partial_t \mathcal{J}_1 = \partial_t \iint u\mathrm{d}x\mathrm{d}y = 0,$$

[2] See Sect. 1.1.1.

$$\partial_t \mathcal{J}_2 = \partial_t \iint u^2 \mathrm{d}x\mathrm{d}y = 0,$$

and

$$\partial_t \mathcal{J}_3 = \partial_t \iint \left[\frac{1}{2}(\partial_x u)^2 - \frac{3}{2}\sigma^2 \left(\int_{-\infty}^{x} \partial_y u \mathrm{d}x \right)^2 - u^3 \right] \mathrm{d}x\mathrm{d}y = 0.$$

The last two integrals correspond to the conservation of the momentum and energy, respectively, in the system described by the KP equation. We can see that for $\sigma = 0$ (i.e., the right hand side of the KP equation is equal to zero) these integrals (accounting for the proper scale transforms) convert to the usual integrals of the KdV equation (1.16).

The *KP equation* has a wide class of exact solutions including the soliton solutions, as well as solutions in the form of *cnoidal waves*. An analytical solution of the Cauchy problem for the KP equation requires the setting and the subsequent solution of the direct and the inverse scattering problems for the *Schrödinger equation* with the potential satisfying the KP equation; this is quite a difficult and cumbersome procedure. We consider this procedure in detail below in Sect. 3.2.1 while presenting here only some basic results.

It is known that unlike the KdV equation where the form of the soliton solution does not depend on the sign of the dispersion term (the change of the dispersion sign to the opposite is merely equivalent to the change $x \to -x$ and $u \to -u$ in the original equation, see Sect. 1.1), the form of the soliton solution of the KP equation is directly determined by the dispersion sign. It was established in 1970 by Kadomtsev and Petviashvili using the Krylov–Bogolyubov method [16] that in the case of the negative dispersion (corresponding to $\sigma^2 = -1$ in (3.9) with $\beta > 0$), the one-dimensional soliton is stable with respect to the perpendicular infinitesimal perturbations, i.e., "bending" of the soliton in the transverse direction results in its "elastic" decremental oscillation (see Fig. 3.1). This is related to the fact that for $\sigma^2 = -1$ the perturbations can be easily transferred from the soliton to the medium and then spread around in all directions.

In the opposite case, the soliton is unstable with respect to its bending, i.e., for $\sigma^2 = 1$ the small perturbations cannot be transferred from the soliton to the medium and therefore they exponentially grow in time, $\sim \exp(\gamma t)$, leading to the increase of the soliton's amplitude. The instability rate is quite large, $\gamma \sim \sqrt{a}$. In the region of the localization of the perturbations, the soliton's velocity differs from that of an unperturbed soliton, leading to even further increase of the amplitude of the perturbation. Thus the KP equation with the positive dispersion does not have stable one-dimensional soliton solutions.

Analytically, the one-dimensional soliton of the KP equation is described by expression (3.2), which is valid in the case when the soliton propagates along the x-axis. A more general expression for the one-dimensional KP solitons propagating obliquely with respect to the x-axis as well as their collisions, can be obtained analytically using the *dressing method*. Section 3.2.1

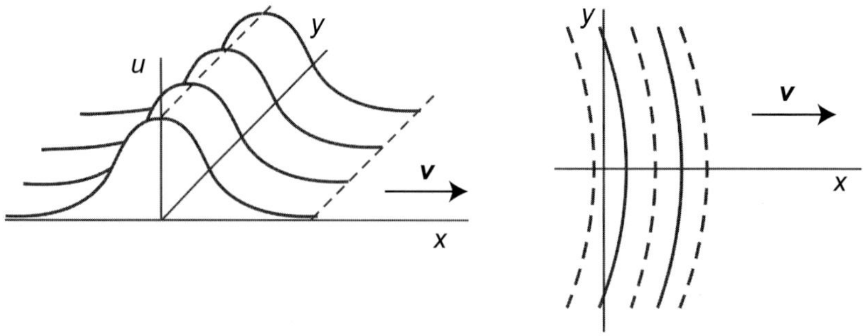

Fig. 3.1. Evolution of the one-dimensional KP soliton in the presence of transverse perturbation

is specifically devoted to the study of this method and its application to the integration of KP equation.

In 1976, Petviashvili demonstrated numerically (by using the *method of stabilizing factor*, see the next section) [31] that in the case of positive dispersion the KP equation has a stable solution in the form of a two-dimensional soliton with the algebraic asymptotics, i.e.,

$$u \sim \left(x^2 + y^2\right)^{-1}$$

when $|x, y| \to \infty$. The soliton in the general form as isolines $u(x, y)$ =const for $t = T$ is shown in Fig. 3.2. The cross-sections of such a two-dimensional soliton along the x- and y-axes are presented in Fig. 3.3. The dressing method

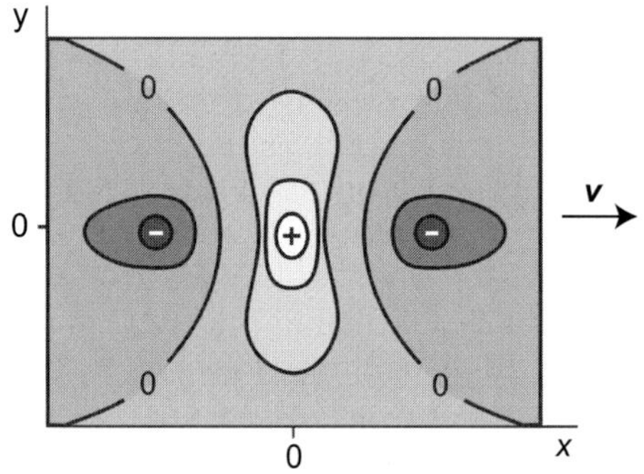

Fig. 3.2. The two-dimensional KP soliton: isolines $u(x, y)$ =const

for the analytical integration of the (1+2)-dimensional KP equation (3.10) with $\sigma^2 = 1$ demonstrates that the two-dimensional *KP soliton* with the algebraic asymptotics is described by

$$u(t, x, y) = 2\partial_x^2 \ln \det B, \tag{3.15}$$

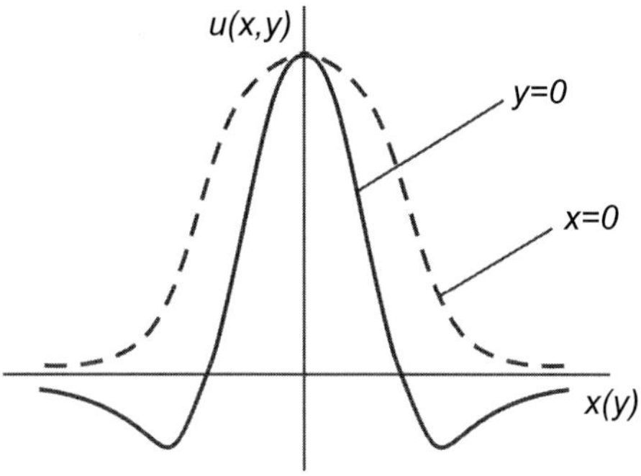

Fig. 3.3. Cross-sections of the two-dimensional KP soliton along the x- and y-axes

where

$$\det B = 4\left(\nu+\nu^*\right)^{-2} + \left|x - \mathrm{i}\nu y - \xi - 3\nu^2 t\right|^2 .$$

Such a two-dimensional soliton has the following parameters [83]:

- Velocity $\mathbf{v} = (v_x, v_y)$, with $v_x = 3|\nu|^2$ and $v_y = -6\mathrm{Im}\nu$
- Amplitude $u_{\max} = (\nu+\nu^*)^2$, $u_{\min} = -u_{\max}/8$
- Coordinates of its center at the moment $t = 0$: $x_c = \mathrm{Re}\xi - \mathrm{Im}\xi\mathrm{Im}\nu$, $y_c = -\mathrm{Re}\nu\mathrm{Im}\xi$
- Orientation (the inclination angle of the big semi-axis to the x-axis) $\alpha = \arctan(-\mathrm{Re}\nu/\mathrm{Im}\nu)$
- Small and big "axes" of the line $u = 0$ (for $\alpha = \pi/2$): $l_s = (2|\alpha|/\pi)^{1/2}$, $l_b = 2l_s$
- "Eccentricity" $\varepsilon = l_s/l_b = |\alpha|/\pi$

The soliton (3.15) is stable in the case of positive dispersion. The stability problem for two-dimensional solitons is discussed in detail below (see Sect. 3.1.3) together with the stability problem of three-dimensional solutions of the (1+3)-dimensional KP equation. When developing an analytical approach to integration of the KP equation (Sect. 3.2.1), we will see that the "dressing" method enables us to obtain not only the one-soliton solution but also the N-soliton solutions; the latter describe collisions of N two-dimensional solitons of the KP equation.

Besides the considered class of exact solutions (solitons), the KP equation allows solutions in the form of stationary periodic *cnoidal waves* [24,182] given by

$$u\left(x - c_0 t\right) = \frac{1}{6}c_0 - 2\wp\left(x - c_0 t + \mathrm{i}\omega' \mid \omega, \omega'\right), \qquad (3.16)$$

where $\wp\left(z|\omega,\omega'\right)$ is the elliptic Weierstrass function with the real (2ω) and imaginary ($2\mathrm{i}\omega$) periods ($\wp(x+\mathrm{i}\omega')$ is real and limited for real x).

The transverse *modulational perturbations* on the background of the periodic waves (within the framework of KP equation) were first considered in [183,184] where the dispersion relation (cubic in the wave frequency) was obtained; it was shown that in the low-frequency limit the *cnoidal waves* are stable/unstable for the same cases when the solitons are stable/unstable.[3] In these papers, the perturbation theory (i.e., direct expansion in a series of quasi-momenta and transverse wave numbers) was used. However, for the nonlinear equations admitting the commutative *Lax representation* (in particular, for KdV and KP equations), the *IST method* [24] enables us to advance significantly in the study of stability of the exact periodic solutions with respect to arbitrary small perturbations [186]. Reference [182] gives a detailed analysis of the stability of cnoidal waves in the KP model. Thus for the case of the negative dispersion, the stability of these (long-wavelength) waves (as well as oscillations with arbitrary quasi-momenta and wave numbers) is strictly proved. For a medium with positive dispersion, when $\sigma^2 > 0$ in the KP equations (3.9)–(3.12), it was shown that for the transverse wave numbers $k < k_{\rm cr}$, an instability of cnoidal waves takes place. The threshold is given by

$$|\sigma|^2 \left(k/q\right)^2_{\rm cr} = \nu^2 \left(s'^2/\lambda\right)^2 ,$$

such that above the threshold the solutions are unstable for $s'^2 \ll 1$, when the real period of the wave (i.e., the distance between two soliton maxima), 2ω, significantly exceeds the imaginary period, $2\omega'$, corresponding to the characteristic scale of the solution change near the solitons' maxima, and showing that the solitons are not overlapped. Here, q is the (small) quasi-momentum, $\nu^2 = \wp(\omega) - \wp(\mathrm{i}\omega')$, $s'^2 = 1 - s^2$ is the modulus of the elliptic Jacobi functions, and $\lambda = E(s)/K(s)$ is the ratio of the full elliptic integrals of the first and the second kind [187].

Note that the cnoidal wave (3.16) can be represented as a sum of the one-dimensional soliton solutions [182]:

$$\wp\left(x + \mathrm{i}\omega'\right) = C - \left(\frac{\pi}{2\omega'}\right)^2 \sum_n \cosh^{-2}\left[\frac{\pi}{2\omega'}\left(x - 2n\omega\right)\right]. \tag{3.17}$$

Generally, if a stationary wave is expressed via the set of elliptic functions $\mathcal{U}(x,h)$, where h is the real period which can take any values, the problem of wave decomposition into solitons is easily solved with the assumption that the wave velocity remains finite for $h \to \infty$ [188].

Reference [188] provides numerous examples demonstrating that one-dimensional stationary solutions of a number of nonlinear equations can be expressed as similar-type linear combinations of solitons of equal amplitude. For the KP equation, this helps us find stationary solutions of a new kind. Thus a superposition of a sequence of two-dimensional KP solitons of equal amplitudes shifted from each other can also result in the formation of a stationary wave but in this case, in contrast to the solutions of (3.16) and (3.17), the dynamics and the cross-sectional structure of the resultant wave change. The *KP equation* given by

$$\partial_x \left(\partial_t u + 2u\partial_x u + \partial_x^3 u\right) = \sigma^2 \partial_y^2 u \tag{3.18}$$

[3] These results were later confirmed in Ref. [185].

has the *rational soliton* [188]

$$u^0(x, y, t) = e\left(x - c_x^0 t, y - c_y^0 t\right),$$
$$e(x, y) = -6\left[\left(x - x_0^0\right)^{-2} + \left(x - x_0^{*0}\right)^{-2}\right],$$

where $x_0^0 = \lambda y + \mathrm{i}r_0$, $r_0 = (\alpha^2 y^2/\mu^2 + 3/\alpha^2)^{1/2}$, $c_x^0 = \lambda^2\mu^2 + \alpha^2$, $c_y = 2\lambda\mu^2$, α and λ are the arbitrary real numbers, and $*$ stands for the complex conjugate. The stationary periodic (in x) solution can then be constructed from the rational solitons as

$$u(t, x, y) = \mathcal{U}(x - c_x t, y - c_y t), \tag{3.19}$$

where

$$\mathcal{U}(x, y) = -6 \sum_{n=-\infty}^{\infty} \left[(x - x_0 - nh)^{-2} + (x - x_0^* - nh)^{-2}\right],$$

$x_0 = \lambda y + \mathrm{i}r$, $r = (h/2)\ln(a\cosh\delta y + \sqrt{a^2\cosh^2\delta y - 1})$, $a = \sqrt{1 + 3\gamma^2/\alpha^2}$, $\delta = \gamma\alpha/\mu$, $c_x = c_x^0 - \gamma^2$, and $\gamma = 2\pi/h$. After summation we obtain

$$\mathcal{U}(x, y) = 6\gamma\frac{\partial}{\partial x}\left[\frac{\sin\gamma(x - \lambda y)}{a\cosh\delta y - \cos\gamma(x - \lambda y)}\right]. \tag{3.20}$$

Equation (3.20) shows that the stationary periodic waves of the KP equation formed by superposition of the rational solitons exponentially decrease (i.e. much faster than each of the constituting solitons) in the perpendicular direction. In Refs. [24,188], more general classes of exact solutions of the KP equation were studied, but expressions of type (3.20) were not obtained. Note, that if we assign pure imaginary values to the parameters α and γ in (3.20), new sets of real stationary solutions of the KP equation can be obtained. Thus the change $x \to x + \pi/\gamma$, $\alpha \to \mathrm{i}\alpha$, and $\gamma \to \mathrm{i}\gamma$ gives [188]

$$\mathcal{U}(x, y) = 6\gamma\frac{\partial}{\partial x}\left[\frac{\sinh\gamma(x - \lambda y)}{a\cosh\delta y + \cosh\gamma(x - \lambda y)}\right]. \tag{3.21}$$

This soliton decreases exponentially in all directions. The analogous soliton exists for $\sigma^2 < 0$ (negative dispersion). In the latter case, it is only necessary to change x and λ, and to keep α real. In addition, to conserve the real character of the function describing the soliton, it is necessary to satisfy the condition $|\gamma|\sqrt{3} < |\alpha|$.

If δ is purely imaginary in (3.21) then the real wave, periodic in the y-direction and exponentially decreasing along the x-axis, can be obtained. The constructed regular solutions can be transformed into the real singular ones by shifting the coordinates x and/or y to the corresponding imaginary values. The above reasons thus enable us to generalize the considered classes of the exact solutions of the KP equation. Now, we proceed to the problem of stability of multidimensional soliton solutions of the KP equation.

3.1.3 Stability of Two- and Three-Dimensional KP Solitons

In the previous section we explained qualitatively the stability of the one-dimensional soliton solutions of the KP equation and presented some considerations on the stability of two-dimensional KP solitons. Note that while the stability problem of the one-dimensional soliton can be solved relatively easy, the problem of stability of the two-dimensional and three-dimensional solutions of the KP equation is far less trivial.

Consider equation (3.11). Taking into account that the *KP equation* is a Hamiltonian one,

$$\partial_t u = \partial_x \left(\delta\mathcal{H}/\delta u\right), \tag{3.22}$$

with the *Hamiltonian*

$$\mathcal{H} = \int \left[\frac{1}{2}\beta\left(\partial_x u\right)^2 + \frac{3}{2}\sigma^2\left(\boldsymbol{\nabla}_\perp w\right)^2 - u^3\right] \mathrm{d}\mathbf{r},$$

and $\partial_x w = u$, we can rewrite it as

$$\delta\left(\mathcal{H} + v\mathcal{P}_x\right) = 0, \tag{3.23}$$

where

$$\mathcal{P}_x = \frac{1}{2}\int u^2 \mathrm{d}\mathbf{r}$$

is the projection of the system's momentum onto the x-axis. This means that all finite solutions of (3.11) are the fixed points of the Hamiltonian $\mathcal{H}$ at the fixed $\mathcal{P}_x$. In this case v is the *Lagrange factor.* For further convenience we use the notations [60]

$$\mathcal{I}_1 = \int \left(\partial_x u\right)^2 \mathrm{d}\mathbf{r}, \qquad \mathcal{I}_2 = \int \left(\boldsymbol{\nabla}_\perp w\right)^2 \mathrm{d}\mathbf{r}, \quad \text{and} \quad \mathcal{I}_3 = \int u^3 \mathrm{d}\mathbf{r}.$$

Then the Hamiltonian of KP equation can be written as

$$\mathcal{H} = \frac{1}{2}\beta\mathcal{I}_1 + \frac{3}{2}\sigma^2\mathcal{I}_2 - \mathcal{I}_3. \tag{3.24}$$

Consider now the *stability problem.* According to the Lyapunov's theorem, the absolutely stable stationary points of a dynamic system correspond to the maximum or the minimum of $\mathcal{H}$, and if the extremum is local then the locally stable solutions are possible. The unstable states correspond to the monotonous dependence of $\mathcal{H}$ on its variables, i.e. to the case when the stationary point is the *saddle point.* Therefore, it is necessary to prove that the Hamiltonian $\mathcal{H}$ is limited from below for a fixed $\mathcal{P}_x$.

Consider the simple scale transforms [59,60] (in the real vector space $\mathbb{R}$)

$$u\left(x, \mathbf{r}_\perp\right) \to \zeta^{-1/2}\eta^{(1-d)/2} u\left(x/\zeta, \mathbf{r}_\perp/\eta\right)$$

conserving $\mathcal{P}_x$ (d is the space dimensionality, and $\zeta, \eta \in \mathbb{R}$), and write the *Hamiltonian* as a function of ζ and η

$$\mathcal{H}(\zeta,\eta)=\frac{\beta}{2\zeta^2}\mathcal{I}_1+\frac{3\zeta^2}{2\eta^2}\sigma^2\mathcal{I}_2-\zeta^{-1/2}\eta^{(1-d)/2}\mathcal{I}_3.$$

For $d=2$, simple analysis shows, that the Hamiltonian $\mathcal{H}$ is limited from below, and with a simple substitution it follows that the minimum of $\mathcal{H}$ is reached on the soliton solution (3.15). In the three-dimensional case, the *focus point* changes to the saddle point and the minimum is absent. It is easy to see that considering lines $\zeta^2=c\eta$ (c =const), where $\mathcal{H}$ varies monotonically, guarantees the absence of the locally stable solutions. The absence of additional integrals of motion for $d=3$ [60] also confirms that the three-dimensional soliton is unstable.

Note that in the two-dimensional case the considered scale transforms do not include all possible deformations of the Hamiltonian and only testify if $\mathcal{H}$ is limited. The exact proof of this fact is given in [60] where the inequality $\mathcal{H}\geq-(1/12)\mathcal{P}_x^3$ was obtained. According to the Lyapunov's theorem this also proves the stability of the two-dimensional soliton with respect to two-dimensional perturbations. The analysis of the problem of stability of the two-dimensional soliton with respect to bending of its front (three-dimensional perturbations), based on the analysis of the KP equation linearized on the background of solution (3.15) using the perturbation theory [60], shows that the two-dimensional soliton is unstable in this case (shifting along the x-axis). Qualitative reasons for the instability are the same here as for the one-dimensional soliton (see the previous section and Refs. [30,31]). Consideration of small three-dimensional perturbations corresponding to the shift along the y-axis demonstrates that in the long-wavelength limit the two-dimensional solitons are stable, however.

3.1.4 Numerical Approaches to Integration

Despite the availability of the well developed analytical technique for integration of nonlinear evolution equations by the inverse scattering transform (IST) method, in a number of cases it fails to obtain solutions (in the closed form) of KP equation as well as other equations of this class using analytical techniques. Thus there is an obvious need for use of computational mathematics. This also applies to solution of many practically important problems related to various applications of the KP model, when the use of unwieldy and rather complicated technique of the IST method is inexpedient. We note here that the two-dimensional soliton solutions of the KP equation were obtained for the first time in 1976 in the numerical simulation [31]. As we demonstrate below, the wide spectrum of applications of the KP model related to nonlinear wave processes in quite different media requires numerical simulations. In this section, on the basis of the "classic" KP equation, we briefly consider two approaches (based on different ideologies) to integration of the multidimensional nonlinear equations of the KP class.

Method of Stabilizing Factor. Historically, perhaps the first numerical method designed and successfully used for solution of the KP equation was the *method of stabilizing factor* [31,189]. Essentially, it consists of the following. First, the transform

$$u = af\left(\xi, \eta\right),$$

with

$$\xi = \sqrt{a}\left(x - at\right), \qquad \eta = ay, \quad \text{and} \quad a = \text{const} > 0,$$

is applied to the KP equation (3.18). We obtain

$$\partial_\xi^4 f - \partial_\xi^2 f - \partial_\eta^2 f = -\partial_\xi^2 \left(f^2\right) \tag{3.25}$$

(i.e., the time derivative is excluded and the equation is written in the reference frame fixed relatively to the soliton). Then, we apply the Fourier transform in ξ and η and reduce (3.25) to

$$F = GA \quad \text{and} \quad G(p,q) = \frac{p^2}{p^4 + p^2 + q^2}, \tag{3.26}$$

where

$$F = \left(2\pi\right)^{-2} \iint\limits_{-\infty}^{\infty} f\left(\xi, \eta\right) \cos p\xi \cos q\eta \mathrm{d}\xi \mathrm{d}\eta$$

and

$$A\left(p,q\right) = \iint\limits_{-\infty}^{\infty} F\left(p', q'\right) F\left(p - p', q - q'\right) \mathrm{d}p' \mathrm{d}q'.$$

When solving (3.26) using the usual iteration method, the resulting iteration series diverges because of the strong instability. This instability can be suppressed if we introduce a stabilizing factor [31]. Instead of (3.26) it is thus necessary to solve the equation

$$F = \left(s_1/s_2\right)^\alpha GA, \qquad \alpha = \text{const} > 0, \tag{3.27}$$

where

$$s_1 = \iint F^2(p,q) \mathrm{d}p\mathrm{d}q \quad \text{and} \quad s_2 = \iint GAF \mathrm{d}p\mathrm{d}q$$

(note that if F satisfies (3.26) then $s_1 = s_2$). Applying the process of iterations to (3.27) (it has the degree of homogeneity $2 - \alpha$, unlike (3.26) where the degree of homogeneity is 2), we obtain a rapidly converging series, at the same time we have $s_1 \to s_2$. This means that reduction of the degree of homogeneity by introduction of the stabilizing factor eliminates the instability and the solution of (3.27) converges to the solution of (3.26).

The formulated method proved to be a good one in practice, also when solving more complicated equations, e.g., the set of equations generalizing (3.25) in the three-dimensional case with an arbitrary nonlinearity

$$\partial_\xi^4 f - \Delta f = -\partial_\xi^2 \left(f^n\right). \tag{3.28}$$

For example, equations for the *Langmuir waves* and the slow extraordinary modes in a *magnetized plasma* under the assumption of a stationary wave packet result in (3.28) with $n = 3$ [189,190].

Method of Iterative Splitting. Note that the method of stabilizing factor considered above also has important deficiency: being "static," it does not allow to solve dynamic problems, when the main analyzed objects are parameters of the solitons' dynamics, collisions, interactions, etc. In this regard, there is a need to develop "dynamic" methods for integration of the KP equation. Here, however, we face essential difficulties. The diffraction term in the KP equations (3.9) and (3.11) is non-local, therefore it is hardly possible to directly apply the known standard methods of numerical solution of multidimensional equations based on the ideology of splitting [191,192]. Using the projection methods involving decomposition of spatial harmonics, is also difficult in the traditional set up. This is because both the group and phase velocities of the harmonics increase with the increase of the scale in the longitudinal coordinate λ_x reaching very large values for large λ_x, as a result of the following [193]. For small perturbations $\delta u \sim \exp(-\mathrm{i}\omega t + \mathrm{i}px - \mathbf{iq}\cdot\mathbf{r})$ the *dispersion law* corresponding to the linear part of (3.11) can be written as $p(\omega + p^3) = -q^2\sigma^2$. It then follows that the group velocity of the perturbations $v_x^{\mathrm{gr}} = \partial\omega/\partial p \sim q^2/p^2 \to \infty$ for $p \to 0$ $(\lambda_x \to \infty)$.

In Ref. [193], the technique of the *iterative splitting* was proposed for numerical integration of the KP equation (3.11) written as a set of equations:

$$\begin{aligned} \partial_t u + 6u\partial_x u + \partial_x^3 u &= v, \\ \partial_x v &= \beta \Delta_\perp u. \end{aligned} \tag{3.29}$$

Here, for the difference approximation of (3.29) in time, the implicit *Crank–Nicholson scheme* with $O(\tau^2)$ approximation was applied. To solve the arising set of equations connecting variables in the different time layers, the following iteration process is built:

$$\begin{aligned} \frac{\mathrm{d}v^{n+1,k}}{\mathrm{d}x} &= \beta\Delta_\perp u^{n+1,k}, \\ \frac{u^{n+1,k+1} - u^n}{\tau} &= \frac{1}{2}\left[u^n \frac{\mathrm{d}u^{n+1,k+1}}{\mathrm{d}x} + u^{n+1,k+1}\frac{\mathrm{d}u^n}{\mathrm{d}x}\right] \\ &+ \frac{1}{2}\left[\frac{\mathrm{d}^3 u^{n+1,k+1}}{\mathrm{d}x^3} + \frac{\mathrm{d}^3 u^n}{\mathrm{d}x^3}\right] = v^{n+1,k}, \end{aligned} \tag{3.30}$$

where n is the number of the time layer, k is the number of the iteration, and $u^{n+1,0} \equiv u^n$. When approximating the spatial derivatives, the scheme of the high order of accuracy, $O(h_x^4, h_\perp^2)$, is used in order to not distort the dispersion law. The convergence of the iterative process (3.31) and, therefore, the stability of the difference scheme for the set (3.29) was proved [193]. This method appeared as the most effective in the problem of the *wave collapse* for (3.11) when $\sigma^2 > 0$ (see Sect. 3.2.3). A drawback of the method of iterative splitting that we can mention here, perhaps, is the relatively large time cost at its computer realization.

Concluding Remarks. Some methods for solution of the Cauchy problem for the class of equations such as

$$\partial_t u + \alpha u \partial_x u + \beta \partial_x^3 u = \mathcal{R} \tag{3.31}$$

were proposed in Refs. [83,98] (here, $\mathcal{R} = \mathcal{R}[u]$ is some linear functional of u). This class of equations describes the propagation of multidimensional nonlinear waves in a medium with weak dispersion and nonlinearity. We can see that for $\mathcal{R} = \sigma^2 \partial_y w$, $\partial_x w = \partial_y u$, and $\mathcal{R} = \sigma^2 \Delta_\perp w$, $\partial_x w = u$, equation (3.31) is transformed into KP equations (3.9) and (3.11), respectively. The corresponding algorithms and numerical (simulation) codes are based on the explicit and implicit finite-difference schemes with $O(\tau^2, h_{\mathbf{r}}^2)$ and $O(\tau^2, h_{\mathbf{r}}^4)$ approximations, as well as on the dynamic spectral method. Below, in Chap. 4, we consider them in detail for KP-class equations with arbitrary constants α, β, and σ at the nonlinear, dispersive, and diffraction terms, respectively, and also discuss their application conditions and compare their characteristics with the methods considered above.

3.2 KP Equation: Analytical Integration and Dynamics of Waves

In this section, we consider the method of analytical integration of the KP equation by "dressing" of *L–A* pairs on an example of the two-dimensional KP equation (Sect. 3.2.1), the method of three-dimensional inverse scattering problem (Sect. 3.2.2), and problems related to the self-influence phenomena, namely, the wave collapse and self-focusing in the three-dimensional KP model (Sect. 3.2.3).

3.2.1 Analytical Integration. "Dressing" Method

"Dressing" of *L–A* Pairs. First, let us recall that the *KP equation*,

$$\partial_x \left(\partial_t u - 6u\partial_x u - \partial_x^3 u \right) = 3\sigma^2 \partial_y^2 u, \tag{3.32}$$

can be written, similar to the KdV equation, in the *Lax representation* [24]

$$\partial_t \hat{\mathrm{L}} = \left[\hat{\mathrm{L}}, \hat{\mathrm{A}} \right] = \hat{\mathrm{L}}\hat{\mathrm{A}} - \hat{\mathrm{A}}\hat{\mathrm{L}}, \tag{3.33}$$

where $\hat{\mathrm{L}}$ and $\hat{\mathrm{A}}$ are the differential operators[4]

[4] The notation $\hat{\mathrm{L}}, \hat{\mathrm{A}}$ is introduced in order to distinguish these operators from the kernels of operators.

$$\begin{aligned} &\hat{\mathrm{L}} = \sigma\partial_y - \hat{\mathrm{M}}, \qquad \hat{\mathrm{M}} = -\partial_x^2 - u(t,x,y), \\ &\hat{\mathrm{A}} = -4\partial_x^3 - 6u\partial_x - 3\partial_x u + 3\sigma w, \qquad \partial_x w = \partial_y u, \end{aligned} \tag{3.34}$$

and

$$w = \int_{-\infty}^{x} \partial_y u \mathrm{d}x.$$

As we know from the previous section, properties of the solutions of the KP equation significantly depend on the sign of σ^2. Without loss of generality, we further assume that $\sigma^2 = \pm 1$.

When $\sigma = 1$, the $\hat{\mathrm{L}}$-operator is the heat conductivity operator, i.e., $\hat{\mathrm{L}} = \partial_y + \partial_x^2 + u$. On the other hand, when $\sigma = \mathrm{i}$ the $\hat{\mathrm{L}}$-operator is the non-stationary *Schrödinger operator*, i.e., $\hat{\mathrm{L}} = \mathrm{i}\partial_y + \partial_x^2 + u$. The solution of the Cauchy problem for equations of type (3.33) requires setting and solving the direct and the inverse scattering problems for the operator $\hat{\mathrm{L}}$. However, to construct wide classes of partial solutions one can use the *dressing method* developed by Zakharov and Shabat [30], see also in [24]. This method is based on the ideas that:

- Linear operators with variable coefficients can be obtained with the help of the transform operators from operators with the constant coefficients (this in fact is the standard way of solution of the inverse spectral problems – see, e.g., Sec. 1.2)
- For simultaneous transformation of two such constant operators with a common spectrum, the condition of combinedness will take the form of a nonlinear equation on the coefficients, i.e., the required equation to be integrated.

The procedure of transformation of an equation with constant coefficients to an equation with variable coefficients can be called "dressing" by using the language of theoretical physics, and the whole method is therefore called the *dressing method.* It is important to note that actual integration of nonlinear differential equations by the IST method requires development of the technique of solution of the corresponding inverse spectral problem; however, this is often quite difficult for many equations (especially multidimensional ones). A great advantage of the dressing method is the possibility of its relatively simple extension to multidimensional problems; and it proved to be (partially) successful in eliminating a very unpleasant question of the setting and solvability of the corresponding inverse spectral problem (i.e., one of the basic difficulties of multidimensional variants of the IST method, see Sect. 3.2.2).

Consider the method of "dressing" of the L–A pairs in detail. We introduce (on the line $-\infty < x < \infty$) the integral *Fredholm operator* $\hat{\mathrm{F}}$ depending on t and y,

$$\hat{\mathrm{F}}\Psi(x) = \int \mathbf{F}(t,x,y,z)\Psi(z)\mathrm{d}z, \tag{3.35}$$

and assume that it allows the triangular factorization, i.e., it can be represented as

$$1+\hat{\mathrm{F}}=\left(1+\hat{\mathrm{K}}^{+}\right)^{-1}\left(1+\hat{\mathrm{K}}^{-}\right), \tag{3.36}$$

where $\hat{\mathrm{K}}^{+}$ and $\hat{\mathrm{K}}^{-}$ are the Volterra operators (or the Volterra factors of the operator $\hat{\mathrm{F}}$)

$$\pm\hat{\mathrm{K}}^{\pm}\Psi=\int_{x}^{\pm\infty}\mathbf{K}^{\pm}(t,x,y,z)\Psi(z)\mathrm{d}z.$$

We are interested, among all the operators $\hat{\mathrm{F}}$, in the operators commutating with the differential operators. Let us introduce the differential operator

$$\hat{\mathrm{L}}_0=\sigma\partial_y-\hat{\mathrm{M}}_0, \tag{3.37}$$

where

$$\hat{\mathrm{L}}_0=\lim_{|x|\to\infty}\hat{\mathrm{L}}, \qquad \hat{\mathrm{M}}_0=\lim_{|x|\to\infty}\hat{\mathrm{M}},$$

$\hat{\mathrm{L}}$ and $\hat{\mathrm{M}}$ are defined by (3.34), and

$$\hat{\mathrm{M}}_0=m_0\partial_x^n+m_1\partial_x^{n-1}+\ldots+m_n=\sum_{k=0}^{n}m_k\frac{\partial^{n-k}}{\partial x^{n-k}}. \tag{3.38}$$

The coefficients m_k of the operator $\hat{\mathrm{M}}_0$ are, in general, the functions of t, x, and y. And let the operator $\hat{\mathrm{F}}$ commute with $\hat{\mathrm{L}}_0$, i.e.,

$$\hat{\mathrm{F}}\hat{\mathrm{L}}_0-\hat{\mathrm{L}}_0\hat{\mathrm{F}}=0. \tag{3.39}$$

Applying this relation to an arbitrary function $\Psi(x)$ and integrating n times by parts we see that the kernel of the operator $\hat{\mathrm{F}}$ satisfies the differential equation

$$\sigma\partial_y\mathbf{F}=\hat{\mathrm{M}}_0\mathbf{F}-\mathbf{F}\hat{\mathrm{M}}_0^{+}, \tag{3.40}$$

where $\mathbf{F}$ is the kernel of the operator $\hat{\mathrm{F}}$ and $\hat{\mathrm{M}}_0^{+}$ is the operator conjugate to $\hat{\mathrm{M}}_0$:

$$\mathbf{F}\hat{\mathrm{M}}_0^{+}=\sum_{k=0}^{n}(-1)^{k+1}\partial_z^{n-k}\mathbf{F}m_k \tag{3.41}$$

(note that the matrix components are multiplied by $\mathbf{F}$ from the right).

It is generally important to also consider operators commutating with $\hat{\mathrm{L}}_0$ since the operator $\hat{\mathrm{L}}$ in (3.34) consists of two parts: the differential operator with the variable coefficients depending on $\hat{\mathrm{K}}^{+}$, and the integral Volterra (from the right) operator. It is possible to find such $\hat{\mathrm{K}}^{+}$ that the integral part of $\hat{\mathrm{L}}$ turns to zero, however.

Consider the transformation of $\hat{\mathrm{L}}_0$ into $\hat{\mathrm{L}}$ with the help of the operators $1+\hat{\mathrm{K}}^{+}$ and $1+\hat{\mathrm{K}}^{-}$, where $\hat{\mathrm{K}}^{+}$ and $\hat{\mathrm{K}}^{-}$ are the "triangular factors" of the operator $\hat{\mathrm{F}}$ given by (3.36). We have

$$\hat{\mathrm{L}}_0 \to \hat{\mathrm{L}} = \left(1+\hat{\mathrm{K}}^+\right)\hat{\mathrm{L}}_0\left(1+\hat{\mathrm{K}}^+\right)^{-1} = \left(1+\hat{\mathrm{K}}^-\right)\hat{\mathrm{L}}_0\left(1+\hat{\mathrm{K}}^-\right)^{-1}. \quad (3.42)$$

It then follows that since the operators $1+\hat{\mathrm{K}}^{\pm}$ transform the operator $\hat{\mathrm{L}}_0$ to one and the same operator $\hat{\mathrm{L}}$, the last is a purely differential operator (because it is of Volterra type simultaneously to both sides, i.e., its integral Volterra components are equal to 0). It has the form (3.38) where $\hat{\mathrm{M}}$, unlike $\hat{\mathrm{M}}_0$ (3.38), is an operator with the variable coefficients of the form

$$\hat{\mathrm{M}} = m_0\partial_x^n + u_1\partial_x^{n-1} + \ldots + u_n,$$

and $\hat{\mathrm{M}} \to \hat{\mathrm{M}}_0$ when $|x| \to \infty$. Here, $u_1, \ldots, u_n$ are the functions of t, x, and y, expressed via the kernel of the operator $\hat{\mathrm{K}}^+$ (or $\hat{\mathrm{K}}^-$). Expressions for these functions can be obtained from (3.42) modified as

$$\hat{\mathrm{L}}\left(1+\hat{\mathrm{K}}^-\right) = \left(1+\hat{\mathrm{K}}^-\right)\hat{\mathrm{L}}_0. \quad (3.43)$$

If we assume that in (3.38) $m_0, m_1 =$const, we obtain for u_1 and u_2 [24]

$$u_1 = [m_0, \xi_0] + m_1$$

and

$$u_2 = (n-1)m_0 d_x\xi_0 + \frac{1}{2}[d_x\xi_0, m_0] + \frac{1}{2}[m_0, \xi_1] + u_1\xi_0 + [m_1, \xi_0],$$

where

$$\xi_i(x,y,t) = (\partial_x - \partial_z)^i \mathbf{K}(x,y,z,t)\,|_{x=z}\,. \quad (3.44)$$

The procedure of obtaining the operator $\hat{\mathrm{L}}$ can be called "dressing" of the operator $\hat{\mathrm{L}}_0$ with the help of the integral Fredholm operator $\hat{\mathrm{F}}$.

Assume now that $\hat{\mathrm{F}}$ commutes, in addition to $\hat{\mathrm{L}}_0$, also with the operator $\partial_t - \hat{\mathrm{A}}_0$, where $\hat{\mathrm{A}}_0$ is a differential (with respect to x) operator. Then the kernel $\mathbf{F}$ of the operator $\hat{\mathrm{F}}$, in addition to (3.40), satisfies also the equation

$$\partial_t\mathbf{F} - \hat{\mathrm{A}}_0\mathbf{F} + \mathbf{F}\hat{\mathrm{A}}_0 = 0, \quad (3.45)$$

and therefore the operators $1+\hat{\mathrm{K}}^{\pm}$, besides the operator $\hat{\mathrm{L}}_0$, dress the operator $\partial_t - \hat{\mathrm{A}}_0$, transforming it to $\partial_t - \hat{\mathrm{A}}$.

Now let the operators $\hat{\mathrm{L}}_0 = \sigma\partial_y - \hat{\mathrm{M}}_0$ and $\partial_t - \hat{\mathrm{A}}_0$ be connected by the Lax relation

$$\left[\hat{\mathrm{L}}_0, \partial_t - \hat{\mathrm{A}}_0\right] = 0. \quad (3.46)$$

Then there exists a class of functions Ψ_0 for which the set of equations

$$\begin{cases} \hat{\mathrm{L}}_0\Psi_0 = 0, \\ \left(\partial_t - \hat{\mathrm{A}}_0\right)\Psi_0 = 0, \end{cases}$$

is a conjugate one. If we apply now relation (3.43) to Ψ_0 and the corresponding relation for the operator $\partial_t - \hat{\mathrm{A}}_0$ we then obtain [24]

$$\begin{cases} \left(\sigma\partial_y - \hat{\mathrm{M}}\right)\Psi = 0, \quad \left(\partial_t - \hat{\mathrm{A}}\right)\Psi = 0, \\ \Psi = \left(1 + \hat{\mathrm{K}}^-\right)\Psi_0. \end{cases} \tag{3.47}$$

This means that the operators $\hat{\mathrm{M}}$ and $\hat{\mathrm{A}}$ also satisfy the Lax relation (3.46), and the coefficients of these operators satisfy the considered nonlinear set of equations (certainly, the coefficients of the operators $\hat{\mathrm{M}}_0$ and $\hat{\mathrm{A}}_0$ also satisfy them).

Thus, the dressing method enables us to "multiply" the number of solutions by constructing wide classes of new solutions on the basis of the known ones. For the particular "dressing," it is necessary to know the kernel of the operator $\hat{\mathrm{K}}^+$. Multiplying equation (3.36) from the right by $1 + \hat{\mathrm{K}}^+$ and assuming $z < x$, we come to the generalization of the *GLM equation*, namely,

$$\mathbf{K}^+ + \mathbf{F} + \int_x^\infty \mathbf{K}^+(x, z', y, t)\mathbf{F}(z', z, y, t)\mathrm{d}z' = 0. \tag{3.48}$$

Solving this equation for the kernel $\mathbf{K}^+$ (taking into account that the kernel $\mathbf{F}$ is defined by the scattering data, i.e., ultimately, by the initial conditions of the Cauchy problem) and finding $\mathbf{K}^+$, we can proceed with dressing of the operators $\hat{\mathrm{M}}_0$ and $\hat{\mathrm{A}}_0$. Consider now application of the dressing method to the integration of the KP equation [24].

Analytical Integration. Following [24], we proceed with the dressing of the trivial solution

$$u_0 = 0 \qquad \text{and} \qquad w_0 = 0,$$

i.e., we "multiply" the trivial solutions by obtaining another (non-trivial) one. Write the *KP equation* as the set

$$\begin{cases} \partial_t u - 6u\partial_x u - \partial_x^3 u - 3\sigma^2\partial_y w = 0, \\ \partial_x w = \partial_y u. \end{cases} \tag{3.49}$$

First, consider the case of the negative dispersion when $\sigma = 1$. Then the kernel $\mathbf{F}$ satisfies the equations [24]

$$\begin{cases} \partial_y \mathbf{F} + \partial_x^2 \mathbf{F} - \partial_z^2 \mathbf{F} = 0, \\ \partial_t \mathbf{F} + 4\left(\partial_x^3 \mathbf{F} + \partial_z^3 \mathbf{F}\right) = 0. \end{cases}$$

Choose a solution of these equations in the form

$$\mathbf{F} = \sum_{n=1}^{N} c_n(t, y)\exp(p_n x + q_n z),$$

where

$$c_n(t, y) = c_n(0) \exp\left[\left(q_n^2 - p_n^2\right) y - 4\left(p_n^3 + q_n^3\right) t\right],$$

and $p_n, q_n, c_n(0) > 0$. Solving the GLM equation (3.48) for the kernel $\mathbf{K}^-(= \mathbf{K})$, we obtain the solution of the KP equation given by

$$u(t, x, y) = -2\partial_x \mathbf{K}\,(t, x, y, x)\,. \tag{3.50}$$

Since the kernel $\mathbf{K}$ of (3.48) is degenerate, the solution of this equation can be easily obtained and written as

$$\mathbf{K}(t, x, y, z)\,|_{z=x} = -\partial_x \ln \det A,$$

where A is the quadratic $N \times N$ matrix with the elements given by

$$A_{nm} = \delta_{nm} + c_n(y, t)\frac{\exp\left[(p_n + q_m)\, x\right]}{p_n + q_m} \tag{3.51}$$

(the Kroneker delta is, as usual, written as $\delta_{nm} = 1$ for $n = m$ and $\delta_{nm} = 0$ for $n \neq m$). Thus the solution of the KP equation for $\sigma = 1$ is

$$u(t, x, y) = 2\partial_x^2 \ln \det A. \tag{3.52}$$

Expressions (3.51) and (3.52) describe the interaction of N one-dimensional solitons. In particular, for $N = 1$ we obtain the one-soliton solution:

$$u = \frac{(p+q)^2}{2} \cosh^{-2}\left\{\frac{1}{2}\left[(p+q)x + (q^2 - p^2)y - 4(p^3 + q^3)t + \ln\frac{c(0)}{p+q}\right]\right\}. \tag{3.53}$$

This formula describes the one-dimensional soliton moving at some angle with respect to the x-axis (for $p \neq q$). A general view of solution (3.53) is shown in Fig. 3.4. In particular, for $p = q$ we obtain the well-known soliton

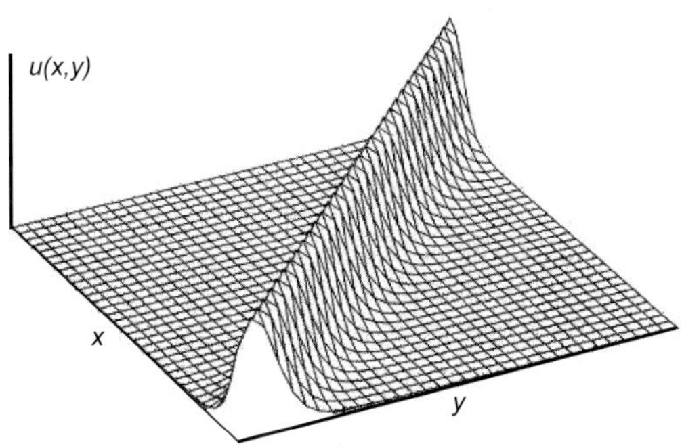

Fig. 3.4. General view of the solution of the KP equation with $\sigma = 1$ and $p \neq q$

of the KdV equation moving along the x-axis.

Solution (3.52) in general describes intersection of ("oblique") solitons traveling at different angles with respect to the x-axis; note that this solution does not decrease asymptotically in the directions

$$x/y = p_n - q_n, \qquad n = 1, 2, \ldots, N,$$

when $|x, y| \to \infty$.

Consider now the case of the positive dispersion when $\sigma^2 < 0$ in the KP equation (3.49). Note that in the KP equation the change of the sign of σ^2 is equivalent to continuation of the solutions from the real y to the purely imaginary ones, i.e., to the change $\sigma^2 \to -\sigma^2$ and $y \to \mathrm{i}y$. Following [24], we obtain explicit solitons for $\sigma^2 < 0$ by the procedure of special degeneration of solution (3.52). When degenerating this solution, we use the notations

$$p_n + q_n = \kappa_n, \qquad p_n - q_n = \nu_n, \quad \text{and} \quad c_n(0) = -a_n \kappa_n.$$

Rewrite the matrix (3.51) as

$$A_{nm} = \exp\left(-\frac{\kappa_n - \kappa_m}{2} x + \frac{\nu_n - \nu_m}{2} x \right) B_{nm}, \tag{3.54}$$

where

$$\begin{aligned} B_{nm} = \delta_{nm} - & \frac{2a_n \kappa_n}{\nu_n - \nu_m + \kappa_n + \kappa_m} \\ & \times \exp\left\{ \kappa_n \left[x - \nu_n y - \left(3\nu_n^2 + \kappa_n^2 t \right) \right] \right\}. \end{aligned} \tag{3.55}$$

Since (3.54) transforms A to B, this is the similarity transform, and we have $\det A = \det B$. Thus instead of (3.52) we obtain

$$u(t, x, y) = 2\partial_x^2 \ln \det B. \tag{3.56}$$

Furthermore we proceed to the limit $\kappa_n \to 0$ and assume

$$a_n = 1 - \xi_n \kappa_n + O(\kappa^2).$$

As a result, we obtain the determinant of the matrix B given by

$$\det B = \prod_n (-\kappa_n) \det \tilde{B} + O\left(\kappa^{N+1}\right),$$

where, after the change $y \to \mathrm{i}y$, the matrix $\tilde{B}$ is

$$\tilde{B}_{nm} = \delta_{nm} \left(x - \mathrm{i}\nu_n y - \xi_n - 3\nu_n^2 t \right) + 2\left(1 - \delta_{nm}\right) / \left(\nu_n - \nu_m\right) \tag{3.57}$$

(it should be taken into account that the last term on the right-hand side is equal to zero for $n = m$ [24]).

Obviously, $u(t, x, y) = 2\partial_x^2 \ln \det \tilde{B} + O(\kappa)$ and, if we assume now that $\kappa = 0$, we have [24]

$$u(t, x, y) = 2\partial_x^2 \ln \det \tilde{B}. \tag{3.58}$$

For a certain choice of the parameters ν and ξ, it is possible to secure in (3.58) both the reality and regularity of the function u for the real values of its arguments. Generally, solution (3.58) is real and not singular if

$$N = 2K, \qquad \nu_{K+n} = -\nu_n^*, \qquad \xi_{K+n} = \xi_n^*, \qquad n = 1, \dots, K, \tag{3.59}$$

and there are no coinciding numbers in the set of complex numbers $\nu_1, \dots, \nu_K$ (these conditions guarantee that the matrix $\tilde{B}$ is positively determined).

Considering the asymptotics of the solution (3.58) with (3.57) and (3.59) at $t \to \pm\infty$ as a function of x_0 and y_0, where

$$x = x_0 + v_x^{(m)} t \quad \text{and} \quad y = y_0 + v_y^{(n)} t,$$

we can see that the corresponding values $\xi_n^\pm$ coincide, i.e., $\xi_n^+ = \xi_n^- = \xi_n$. Thus we note that for the solution (3.58) with (3.57) and (3.59) describing the collision of K two-dimensional solitons, the phase shifts of the solitons after the collision are identically zero, in contrast to the well-known effect in the one-dimensional problem. Each of the colliding two-dimensional solitons described by the constructed solution is defined by (3.57)–(3.59) for $N = 2$, i.e., when $\nu_2 = -\nu_1^*$, $\xi_2 = \xi_1^*$, and

$$\det \tilde{B} = 4\left(\nu_1 + \nu_1^*\right)^{-2} + \left|x - \mathrm{i}\nu_1 y - \xi_1 - 3\nu_1^2 t\right|^2 . \tag{3.60}$$

Formulae (3.58) and (3.60) give the one-soliton solution. A general view of such a *rational soliton* is shown in Fig. 3.5. Thus, by proceeding from a triv-

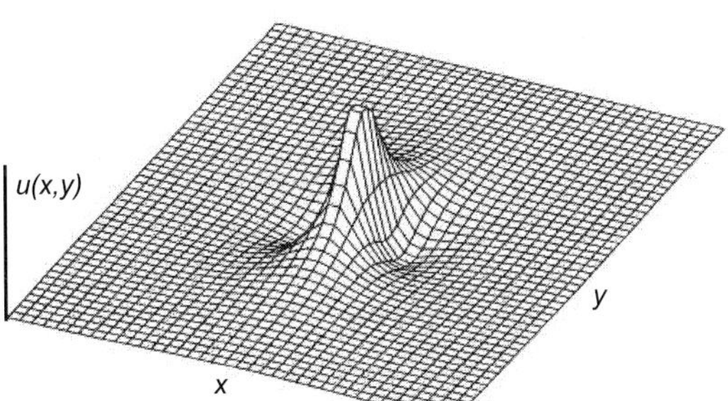

Fig. 3.5. General view of the solution of the KP equation with $\sigma = i$ (two-dimensional rational soliton)

ial (zero) solution with the technique developed in Ref. [24], we analytically obtained the one-dimensional and two-dimensional soliton solutions of the KP equation using the dressing method ("multiplying" or dressing the corresponding operators).

3.2.2 Three-Dimensional Inverse Scattering Problem

The dressing method considered in the previous subsection is a sort of two-dimensional generalization of the inverse scattering problem for the KdV equation considered in Sect. 1.2 in detail. Having in mind the possibility of extension of the IST method for a wider class of multidimensional nonlinear evolution systems allowing soliton-type solutions important for a number applications of the nonlinear wave theory in the physics of particular media, we consider here how the inverse scattering problem can be formulated in the general three-dimensional case.

Introductory Notes. The three-dimensional stationary theory of non-relativistic scattering deals with the *Schrödinger equation*

$$-\Delta\Psi + V(\mathbf{r})\Psi = E\Psi, \tag{3.61}$$

where Δ is the three-dimensional Laplace operator and Ψ is the wave function. The properties of the potential are such that the asymptotics of the wave function is given by

$$\Psi(\mathbf{k},\mathbf{r}) = \mathrm{e}^{\mathrm{i}\mathbf{k}\cdot\mathbf{r}} + r^{-1}\mathrm{e}^{ikr}A(\hat{\mathbf{r}},\mathbf{k}) + O(r^{-1}). \tag{3.62}$$

Furthermore, E is the full energy and $V(\mathbf{r})$ is the potential energy. Here and further we use the following notations:

- The absolute value of any vector $\mathbf{v}$ is written as $v \equiv |\mathbf{v}|$,
- The unit vector of its direction is $\hat{\mathbf{v}} \equiv \mathbf{v}/v$,
- In the scattering amplitude $\mathbf{k}' \equiv k\hat{\mathbf{r}}$,
- In a more general case, $\mathbf{k}'$ is the vector of the length k but not necessarily coinciding with the direction of the vector $\mathbf{k}$.

We assume that the amplitude $A(\hat{\mathbf{r}},\mathbf{k})$ is calculated on the basis of the measured values. The inverse problem is to obtain the potential $V(\mathbf{r})$ via the known $A(\hat{\mathbf{r}},\mathbf{k})$.

Since the amplitude $A(\hat{\mathbf{r}},\mathbf{k})$ depends on the five variables (x, y, z, r, k), and the potential $V(\mathbf{r})$ depends only on the three variables (x, y, z), the problem of existence of the solution inevitably arises. In other words, the existence of a local potential corresponding to the initial data is accompanied by strong limitations on the *scattering amplitude* $A(\hat{\mathbf{r}},\mathbf{k})$. However, if we proceed from the scattering amplitude corresponding to some local potential (for example, $A(\hat{\mathbf{r}},\mathbf{k})$ is obtained as a result of the direct problem and corresponds to the initial condition of a three-dimensional evolution equation of a "soliton type") and want to restore such a local potential, then the problem of the solution's existence does not arise.

The problem of uniqueness of the solution can be solved quite easily because for any "reasonable" potential the scattering amplitude at high energies should tend to its limit in the Born approximation:

$$A \approx -(4\pi)^{-1} \int \mathrm{d}\mathbf{r}' V(\mathbf{r}') \mathrm{e}^{\mathrm{i}\boldsymbol{\tau}\cdot\mathbf{r}'},$$

where $\boldsymbol{\tau} = \mathbf{k} - \mathbf{k}'$. Since $\tau_{\max} = 2k$ (which takes place for the anti-parallel vectors $\mathbf{k}$ and $\mathbf{k}'$), the amplitude A approaches the full Fourier image of the potential V. However, the high-energy limit of the amplitude is unknown, and, in addition, the Schrödinger equation (3.61) is useless for the actual description of the particles' behavior at high energies. It is therefore necessary to find such a solution of the IST problem which should not be too dependable on the high-energy data. In the language of nonlinear evolution equations, this means that we do not consider solutions of these equations with very high amplitudes, nor do we consider their singular solutions.

If the scattering amplitude unequivocally determines the potential, then it has to define bound states, i.e., the discrete spectrum of the *Schrödinger operator*,

$$\hat{\mathrm{H}} = -\Delta + V(\mathbf{r}).$$

If the potential satisfies the condition

$$|V(\mathbf{r})| \leq C\,|1 + r|^{-3-\varepsilon},$$

we can obtain the discrete spectrum, e.g., via the amplitude of the forward scattering $A(k) \equiv A(\hat{\mathbf{k}}, \mathbf{k})$. The value $A(k)$ appears on the real axis as the limit of an analytic function, regular in the whole upper half-plane except for the poles on the imaginary axis $k = \mathrm{i}K_n$ where $-K_n^2$ are the eigenvalues of the operator $\hat{\mathrm{H}}$ (compare with the one-dimensional case, Sect. 1.2), and $A(k)$ tends to a constant when $|k| \to \infty$ ($\mathrm{Im}k \geq 0$). These properties enable us to obtain the positions of the poles by knowing the function A on the real axis and analytically continuing it on the upper half-plane. Thus we obtain the discrete spectrum. The full *scattering amplitude* (similar to the one-dimensional case, Sect. 1.2) is related to the discrete spectrum, and it is impossible to present them independently. Having in mind these remarks we can now proceed directly to study of the three-dimensional inverse problem.

Outline of the Method. Studies of the three-dimensional inverse scattering problem (*3-ISP*) were first (independently) done by Faddeev (1971) and Newton (1974, 1980) for the case of a local potential. The general scheme of the 3-ISP solution can be represented by the diagram shown in Fig. 3.6. The kernel of the transformation and the symmetric kernel can be obtained in two different ways:

1. The kernels are investigated as solutions of some hyperbolic partial differential equations (not the Schrödinger equation).
2. The kernels are obtained from the relations of completeness for the set of solutions of the Schrödinger equation.

Both these ways are shown in the diagram Fig. 3.6. The second procedure also answers the questions on the existence and uniqueness of the solution and

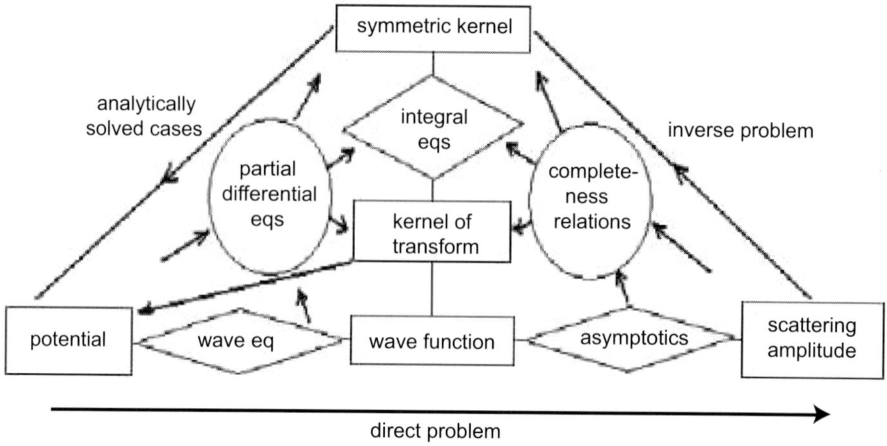

Fig. 3.6. General scheme of the solution of the three-dimensional inverse scattering problem

about the algorithm to construct the potential $V(\mathbf{r})$. In this case, a general way of solution is similar to the GLM method studied in Sect. 1.2 for the one-dimensional problem on the whole axis $-\infty \leq x \leq \infty$. Consider therefore the second procedure, namely, obtaining the kernel of the transform and the symmetric kernel from the completeness relations. The procedure can be divided into three steps according to the corresponding parts of the diagram Fig. 3.6. Here, we consider these steps on an example of the *Gelfand–Levitan–Marchenko method* (*GLM method*) for the S-wave.[5]

The "wave function" step

The "wave function" is the solution of the wave equation, this solution is central in our investigation (see Fig. 3.7). Generally speaking, it is not nec-

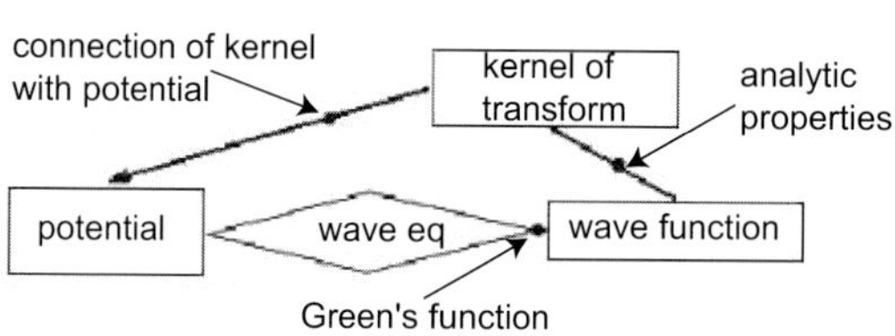

Fig. 3.7. Scheme of the solution on the "wave function" step

essarily a physical wave function (for example, the Jost solutions in the one-dimensional ISP on the axis $-\infty \leq x \leq \infty$). However, in the *GLM method*

[5] The S-wave means the partial wave with the zeroth angular momentum $l = 0$. The choice of S-wave is motivated by considerable simplification of the relations and clarity of the results obtained.

the wave function is indeed the physical wave function, with the accuracy up to its normalization, i.e., it is a solution of the *Schrödinger equation*

$$\hat{\mathrm{H}}\varphi(k,r) \equiv \Delta\varphi(k,r) - V\varphi(k,r) = -k^2\varphi(k,r). \tag{3.63}$$

It is equal to zero at $r = 0$, and such a choice is equivalent to the choice of the Green's function:

$$\varphi(k,r) = \varphi_0(k,r) + \int_0^r G_0(k;r,\rho)V(\rho)\varphi_0(k,\rho)\mathrm{d}\rho, \tag{3.64}$$

where

$$\varphi_0(k,r) = \frac{\sin kr}{k} \quad \text{and} \quad G_0(k;r,\rho) = \frac{\sin k(r-\rho)}{k}. \tag{3.65}$$

We can see from relations (3.64) and (3.65) that $\varphi(k,r)$ is the even integer function of the variable k and in the complex k-plane:

$$\varphi(k,r) - \frac{\sin kr}{k} = O(\frac{\sin kr}{k}), \qquad |k| \to \infty. \tag{3.66}$$

It follows straightforwardly from these analytical properties that there exists the triangular kernel $\mathbf{K}(r,r')$ with the following properties (*Peli–Winner theorem*):

Theorem.
(1) $\mathbf{K}(r,r')$ *does not depend on* k*;*
(2) $\mathbf{K}(r,r') = 0$ *for* $r' \geq r$*;*
(3)

$$\varphi(k,r) = \varphi_0(k,r) + \int_0^r \mathbf{K}(r,r')\varphi_0(k,r')\mathrm{d}r'. \tag{3.67}$$

Substituting (3.67) into (3.64) we obtain the relation between the potential $V(r)$ and the kernel of the transformation $\mathbf{K}$, namely,

$$\mathbf{K}(r,r) = \frac{1}{2}\int_0^r V(\rho)\mathrm{d}\rho. \tag{3.68}$$

The "asymptotics and completeness" step (Fig. 3.8)
The *Jost function* $F(k)$ can be obtained in two different ways:

1. On the basis of the asymptotic behavior of the function $\varphi(k,r)$ for $r \to \infty$

$$2\mathrm{i}k\varphi(k,r) = \mathrm{e}^{\mathrm{i}kr}F(-k) - \mathrm{e}^{-\mathrm{i}kr}F(k) + O(1) \tag{3.69}$$

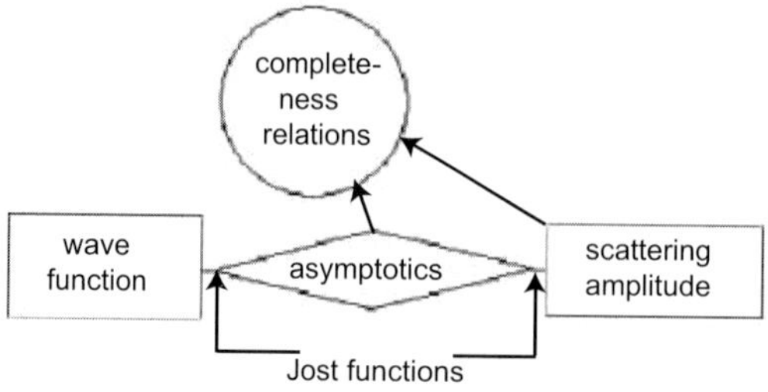

Fig. 3.8. Scheme of the solution on the "asymptotics and completeness" step

2. On the basis of the regular solution of the integral *Lippman–Swinger equation*

$$\psi(k,r) = \varphi_0(k,r) + \int_0^\infty g(k,r,\rho)V(\rho)\psi(k,\rho)\mathrm{d}\rho, \tag{3.70}$$

where

$$g(k,r,\rho) = -k^{-1}\exp(\mathrm{i}kr_>)\sin(kr_<)$$

and

$$\begin{cases} r_> = \max(r,r'), \\ r_< = \min(r,r'). \end{cases}$$

This gives the formula

$$\varphi(k,r) = F(k)\psi(k,r), \tag{3.71}$$

i.e., the function $\varphi(k,r)$ is directly proportional to the solution of the integral *Fredholm equation* (3.70), with the factor being the *Jost function* $F(k)$. Moreover, one can prove that $F(k)$ is the Fredholm determinant of (3.70). It then follows that the zeros of the function $F(k)$ in the upper half-plane are exactly at the same values $k = \mathrm{i}K_n$ where the wave $-K_n^2$ is the eigenvalue of the operator $\hat{\mathrm{H}}$, and the multiplicity of every zero is the degeneration degree of the corresponding eigenvalue.

It is possible to prove that the completeness relation of the functions $\Psi(k,r)$, defined above as solutions of the integral Fredholm equation, together with the eigenfunctions of the operator $\hat{\mathrm{H}}$, is fulfilled, namely,

$$\frac{2}{\pi}\int_0^\infty \psi(k,r)\psi^*(k,r')k^2\mathrm{d}k + \sum_n \psi^{(n)}(r)\psi^{(n)}(r') = \delta\left(r-r'\right), \tag{3.72}$$

where $\Psi^{(n)}(r)$ is the n-th normalized eigenfunction of the *Schrödinger operator* $\hat{\mathrm{H}}$

$$\int_0^\infty \left|\psi^{(n)}(r)\right|^2 \mathrm{d}r = 1. \tag{3.73}$$

We can write on the basis of these relations and equality (3.71) the completeness relation expressed via the wave function (the solution of the Schrödinger equation) φ as the *Stieltjes integral*:

$$\int \varphi(k,r)\varphi(k,r')\mathrm{d}\rho(E) = \delta\left(r - r'\right), \tag{3.74}$$

where $\rho(E)$ is the spectral function defined by

$$\frac{\mathrm{d}\rho}{\mathrm{d}E} = \frac{k}{\pi}|F(k)|^{-2}, \qquad E \geq 0,$$

and

$$\frac{\mathrm{d}\rho}{\mathrm{d}E} = \sum_n \left[\int_0^\infty \varphi^2(\mathrm{i}K_n, r)\mathrm{d}r\right]^{-1} \delta(E - E_n), \qquad E < 0.$$

Certainly, there is similar completeness relation for the function φ_0 included in (3.64) and (3.70) together with the spectral function ρ_0, where $\mathrm{d}\rho_0/\mathrm{d}E = 0$ for $E < 0$ and $\mathrm{d}\rho_0/\mathrm{d}E = k\pi$ for $E \geq 0$.

The "integral equation" step (Fig. 3.9)

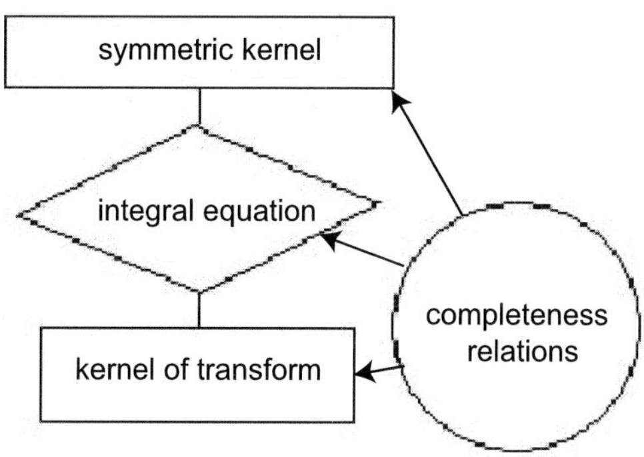

Fig. 3.9. Scheme of the solution on the "integral equation" step

1. First, we note the following. Let $\mathbb{C}^*$ be the linear space containing all continuous functions on the set $\mathbb{R}^+$ as well as all δ-functions in all points of the set $\mathbb{R}$. For the linear operator $\hat{\mathrm{T}}$ transforming $\mathbb{C}^*$ to $\mathbb{C}^*$, we define the conjugated operator $\hat{\mathrm{T}}^*$ by the equality

$$\int \overline{\hat{\mathrm{T}}a(\rho)b(\rho)}\mathrm{d}\rho = \int a(\rho)\hat{\mathrm{T}}^* b(\rho)\mathrm{d}\rho, \tag{3.75}$$

where the overline means the value of the corresponding function for $\rho' \neq \rho$. Let this be true for every combination of a and b from the space $\mathbb{C}^* \times \mathbb{C}^*$ for which one of the sides of (3.75) holds true.

If $\hat{\mathrm{U}}$ is the operator of the form

$$\hat{\mathrm{U}} = \delta(r-\rho) + \mathbf{K}(r,\rho)\Theta(r-\rho)\,\Theta(\rho)\,,$$

then the conjugated operator is given by

$$\hat{\mathrm{U}}^* = \delta(r-\rho) + \mathbf{K}^*(r,\rho)\Theta(\rho-r)\,\Theta(r)\,.$$

Here,

$$\mathbf{K}^*(r,\rho) = \mathbf{K}(\rho,r) \qquad (r<\rho).$$

Thus the kernel conjugated to the triangular kernel is the triangular kernel. One can rewrite (3.75) as

$$\overline{\hat{\mathrm{T}}a(\rho)} = a(\rho)\hat{\mathrm{T}}^*.$$

Obviously, the δ-function $\delta(r-\rho)$ is the unit operator, and we use the notation $\hat{1}$ for it.

2. Consider a formal solution of the problem on the "integral equation" step. Expression (3.67) (the third property of the triangular kernel $\mathbf{K}(r,r')$) can now be written in the operator form as

$$\varphi_k = \hat{\mathrm{U}}\varphi_{0k}. \tag{3.76}$$

Since it is possible to expand any vector from the space $\mathbb{L}^2$ in a series of functions φ_{0k}, (3.76) means that $\hat{\mathrm{U}}$ is the "interlacing" operator, i.e., $\hat{\mathrm{H}}\hat{\mathrm{U}} = \hat{\mathrm{U}}\hat{\mathrm{H}}_0$. If we write the completeness relation (3.74) for the functions φ_k and φ_{0k} symbolically as

$$\int \mathrm{d}\rho\varphi_k\overline{\varphi}_k = \hat{1} \quad \text{and} \quad \int \mathrm{d}\rho_0\varphi_{0k}\overline{\varphi}_{0k} = \hat{1},$$

and substitute (3.76) in both, we then obtain

$$\hat{\mathrm{U}}\mathbf{g}\hat{\mathrm{U}}^* = \hat{1} - \hat{\mathrm{U}}\hat{\mathrm{U}}^*, \tag{3.77}$$

where

$$\mathbf{g} = \int \mathrm{d}(\rho-\rho_0)\varphi_{0k}\overline{\varphi}_{0k}$$

is the symmetric kernel.

Note that it is possible to proceed in this way with any operator of the transformation. Furthermore, since $\hat{\mathrm{K}}$ is the triangular operator, it is possible to construct the inverse operator

$$\hat{\mathrm{U}}^{-1} = \hat{1} + \hat{\mathrm{K}}', \tag{3.78}$$

and the operator $\hat{\mathrm{K}}'$ is also triangular. Therefore, dividing (3.77) by $\hat{\mathrm{U}}^*$ and using (3.78), we obtain

$$\hat{\mathrm{U}}\mathbf{g} = \hat{\mathrm{K}}'^* - \hat{\mathrm{K}}.$$

For $r' < r$ this gives

$$\mathbf{K}(r,r') + \mathbf{g}(r,r') + \int \mathbf{K}(r,\rho)\mathbf{g}(\rho,r')\mathrm{d}\rho = 0, \qquad (3.79)$$

i.e. the three-dimensional integral *GLM equation.*

Thus, we have considered here the example of the S-wave (i.e., the partial wave with the angular momentum $l = 0$) the scheme of solution of the three-dimensional inverse scattering problem. Every step of the investigation is divided into a number of very complicated "substeps." For example, on the first (the "wave function") step, it is necessary to choose a solution of the Schrödinger equation such that the whole method is applicable. This means that it is necessary to choose the proper Green's function.[6] Generally speaking, this itself is a complicated problem, and its detailed consideration is beyond the scope of this book.

To conclude, we note that we considered the three-dimensional inverse scattering problem in a general form, quite schematically, abstaining from the question about the appropriate nonlinear evolution equation, whose solution is the potential $V(\mathbf{r})$ in the *Schrödinger equation.* The theory of *3-ISP* proves that if a solution of the inverse problem exists then it is unique, but at present there is no success yet in proving that the solution exists in a general case. However, if we have a three-dimensional equation with the solution in the form of a local potential, then the problem of the existence of the solution no longer arises. As an example of the model for which the 3-ISP method can be applied, we can refer to the mathematical model for the elastic scattering of two particles in the non-relativistic limit

$$\mathrm{d}_r^2\varphi_l + \left[k^2 - V(r) - l(l+1)r^{-2}\right]\varphi_l = 0$$

(this equation is written for the partial waves with the angular momentum l). Finally, we note that the problem of integration using the IST method of the three-dimensional "soliton" equations yielding stable solutions in the form of three-dimensional solitons, in particular, represented by the three-dimensional *generalized KP equation* (*GKP equation*, see Sect. 4.1) is not yet solved.

3.2.3 Dynamics of Three-Dimensional Nonlinear Waves in the KP Model. Wave Collapse and Self-Focusing

In Sect. 3.1, on the basis of analysis of deformations of the KP Hamiltonian $\mathcal{H}$ we demonstrated that the KP equation has in the (1+2)-case the following features: stable one-dimensional soliton solutions for negative dispersion, and two-dimensional solitons solutions with the algebraic asymptotics for positive dispersion. The particular form of these solutions was obtained in Sect. 3.2.1

[6] This problem was first solved by Faddeev in 1966 (see, e.g., Ref. [21]).

using the dressing method. In the (1+3)-dimensional case we found that the KP equation with any sign of the dispersion does not have three-dimensional stable solutions because $\mathcal{H}$ turns out to be unlimited from below at its deformations. In this regard it is interesting to study the dynamics of three-dimensional nonlinear waves of the soliton type, i.e., to investigate nonlinear effects associated with the instability.

At present, no analytical three-dimensional solutions of the KP equation have been obtained, therefore, almost all investigations connected with this problem (with the exception of, perhaps, the stability and asymptotics analyses [83,112,194]) are based on numerical experiments (computer simulations). Such investigations were conducted in the 1980s and 1990s in Russia by Kuznetsov and Musher [59,63,64], Petviashvili et al [86,189], and Belashov and Karpman [81,83,148,196]. Since the (1+3)-dimensional KP equation has direct applications in plasma physics (including space plasma, and plasma of the Earth's ionosphere and magnetosphere), all these investigations, apart from studies of the fundamental problem of dynamics of three-dimensional wave structures per se, were related to studies of the nonlinear self-influence of waves in a plasma, namely, the *wave collapse* of sonic waves and the *self-focusing* of the beams of the *fast magnetosonic waves* (*FMS waves*) propagating in a magnetized plasma.

Wave Collapse. Consider first the main results on the wave collapse obtained by numerical integration of the *KP equation* written (in its standard form) as

$$\partial_x \left(\partial_t u + 6u\partial_x u + \partial_x^3 u\right) = \kappa \Delta_\perp u. \tag{3.80}$$

The numerical simulations for (3.80) was done in Refs. [59,63,64] and, independently, in Refs. [81,83,148,196] in the axially symmetric geometry, i.e., when the operator on the right-hand side of KP equation is given by $\Delta_\perp = \partial_\rho^2 + (1/\rho)\partial_\rho$, $\rho^2 = y^2 + z^2$.

The diffraction term in (3.80) is nonlocal, that means, as we already mentioned in Sect. 3.1.4, that the standard methods based on the ideology of splitting of the problem into a number of one-dimensional problems, appear inapplicable. The *method of stabilizing factor* [31], being "static," is suitable only for the search of stationary solutions. Therefore, the Cauchy problem for (3.80) requires use of special "dynamic" methods, some of which, developed by Belashov, are specially considered in detail in Sect. 4.3. Musher has also developed [193] a special difference technique using the *iterative splitting* considered above in Sect. 3.1.4, and constructed the difference scheme with a high order of accuracy, $O(\tau^2, h_x^4, h_\perp^2)$. Below, we present the results of numerical simulations using our "dynamic" methods and test them by comparing the results with the results obtained in [59,63].

In all the above studies, the conditions of a total absorption on the boundaries were used as the boundary conditions. They do not conserve the integrals of motion of KP equation $\mathcal{P}_x$ and $\mathcal{H}$ (i.e., the projection of the system's momentum on the x-axis and the energy, respectively[7]). To control the calculations in our simulations, we calculated the flows of $\mathcal{P}_x$ and $\mathcal{H}$ onto the boundaries of the simulation region and thus checked these invariants. As the initial conditions, the solitary soliton-like three-dimensional axially symmetric pulses given by

$$\begin{aligned} u(0,x,\rho) &= 2\partial_x^2 \ln \det B, \\ \det B &= 4\left(\nu+\nu^*\right)^{-2} + \left|x - \mathrm{i}\nu\rho\right|^2 . \end{aligned} \tag{3.81}$$

were chosen in our numerical simulations.

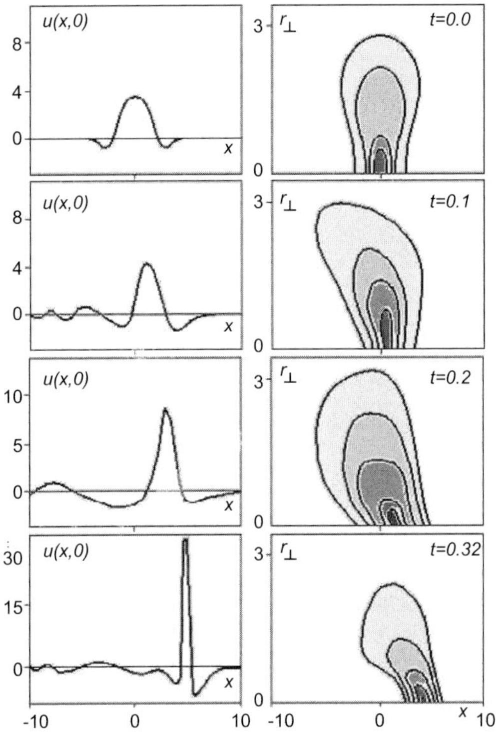

Fig. 3.10. Evolution of the three-dimensional axially symmetric wave pulse in a medium with positive dispersion

First, consider evolution of the three-dimensional nonlinear axially-symmetric pulse (3.81) in a medium with positive dispersion ($\kappa = 1$ in KP equation) for the initial Hamiltonian $\mathcal{H}$ less than its critical value, which on the soliton solution is

$$\mathcal{H} = \upsilon \mathcal{P}_x > 0,$$

where υ is the *Lagrange factor* in the variational equation (3.23) considered in Sect. 3.1.3. Our simulation results show that for $\kappa = 1$, the *wave collapse*

[7] Note that if the KP equation is considered in an open domain $\mathbb{D}$ of the real space R^3 then we have $\partial_t \mathcal{P}_x = \partial_t \mathcal{H} = 0$

is observed as a result of the time evolution of a three-dimensional initial wave pulse (see Fig. 3.10). Since the regions with larger k have larger phase velocity in the case of positive dispersion, the "wings" of the pulse lag behind its center during the evolution, and the distinctive U-type profile is formed, i.e., the *self-focusing* instability is developed. The amplitude of the wave field in the forming cavity ($u \geq 0$) within the time period ~ 0.3 is increased by an order of magnitude. The radiation of the waves from the whole system as well as from the cavity is observed. These results fully correspond to those obtained in Refs. [59,63].

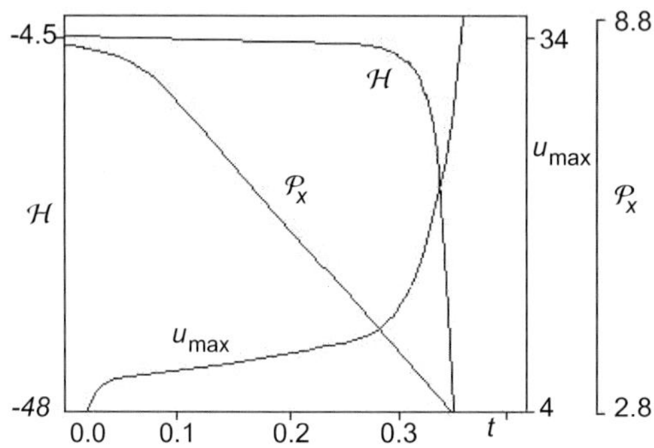

Fig. 3.11. Change of the wave amplitude $u_{\max}$, momentum $\mathcal{P}_x$, and energy $\mathcal{H}$ in the process of time evolution of an axially symmetric wave pulse in a medium with positive dispersion

On the next stage of the evolution, after $t \sim 0.3$, the tendency to the self-similar time behavior of $u_{\max}$, $\mathcal{P}_x$, and $\mathcal{H}$ takes place as shown in Fig. 3.11. We can see that the strongest changes of the wave amplitude, as well as those of the energy and x-projection of the system's momentum are in the cavity. The simulation data show that, for example, the dependence of the amplitude of the wave field on the axis of the cavity is proportional to $(t - t_0)^{-2/3}$; this means, according to the results of Ref. [64], that the regime of the wave collapse with the maximum radiation is realized in this case. The wave collapse is finished by the depletion of the cavern, with transition of the energy into the radiation propagating out of the cavern.

It follows from the results obtained in Sect. 3.1.3 by the analysis of deformations of the Hamiltonian $\mathcal{H}$, that for $\kappa < 0$ in (3.80), i.e., in the case of the negative dispersion in the medium, the KP equation has also no stable three-dimensional soliton solutions. In this regard, there is a problem of the asymptotic behavior of such solutions for $t \to \infty$. Different signs of the quadratic terms of the *Hamiltonian*

$$\mathcal{H} = \int \left[\frac{1}{2} (\partial_x u)^2 + \frac{\kappa}{2} (\mathbf{\nabla}_\perp w)^2 - u^3 \right] \mathrm{d}\mathbf{r}$$

for $\kappa < 0$ and $c = \int u^3 \mathrm{d}\mathbf{r} > 0$ correspond to attraction in the longitudinal coordinate direction and repulsion the transverse one (for $c < 0$ there is the interchange of the directions). In principle, in this situation with a sufficiently large wave amplitude, the time evolution can result in the formation of singularities such as caustics [59]. Numerical simulations [63] as well as the results

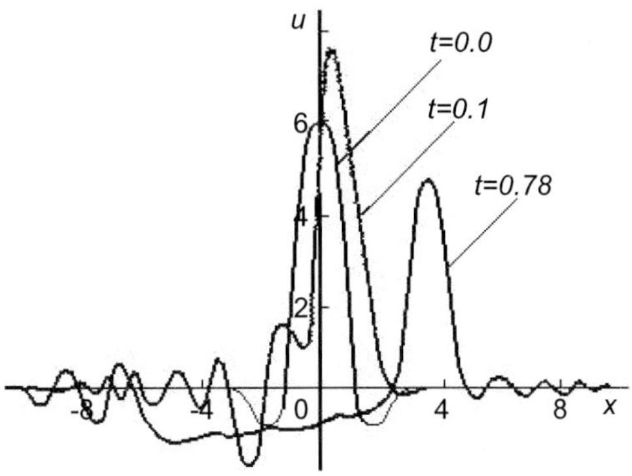

Fig. 3.12. Evolution of the three-dimensional axially symmetric wave pulse in a medium with negative dispersion

of our simulations demonstrate, however, that for negative dispersion of the medium, the wave collapse does not develop. At the initial stage of the time evolution the sub-focusing of the wave field is observed and the wave pulse amplitude grows (see Fig. 3.12). The peripherals ("wings") of the wave pulse lag behind its central part and the pattern of the evolution is qualitatively similar to the case $\kappa > 0$. At the later stage, the compression of the wave packet stops and, for all cases, converts into the regime of the wave defocusing, leading in the end to the spreading of the wave packet. It is interesting that such a pattern of the evolution of sonic waves in a plasma with negative dispersion is similar to the self-influence of electromagnetic waves in those media where the derivatives $\partial^2\omega/\partial k_x^2$ and $\partial^2\omega/\partial k_y^2$ have different signs (e.g., ion-cyclotron waves and whistlers observed in the Earth's magnetosphere and ionosphere) [197].

Self-Focusing of a Three-Dimensional Wave Beam. Consider now another nonlinear effect, typical for waves in a plasma: the *self-focusing* of a beam of *FMS waves.* It is well known that in a magnetized plasma with the small ratio of the kinetic pressure to the magnetic pressure ($\beta = 4\pi nT/B^2 \ll 1$), the FMS waves can easily be excited in the frequency range $\omega \ll \omega_{Bi}$, where $\omega_{Bi} = eB/m_i c$ is the ion-cyclotron frequency. Their *dispersion law* is given by

$$\omega = v_A k\left[1 + \chi(\theta)k^2 D^2\right]. \tag{3.82}$$

Here, $v_A = B^2/4\pi n_i m_i$ is the Alfvén velocity, $D^2 = c^2/2\omega_{pi}^2$ is the (squared) wave dispersion length, $\chi(\theta) = \cot^2\theta - m_e/m_i$, and θ is the angle between $\mathbf{k}$ and $\mathbf{B}$. When $kD \ll 1$, $k_x^2 \gg k_\perp^2$, and $v_x \ll v_A$, the dispersion law (3.82) is reduced to [196]

$$\omega = v_A k_x \left[1 + \frac{k_\perp^2}{k_x^2} + \chi(\theta) D^2 k_x^2\right]. \qquad (3.83)$$

This (weak) dispersion implies that the principal nonlinear process for small-amplitude waves is the three-wave interaction [59,196]. The (weak) nonlinearity determines the small angle between the interacting waves. The character of FMS waves depends on the dispersion sign: for almost transverse propagation, when

$$\left|\frac{\pi}{2} - \theta\right| \leq \left(\frac{m_e}{m_i}\right)^{1/2},$$

the dispersion of such FMS waves is negative (the phase velocity decreases with the increase of the wave number k); outside this cone the wave dispersion is positive.

Consider only the range of the propagation angles, where the inequality $\chi(\theta) > 0$ holds. In this case, the small amplitude FMS waves are described by the *KP equation* [81,83]:

$$\partial_t h + \alpha h \partial_x h - v_A \chi(\theta) D^2 \partial_x^3 h = -\frac{v_A}{2} \int_{-\infty}^{x} \Delta_\perp h dx. \qquad (3.84)$$

The nonlinear term (with $\alpha = (3/2) v_A \sin\theta$) is a consequence of renormalization of the sound velocity and reflects a small probability of other nonlinear processes that can be caused by the vector nonlinearity. Furthermore in (3.84), $h = B_\sim / B_0$ is the dimensionless amplitude of the FMS wave. According to the results known from Sect. 3.1, equation (3.84) can have:

- One-dimensional soliton solutions for angles $|(\pi/2) - \theta| \leq (m_e/m_i)^{1/2}$
- Two-dimensional soliton solutions with the algebraic asymptotics in the range, where the wave dispersion is positive (for FMS waves this takes place for sufficiently high ion temperature T_i, i.e. when $\beta > m_e/m_i$)

The three-dimensional wave packet of FMS waves with $\beta > m_e/m_i$ is not stable and collapses or spreads (for the positive and negative dispersion, respectively, see above). Thus in the case of a sufficiently intensive beam of FMS waves limited in the $\mathbf{k}_\perp$ direction, we can expect that the self-focusing occurs. Such idea was proposed for the first time by Manin and Petviashvili, and the problem of *self-focusing* of the FMS wave beam for the KP equation model was solved [86] by averaging initial equations and solving numerically the obtained expressions.[8] Similar studies were done in 1986-87 by Belashov (together with tests of numerical schemes for the generalized KP equation) as a particular case of nonlinear dynamics of FMS wave beam in the GKP equation model (see Sect. 4.6.2).

[8] The angle $\theta = \pi/2$, i.e. $\mathbf{B} \parallel \mathbf{k}_\perp$ was wrongly indicated in Ref. [86]. Since the wave dispersion is negative in this case, the self-focusing effect cannot be observed as shown by our results and [81].

Thus consider a stationary three-dimensional beam of FMS waves propagating in a plasma at the angle θ with respect to the magnetic field $\mathbf{B}$,

$$\left|\frac{\pi}{2} - \theta\right| \le \left(\frac{m_e}{m_i}\right).$$

For such a wave beam, taking into account the above conditions on the frequency range $\omega < \omega_{Bi}$ and on the projection of the wave vector, we convert (3.84) using the new variables

$$x \to x/r_B, \qquad y \to y/r_B, \qquad z \to z/r_B, \quad \text{and} \quad t = x/r_B - \omega_{Bi}t,$$

where $r_B^2 = 2D^2$, and obtain

$$\partial_x h + 2h\partial_t h + \partial_t^3 h = \int_{-\infty}^{t} \Delta_\perp h dt. \tag{3.85}$$

This KP equation describes the propagation of an FMS wave beam along the x-axis from the boundary $x = 0$. Thus we in fact converted the Cauchy problem for the KP equation (3.84) to the boundary problem for (3.85).

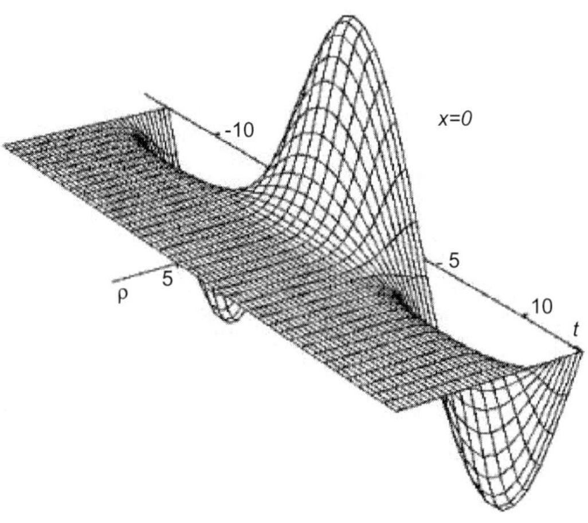

Fig. 3.13. The boundary condition at $x = 0$ for KP equation (3.85)

We solved equation (3.85) in the axially symmetric geometry

$$\Delta_\perp = \partial_\rho^2 + (1/\rho)\partial_\rho, \quad \text{where} \quad \rho^2 = y^2 + z^2$$

by using the implicit scheme (1.84) with approximation of the integral on the right-hand side by the *Newton–Cotes formula*[9], i.e., with the $O(\tau^2, h_{x,\rho}^4)$ accuracy. At the boundaries of the integration region, $t = \pm T$, $\rho = \pm P$, the

[9] See Sect. 4.3 for details of the numerical integration method.

condition of the total absorption was applied. At the boundary $x = 0$, the function

$$h_0 = h(t, 0, \rho) = a \cos(mt) \exp\left(-\rho^2\right), \tag{3.86}$$

i.e., the wave beam localized in the (y, z)-plane and periodic in time, was assumed (see Fig. 3.13).

We now move on to discuss results of numerical simulations of the FMS wave beam self-focusing for various beam intensities (amplitude a in (3.86)) at the boundary $x = 0$. A series of numerical experiments allows us to state:

1. For a small beam intensity h_0, the self-focusing phenomenon is not observed and beam scattering, i.e., decrease of intensity along the direction of its propagation takes place.
2. Above some threshold $h_{0,\mathrm{cr}}$, the self-focusing of the wave beam into a point is observed, i.e., the wave beam compression in the ρ-direction with propagation along the x-axis together with the fast increase of the beam intensity at the axis (see Fig. 3.14). The results of numerical simulations demonstrate

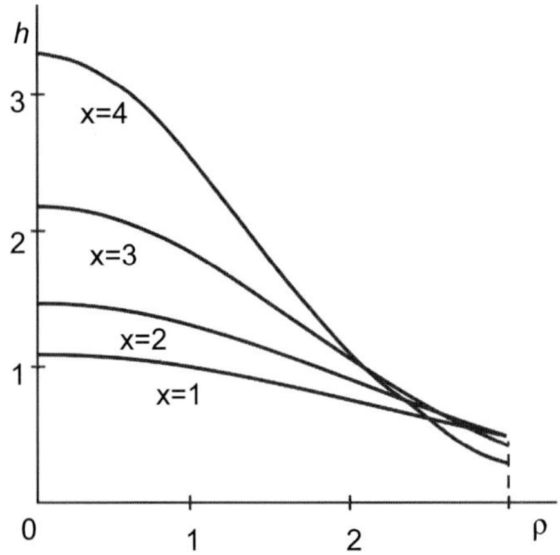

Fig. 3.14. Self-focusing of the three-dimensional axially symmetric FMS wave beam in a medium with the positive dispersion for $h_0 > h_{0,\mathrm{cr}}$

that for the central part of the wave beam the function $h(0, x, 0)$ can be well approximated by

$$h(0, x, 0) \sim h_0 \left[1 + (3.59x)^{2.3}\right]. \tag{3.87}$$

It is possible to estimate the compression of the wave beam on the ρ-axis. We have $l_\rho(x) \sim l_\rho(0) h_0 / h(x)$, where $l_\rho(0)$ is the characteristic transverse size of the wave beam at the boundary $x = 0$.
3. For the angles where the wave dispersion is negative, i.e., when $|(\pi/2) - \theta| \leq (m_e/m_i)^{1/2}$, the beam self-focusing does not occur; a subfocusing of the beam observed in the beginning of its evolution then transforms (with increasing x) into the defocusing regime.

Note that similar *self-focusing* phenomena were obtained in the studies of *acoustic waves in antiferromagnetics*, taking into account the crystal anisotropy [87]. This indicates the wide area of applications of the KP equation model in physics. Note also, that with accounting of finer dispersion effects appearing near the cone of the angles

$$\theta = \arctan\left(m_i/m_e\right)^{1/2} \tag{3.88}$$

in the KP equation (3.85), *higher order dispersion* corrections (of the fifth order of the x-derivative) appear and the range of possible wave beam evolutions can be significantly extended. Thus, for the beam propagation at the angles close to (3.88), for example, formation of a stationary wave beam after the stages of sub-focusing and defocusing is possible. This problem was investigated by Belashov [81,83,148,196] and is considered in detail in Sect. 4.6.2.

4. Generalized Two- and Three-Dimensional Models and Their Applications

4.1 Basic Dynamic Equations in the Long-Wavelength Approximation, Their Generalizations, and Solutions

In this section, we generalize the classic KP equation by introducing the higher order dispersion correction, the terms describing dissipation of the viscous type, as well as an instability and stochastic fluctuations of the wave field. Thus we derive the generalized KP (GKP) equation. We then reduce this equation to a simplified form, allowing its subsequent analysis (Sect. 4.1.1). Furthermore, in Sect. 4.1.2, we derive the three-dimensional derivative nonlinear Schrödinger (3-DNLS) equation from the full set of the plasma one-fluid magnetohydrodynamic (MHD) equations, and then, using the scale transforms, reduce it to a dimensionless form convenient for further analysis. Also, a generalization of 3-DNLS equation in the presence of dissipation in a medium is considered. Finally, in Sect. 4.1.3, we study in detail the stability of two- and three-dimensional solutions of the GKP and 3-DNLS equations.

4.1.1 Generalized KP Equation

In Sect. 1.1 we derived the KdV equation from the full set of the hydrodynamic equations, and in Sect. 3.1 we generalized this type of derivation for the systems described by the classic two-dimensional and three-dimensional KP equations. However, as we already noted in the Introduction, for some cases the coefficient at the third-order derivative in the KP-class equations (0.3) can be negligible or even exactly equal to zero (this takes place, for example, for the *gravity-capillary waves* in *shallow water* when $H \to (3\sigma/\rho g)^{1/2}$, and for the *FMS waves* propagating at the angles close to $\Theta = \arctan(m_i/m_e)^{1/2}$ with respect to an external magnetic field $\mathbf{B}$, see (0.8) and (0.9)). Nevertheless, this does not mean total disappearance of the effects of the medium's dispersion: the equilibrium between the nonlinear and dispersive processes in this case can still be recovered by invoking higher order terms in the expansion of the full dispersion equation in the powers of the wavenumber k. As a result, for equations of class (0.3) in the expression for $\mathcal{R}[u]$, dispersive correction terms proportional to the fifth derivative, $-\gamma\partial^5 u$, appear where

$\gamma > 0$[1], often playing the decisive role in the dynamics of multidimensional solitons (see Refs. [83,113,114,148,196]).

When dissipation cannot be neglected, equations of the class (0.3) should be supplemented by the corresponding terms. Since (as, e.g., in the case of nonlinear plasma waves) we consider here the hydrodynamic approximation for $\omega \ll \omega_{pe} = (4\pi n_e e^2/m_e)^{1/2}$, we are limited in our study with the effects of dissipative processes of the so-called viscous type (assuming that the *Landau damping* can be neglected) on the structure and evolution of nonlinear waves and solitons. In this case, for the ion plasma oscillations when the wave frequency (and the characteristic times of the processes) is significantly less than the electron plasma frequency, $\omega \sim \tau^{-1} \ll \omega_{pe}$, the dissipative effects associated with the processes of relaxation in the medium lead to the imaginary term of the type $-\mathrm{i}\nu k_x^2$ in the dispersion equation (0.5), and, accordingly, to the Burgers term [15] on the right-hand side of equation (0.3). Thus the *KP equation* takes the form

$$\partial_x \left(\partial_t u + \alpha u \partial_x u - \nu \partial_x^2 u + \beta \partial_x^3 u + \gamma \partial_x^5 u\right) = \kappa \Delta_\perp u, \tag{4.1}$$

where $\kappa = -c_0/2$ and it has the same degree of universality, as the standard KdV and KP equations in the sense that it is valid always when the *dispersion law* is given by

$$\omega \approx c_0 k_x \left(1 + \frac{k_\perp^2}{2k_x^2} - \frac{\mathrm{i}\nu k_x}{c_0} + \frac{-\beta k_x^2 + \gamma k_x^4}{c_0}\right). \tag{4.2}$$

To account for various instabilities (determined by the medium and the character of the wave propagation), leading to rapid increase of wave perturbations with the formation of the chaotic turbulent state and transfer of the energy stored in the oscillations to other degrees of freedom, we can introduce into the left-hand side of (0.22) the term proportional to the fourth space derivative, $\delta \partial_x^4 u$, corresponding to the term $-\mathrm{i}\delta k_x^4/c_0$ which appears in this case in the dispersion relation (0.23). With this, the generalization of the KP equation is given by

$$\partial_x \left(\partial_t u + \alpha u \partial_x u - \nu \partial_x^2 u + \beta \partial_x^3 u + \delta \partial_x^4 u + \gamma \partial_x^5 u\right) = \kappa \Delta_\perp u, \tag{4.3}$$

and we call it, as well as equation (0.22), the *generalized KP equation* or the *GKP equation.*

The three-dimensional equation (4.3) has wide applications in the physics of nonlinear dispersive waves. However, it was investigated in detail only recently in Refs. [81,83,113,148,198] where the two-dimensional problems mentioned above in the Introduction were effectively solved, and the specific processes of self-influence of the waves, described earlier by the standard three-dimensional KP equation (0.10) [18,59,86,87], were studied within the

[1] Explicit expressions for the coefficient γ in the cases of the gravity-capillary and the fast magnetosonic waves are given in the Introduction (see (0.18) and (0.21)).

framework of generalization (0.18) representing the approach which is more adequate for real physical systems.

For problems where the question might appear on the appropriate model assumed for the GKP equation, we present here the results obtained for the KP-class equations in the form (4.3). For convenience of further analysis when $\delta = 0$, we transform equation (4.1) using $x \to -sx$, $y \to -s\kappa^{1/2}y$, $z \to -s\kappa^{1/2}z$, $t \to st$, and $u \to -(6/\alpha)u$, where $s = |\gamma|^{1/4}$. Thus the basic equation in this case can be written as

$$\partial_x \left(\partial_t u + 6u\partial_x u - \mu\partial_x^2 u - \varepsilon\partial_x^3 u - \lambda\partial_x^5 u\right) = \Delta_\perp u, \tag{4.4}$$

where $\mu = \nu s^{-1}$, $\varepsilon = \beta s^{-2}$, and $\lambda = \mathrm{sgn}(\gamma)$. In Sect. 4.1.3, we study analytical approaches to the problem of stability of multidimensional solitons and nonlinear wave packets described by equations of the KP class in the form (4.4). In Sect. 4.2 we investigate the classes of possible solutions and their asymptotics employing the methods of qualitative analysis (usually used in the theory of dynamic systems) as well as the asymptotic analysis (when $|x| \to \infty$) of the structure of the solutions.

4.1.2 3-DNLS Equation

We introduced in Sect. 2.4 the derivative nonlinear Schrödinger (DNLS) equation, omitting its detailed derivation, and considered it as an integrability condition for two linear differential equations. Since we investigate multidimensional systems (and would like to emphasize the physics of phenomena described by this equation more clearly) here, we present a brief derivation of the three-dimensional DNLS (3-DNLS) equation. Thus using the ideas and technique detailed also in Refs. [7,37,38], we obtain the *3-DNLS equation* in the form given by (0.24).

Here, we write the full set of the one-fluid *MHD equations* assuming that $k_\perp^2 \ll k_x^2$ (in this case we can change $\nabla \to \hat{\mathbf{x}}\partial_x$) [7]:

$$\left(\partial_t + v_x\partial_x\right)\mathbf{v}_\perp = v_A^2\partial_x\mathbf{h}, \tag{4.5}$$

$$\left(\partial_t + v_x\partial_x\right)v_x = -\frac{v_A^2}{2}\partial_x\mathbf{h}^2, \tag{4.6}$$

$$\partial_t\mathbf{h} = \partial_x\left(\mathbf{v}_\perp - v_x\mathbf{h}\right), \tag{4.7}$$

$$\partial_t\rho + \partial_x\left(v_x\rho\right) = 0, \tag{4.8}$$

where $\mathbf{h} = \mathbf{B}_\perp/B_0$ is the dimensionless perturbation of the perpendicular magnetic field. Following Ref. [7], consider dependence of all functions in these equations on t and x in the form $f = f(t, x - v_A t)$, where the first argument is due to nonlinear effects. Since perturbations of the density ρ and the x-component of the velocity v_x are also stipulated by the presence of the nonlinearity of the medium, it is possible to approximate $\partial_t \cong -v_A\partial_x$ [37]. Thus integrating (4.8) and taking into account that $\lim\limits_{|x|\to\infty} \rho = \rho_0$, we find

$$\rho = \rho_0 \left(1 + \frac{v_x}{v_A}\right). \tag{4.9}$$

Furthermore, by neglecting the small term on the order of v_x^2/v_A^2 on the left-hand side of (4.6), and applying the same assumptions as above, we obtain from (4.6)

$$v_x = \frac{v_A}{2}\mathbf{h}^2. \tag{4.10}$$

According to the relation (4.10), the plasma is pushed away by the wave field in the direction of the wave propagation thus forming the "Alfvén wind" [37] with the effect of the ponderomotive force [157].

Substituting relations (4.9) and (4.10) into (4.5) and (4.7), taking into account the parabolic equation for the magnetic field **B** (see, e.g., Refs. [7,37]), and retaining the dispersion effects, we find that the equation for the dimensionless perpendicular magnetic field $h = h_x + \mathrm{i}h_y$ is given by

$$\partial_x \left[(2/\omega_{pi})\, \partial_t h + r_A \partial_x \left(|h|^2 h \right) + \mathrm{i} r_A^2 \partial_x^2 h \right] = -\frac{r_A}{2} \nabla_\perp^2 h, \tag{4.11}$$

where $r_A = v_A/\omega_{Bi}$. Equation (4.11) describes the left-circularly polarized wave mode; for the right-circularly polarized mode we have the "minus" sign in front of the dispersion term. It is possible to incorporate the sign of the nonlinearity by the factor $s = \pm 1$ in front of the nonlinear term. Thus after introducing $\kappa = -r_A/2$, converting to the dimensionless variables $t \to \omega_{Bi} t/2$, $x \to x/r_A$, $\mathbf{r}_\perp \to \mathbf{r}_\perp \sqrt{2}/r_A$, and integrating in x, we find that the *3-DNLS equation* in the reference frame moving in the positive direction of the x-axis with the Alfvén velocity (0.16) can be written as

$$\partial_t h + s \partial_x \left(|h|^2 h \right) - \mathrm{i}\lambda \partial_x^2 h = \kappa \int_{-\infty}^{x} \Delta_\perp h \mathrm{d}x, \tag{4.12}$$

where the upper(lower) sign of the factor $\lambda = \pm 1$ corresponds to the right(left)-circularly polarized wave mode, respectively.

When the dissipation effects cannot be neglected, (4.12) should be supplemented by the proper term. Taking into account the hydrodynamic approximation considered in this section (when $\omega \ll \omega_{pe}$) it is thus sufficient to limit the study (of the influence of dissipation on the structure and evolution of nonlinear waves) to only processes of the "viscous" type (e.g., taking place in a plasma for the ion oscillations), with the inverse times being much less than the electron plasma frequency, i.e., $\tau^{-1} \ll (4\pi n_e e^2/m_e)^{1/2}$ (in this case, for $T_e \gg T_i$, the *Landau damping* is small). Thus, dissipative effects associated with such type of relaxation process lead to appearance of the imaginary term $-\mathrm{i}\nu k_x^2$ in the dispersion equation. Accordingly, the Burgers-type term $\nu \partial_x^2 u$ [15] has to be included into the left-hand side of (4.12). In this case, the coefficient

$$\nu = \frac{\rho_0}{2\rho}\left(c_\infty^2 - c_0^2\right)\tau \int_0^\infty \xi\varphi(\xi)\mathrm{d}\xi \tag{4.13}$$

defines the logarithmic damping rate and, as it is shown in Ref. [3], is the characteristic rate of the relaxation damping of the "sound" wave. Here, c_∞ and c_0 are the velocities of the high- and low-frequency "sound" mode (the last one coincides with $c_s = (T_e/m_i)^{1/2}$) and $\varphi(t,\tau)$ is the function defining the relaxation process. Thus the *3-DNLS equation* generalized by the viscous-type dissipative term can be written as

$$\partial_t h + s\partial_x\left(|h|^2 h\right) - \mathrm{i}\lambda\partial_x^2 h - \nu\partial_x^2 h = \kappa \int_{-\infty}^{x} \Delta_\perp h \mathrm{d}x. \tag{4.14}$$

For this equation, taking into account $\kappa = -c_0/2$, $c_0 = v_A$, and (formally) $\beta k_x = \lambda$, $\gamma = 0$, $s = c_0$, the *dispersion relation* of the type (0.23) is also valid, namely

$$\omega \approx v_A k_x \left(1 + \frac{k_\perp^2}{2k_x^2} - \frac{\mathrm{i}\nu k_x}{v_A} - \frac{\lambda k_x}{v_A}\right). \tag{4.15}$$

We note here that the three-dimensional equation (4.14) (as well as equation (4.12)) is not completely integrable and for its solution it is necessary to use numerical methods [50,65–70]. It is also necessary to take into account, that even for the one-dimensional equation (1-DNLS equation), the solutions can not always be obtained analytically in the closed form, since the use of the IST procedure requires rather strong limits on the initial and boundary conditions (first of all, on the localization of the potential $h(x,t)$ at any time moment and $|h(x,0)| \to 0$ when $|x| \to \infty$). Thus development of numerical codes for the integration of the DNLS-type equations is an important and actual issue.

4.1.3 Stability of Two-Dimensional and Three-Dimensional Solutions of GKP and 3-DNLS Equations

We here investigate analytical approaches to the problem of stability of multidimensional solitons and nonlinear wave packets. Under the assumption of negligible dissipative effects, these solutions coincide with those of the GKP-class equations in the form (4.4) with $\mu = 0$ and the 3-DNLS equation (4.12). New results on the problem of the soliton stability are obtained. In particular, in the first part of this section we present analytical estimates and formulate the sufficient conditions for the stability of solutions of GKP equation in the two-dimensional and three-dimensional cases, based on the transformational properties of the system's Hamiltonian for the whole range of the dispersion coefficients. Then an analogous problem for the 3-DNLS equation in the three-dimensional geometry is studied. Despite the fact that the considered classes of the Hamiltonian's deformations for both equations do not include all possible deformations of $\mathcal{H}$, the obtained results clearly demonstrate the stability of the solutions if some (found and formulated) conditions are satisfied and can at least be considered as the necessary conditions of the stability of the multi-dimesnional solutions.

Stability of Two-Dimensional and Three-Dimensional Solutions of the GKP Equation. Here we suppose that the dissipation is absent in the medium, i.e., $\mu = 0$ and the *GKP equation* (4.4) is written as

$$\partial_x \left(\partial_t u + 6u\partial_x u - \varepsilon\partial_x^3 u - \lambda\partial_x^5 u\right) = \Delta_\perp u. \tag{4.16}$$

Note that (4.16) is now the Hamiltonian equation. Rewriting it into the form

$$\partial_t u = \partial_x \left(\delta\mathcal{H}/\delta u\right) = -\int_{-\infty}^{\infty} \delta'(x - x') \left(\delta\mathcal{H}/\delta u\right) \mathrm{d}x', \tag{4.17}$$

where

$$\mathcal{H} = \int \left[-\frac{\varepsilon}{2}\left(\partial_x u\right)^2 + \frac{\lambda}{2}\left(\partial_x^2 u\right)^2 + \frac{1}{2}\left(\nabla_\perp \partial_x v\right)^2 - u^3\right] \mathrm{d}\mathbf{r} \tag{4.18}$$

and $\partial_x^2 v = u$, we obtain the Hamiltonian equation where the continuum of values, $u \in \mathbb{M}$, plays the role of the point coordinates in the phase space $\mathbb{M}$, the matrix $\omega(x, x') = \delta'(x - x')$ is skew-symmetric and, because of the inversibility of the operator ∂_x on the decreasing functions for $|x| \to \infty$, is a non-degenerate one on u. Thus the Hamiltonian structure can be represented by the Poisson bracket [24],

$$\{S, R\} = \int_{-\infty}^{\infty} (\delta S/\delta u)\partial_x(\delta R/\delta u)\mathrm{d}x, \tag{4.19}$$

with $S, R \in \mathbb{M}$, which satisfies the Jacobi's identity since ω does not depend on the point u in the space $\mathbb{M}$.

The problem of the stability of the soliton-like solutions of (4.17) was studied before [195] on the basis of an analysis of transformational properties of the *Hamiltonian* (4.18) in the two- and three-dimensional geometry ($\partial_z = 0$ and $\partial_{yz} \neq 0$, respectively) for $\lambda = \pm 1$ and $\varepsilon > 0$ and $\varepsilon < 0$ (corresponding to different types of the medium). The stationary solutions of (4.17) are then defined from the *variational equation*,

$$\delta\left(\mathcal{H} + v\mathcal{P}_x\right) = 0, \tag{4.20}$$

where $\mathcal{P}_x = (1/2)\int u^2 \mathrm{d}\mathbf{r}$ is the momentum projection onto the x-axis and v is the *Lagrange factor*. Equation (4.20) illustrates the fact that all finite solutions of (4.17) are the stationary points of the Hamiltonian $\mathcal{H}$ for the fixed momentum $\mathcal{P}_x$.

Consider now the problem of stability. In a dynamic system, according to the *Lyapunov's theorem*, the stationary points corresponding to the maximum or minimum of the Hamiltonian $\mathcal{H}$ are absolutely stable. If an extremum is local then the locally stable solutions are possible. The unstable states correspond to the monotonous dependence of $\mathcal{H}$ on its variables, i.e., those cases when the stationary point is the *saddle point*. According to that, all we need is to prove that the Hamiltonian $\mathcal{H}$ is limited from below for the fixed $\mathcal{P}_x$.

Similar to what was done for the classic KP equation in Chap. 3 [60], we consider the scale transformations in the real vector space $\mathbb{R}$,

$$u(x, \mathbf{r}_\perp) \to \zeta^{-1/2}\eta^{(1-d)/2}u(x/\zeta, \mathbf{r}_\perp/\eta), \tag{4.21}$$

(where d is the dimension of the problem, and $\zeta, \eta \in \mathbb{R}$) which conserves the momentum $\mathcal{P}_x$. The *Hamiltonian* as a function of the parameters ζ and η now takes the form

$$\mathcal{H}(\zeta, \eta) = a\zeta^2 + b\zeta^2\eta^{-2} - c\zeta^{-1/2}\eta^{(1-d)/2} + e\zeta^{-4}, \tag{4.22}$$

where

$$\begin{aligned} a &= -\frac{\varepsilon}{2}\int (\partial_x u)^2\, d\mathbf{r}, \\ b &= \frac{1}{2}\int (\nabla_\perp \partial_x v)^2\, d\mathbf{r}, \\ c &= \int u^3 d\mathbf{r}, \\ e &= \frac{\lambda}{2}\int \left(\partial_x^2 u\right)^2 d\mathbf{r}. \end{aligned}$$

The necessary conditions for the existence of the Hamiltonian's extremum are given by

$$\partial_\zeta \mathcal{H} = 0 \qquad \text{and} \qquad \partial_\eta \mathcal{H} = 0. \tag{4.23}$$

The latter enables us to obtain the extremum's coordinates, (ζ_i, η_i), if it exists. Holding the inequalities

$$\begin{vmatrix} \partial_\zeta^2 \mathcal{H}(\zeta_i, \eta_j) & \partial_{\zeta\eta}^2 \mathcal{H}(\zeta_i, \eta_j) \\ \partial_{\eta\zeta}^2 \mathcal{H}(\zeta_i, \eta_j) & \partial_\eta^2 \mathcal{H}(\zeta_i, \eta_j) \end{vmatrix} > 0 \quad \text{and} \quad \partial_\zeta^2 \mathcal{H}(\zeta_i, \eta_j) > 0 \tag{4.24}$$

guarantees that the corresponding quadratic form is the positively definite one and therefore these inequalities give the sufficient condition of the existence of the (local) minimum at the point (ζ_i, η_j).

Consider equation (4.17) for $d = 2$, i.e., with $\partial_z = 0$. In this case equations (4.23) form the following set:

$$
\begin{aligned}
G &\equiv \left(c^4/32b\right) t^4 - (at+2e)^3 = 0, \\
t &= \zeta^2, \\
\eta &= \left[(4b/c)^2 \zeta^5\right]^{1/3}.
\end{aligned}
\tag{4.25}
$$

Analysis of (4.25) shows (see Appendix 2) that it has one positive root, $t \in \mathbb{R}$, for every quadruple of the functions $a, b, c, e \in \mathbb{R}$ in the case $e > 0$ and any a; two positive roots, $t_{1,2} \in \mathbb{R}$, for $e < 0$ and $a > 0$; and in the case $e < 0$ and $a \leq 0$ we have $t \notin \mathbb{R}$.

Inequalities (4.24) for $d = 2$, taking into account expressions (4.25), lead to

$$
G - (C_{11}a^3t^3 + C_{12}a^2et^2 + C_{13}ae^2t + C_{14}e^3) < 0, \tag{4.26}
$$

and

$$
G - (C_{21}a^3t^3 + C_{22}a^2et^2 + C_{23}ae^2t + C_{24}e^3) < 0, \tag{4.27}
$$

where $C_{nm} > 0$ are constants. It follows that conditions (4.24) are fulfilled on the set $S_t \subset \mathbb{R}$ of solutions of the set (4.25) for $e > 0$ and $a \geq 0$, and, consequently, the Hamiltonian $\mathcal{H}(\zeta, \eta)$ is bounded from below. Solving (4.26) and (4.27) in the $\mathbb{R}$-space for $e > 0$ and $a < 0$, we obtain for $S_t^{(4.26)} \cap S_t^{(4.27)} = A_t \subset \mathbb{R}$ that $\sup A_t = (3C_{11})^{-1} \times [2C_1 \cos(\varphi_1/3) - C_{12}]\, e/a$ and $\inf A_t = 0$ ($t = 0$ is not the root of the set (4.25) and we therefore discard it). Here, $C_1 = (C_{12}^2 - 3C_{11}C_{13})^{1/2}$ and $\varphi_1 = \operatorname{Arccos}\left\{\left[C_{12}\left(C_{12}^2 - 3C_1^2\right) - 27C_{11}^2C_{14}\right] / \left(2C_1^3\right)\right\}$. Taking into account (5.8) (see Appendix 2; note that $S_t \cap A_t \neq \oslash$), we conclude that for $e > 0$ and $a < 0$, the sufficient condition of the existence of the local minimum of $\mathcal{H}(\zeta, \eta)$ is the relation $S_t \subseteq A_t$, i.e.,

$$
\left(\frac{a^4 b}{c^4 e}\right)^{1/4} \geq \frac{1}{6C_{11}} \left[C_1 \cos\left(\frac{\varphi_1}{3}\right) - \frac{C_{12}}{2}\right]. \tag{4.28}
$$

Analogously, considering inequalities (4.26) and (4.27) for the case $e < 0$ and $a > 0$, we obtain that

$$
\begin{aligned}
\inf B_t^{(1)} &= \frac{1}{3C_{21}} \left[2C_2 \cosh\left(\frac{\varphi_2}{3}\right) - C_{22}\right] \left(\frac{e}{a}\right), \\
\sup B_t^{(1)} &= \frac{1}{3C_{11}} \left[2C_1 \cos\left(\frac{\varphi_1}{3} + \frac{4\pi}{3}\right) - C_{12}\right] \left(\frac{e}{a}\right), \\
\inf B_t^{(2)} &= \frac{1}{3C_{11}} \left[2C_1 \cos\left(\frac{\varphi_1}{3} + \frac{2\pi}{3}\right) - C_{12}\right] \left(\frac{e}{a}\right),
\end{aligned}
$$

where

$$
\begin{aligned}
&B_t^{(1)} \cup B_t^{(2)} = S_t^{(4.26)} \cap S_t^{(4.27)} = A_t \subset \mathbb{R}, \\
&C_2 = \left(C_{22}^2 - 3C_{21}C_{23}\right)^{1/2}, \\
&\varphi_2 = \operatorname{Arccosh}\left\{\left[C_{22}\left(C_{22}^2 - 3C_2^2\right) - 27C_{21}^2C_{24}\right] / 2C_2^3\right\}.
\end{aligned}
$$

Taking into account equalities (5.12) we obtain that $B_t^{(1)} \subset S_t \Longrightarrow B_t^{(1)} \cap S_t = B_t^{(1)}$ and $B_t^{(2)} \cap S_t = \oslash$. Then in (5.9), by changing $(a^4 b/c^4 e) \leq -2^4 \cdot 3^{-3} \cdot Q^{-1}$ with $Q > 1$ and using inequalities (5.11), we obtain $Q = -2^8 \cdot 3^{-3}(T+2)/T^2$ with $T = \inf B_t^{(1)} a/e$. This corresponds to the sufficient condition of the existence of the local minimum of the Hamiltonian $\mathcal{H}(\zeta, \eta)$, namely $\inf S_t = \inf B_t^{(1)}$, which can now be rewritten as

$$\frac{a^4 b}{c^4 e} \leq \frac{T^2}{2^4(T+2)}. \tag{4.29}$$

Fig. 4.1 shows the change of the Hamiltonian $\mathcal{H}(\zeta, \eta)$ for the test values of

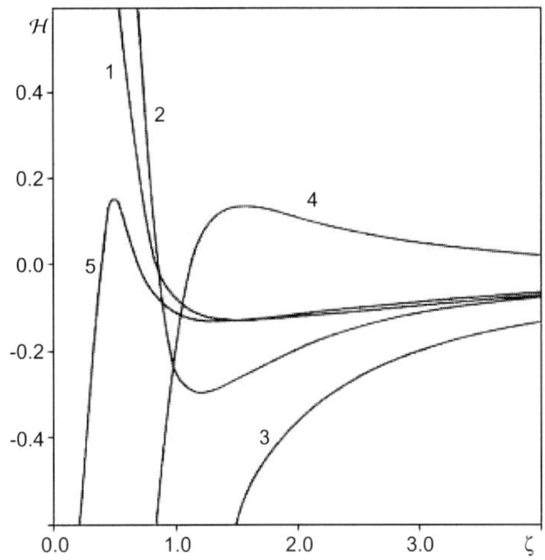

Fig. 4.1. Change of the Hamiltonian $\mathcal{H}(\zeta, \eta)$ in the two-dimensional case ($d = 2$) along lines $\eta = [(4b/c)^2 \zeta^5]^{1/3}$ for the test values: (1) $a = 0.5$, $b = 0.5$, $c = 1.0$, $e = 0.02$; (2) $a = -0.5$, $b = 0.5$, $c = 0.5$, $e = 0.5$; (3) $a = -0.5$, $b = 0.5$, $c = 1.0$, $e = -0.02$; (4) $a = 1.0$, $b = 1.0$, $c = 0.5$, $e = -1.0$; (5) $a = 0.5$, $b = 0.5$, $c = 1.0$, $e = -0.02$

the integrals $a, b, c,$ and e for $d = 2$, $\lambda = \pm 1$, and $|\varepsilon| > 0$.

Consider now equation (4.17) for $d = 3$ ($\partial_{yz} \neq 0$). In this case, for every quadruple $a, b, c, e \in \mathbb{R}$, with $a \neq 0$, we immediately obtain from (4.23)

$$\begin{aligned} \zeta_i &= \frac{1}{16ab}\left(3c^2 \pm \sqrt{9c^4 - 512ab^2 e}\right), \\ \eta_j &= \frac{2b}{c}\zeta_i^{5/2}, \end{aligned} \tag{4.30}$$

where $i = 1, 2$ and $j = 1, 2, 3, 4$. We note here that $(\zeta_i, \eta_j) \notin \mathbb{R}$ for $\zeta_i < 0$, and therefore we consider below only the roots $\zeta_i > 0$ (we map out equality $\zeta_i = 0$, taking into account $e \neq 0$, otherwise (4.17) degenerates into the standard KP equation). Inequalities (4.24) taking into account (4.30) are now given by

$$a\zeta^2 - (c^2/2b)\zeta + 10e/3 > 0, \tag{4.31}$$

and

$$a\zeta^2 + (c^2/48b)\zeta + 10e/3 > 0. \tag{4.32}$$

In the case $e > 0$ and $a > 0$, the condition $\zeta_i \in \mathbb{R}$, i.e.,

$$c^4 \geq (512/9)ab^2 e, \tag{4.33}$$

gives $\zeta_{1,2} > 0$. Elementary analysis then shows that $S_{\zeta_1}^{(4.30)} \cap S_{\zeta_1}^{(4.31)} = \oslash$ and, if strict inequality (4.33) holds, $S_{\zeta_2}^{(4.30)} \subset S_{\zeta_2}^{(4.31),(4.32)}$. Thus, for the existence of the local minimum of the Hamiltonian $\mathcal{H}(\zeta,\eta)$ for $e > 0$ and $a > 0$, it is sufficient to have

$$\frac{ab^2 e}{c^4} < \frac{9}{512}. \tag{4.34}$$

When $e > 0$ and $a < 0$, for each quadruple of $a, b, c, e \in \mathbb{R}$ we have from the first equality of (4.30) $\zeta_1 < 0$ and, therefore, $S_{\eta_{1,2}}^{(4.30)} \cap \mathbb{R} = \oslash$. For $S_{\zeta_2}^{(4.30)}$, elementary analysis of inequalities (4.31) and (4.32) gives us $S_{\zeta_2}^{(4.30)} \subset S_{\zeta_2}^{(4.31),(4.32)}$. Thus for any $e > 0$ and $a < 0$ the function $\mathcal{H}(\zeta,\eta)$ is limited from below.

Analogous consideration in the case $e < 0$ shows that for $a < 0$, when condition (4.33) is satisfied for every quadruple $a, b, c, e \in \mathbb{R}$, we have $\zeta_{1,2} < 0$ and, therefore, $S_{\eta_{1,2,3,4}}^{(4.30)} \cap \mathbb{R} = \oslash$; for $a > 0$ we have $\zeta_2 < 0 \Longrightarrow S_{\eta_{3,4}}^{(4.30)} \cap \mathbb{R} = \oslash$, $\zeta_1 > 0$ but $S_{\zeta_1}^{(4.30)} \cap S_{\zeta_1}^{(4.31)} = \oslash$. For $a = 0$ and $|e| > 0$ (i.e., $e \neq 0$), instead of (4.30) we have

$$\zeta_i = \frac{16be}{3c^2}, \quad \text{and} \quad \eta_j = \frac{2b}{c}\zeta_i^{5/2}, \tag{4.35}$$

where $i = 1$ and $j = 1, 2$, for every triplet of functions $b, c, e \in \mathbb{R}$. For $e < 0$

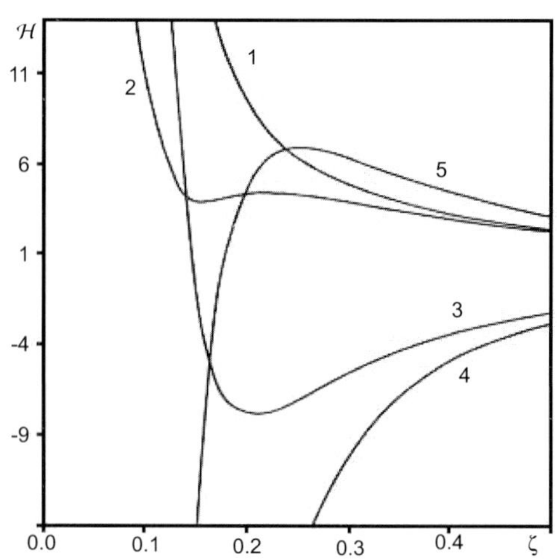

Fig. 4.2. Change of the Hamiltonian $\mathcal{H}(\zeta,\eta)$ for $d = 3$ along the lines $\eta = (2b/c)\zeta^{5/2}$ for the test values: (1) $a = 1.0$, $b = 1.0$, $c = 1.0$, $e = 0.025$; (2) $a = 1.0$, $b = 1.0$, $c = 1.0$, $e = 0.017$; (3) $a = -0.5$, $b = 1.0$, $c = 0.5$, $e = 0.02$; (4) $a = -0.5$, $b = 1.0$, $c = 0.5$, $e = -0.02$; (5) $a = 1.0$, $b = 1.0$, $c = 0.5$, $e = -0.02$

it immediately follows that $S_{\eta_j} \cap \mathbb{R} = \oslash$. For $e > 0$, it is not difficult to show that $S_\zeta \subset S_\zeta^{(4.31),(4.32)}$. Fig. 4.2 shows the change of the Hamiltonian $\mathcal{H}(\zeta,\eta)$ for the test values of the integrals a, b, c, e for $d = 3$, $\lambda = \pm 1$, and $|\varepsilon| > 0$.

To sum up the above results, we conclude the following. In the two-dimensional case the Hamiltonian (4.18) of the equation (4.17) is limited from below at the fixed projection of the momentum $\mathcal{P}_x$ for the integral values $e > 0$ and $a \geq 0$ (i.e. when $\lambda = 1, \varepsilon \leq 0$ in expression (4.18)) and has the local minima for $e > 0$ and $a < 0$ ($\lambda = 1, \varepsilon > 0$) and $e < 0$ and $a > 0$ ($\lambda = -1, \varepsilon < 0$) when the conditions (4.28) and (4.29), respectively, are satisfied. In the three-dimensional case $\mathcal{H}$ has a local minimum for $e > 0$ and $a \geq 0$ (i.e. when $\lambda = 1, \varepsilon \leq 0$ in (4.18)) if the condition (4.34) is satisfied, and it is limited from below for $e > 0$ and $a < 0$ ($\lambda = 1, \varepsilon > 0$). Note that the class of scale transformations (4.21) of course does not include all possible deformations of the Hamiltonian $\mathcal{H}$ but the estimations obtained above justify that it is limited for the cases considered when, according to the Lyapunov's theorem, absolutely and locally stable soliton solutions should exist. Analysis of the boundedness of $\mathcal{H}$ on the numerical solutions of (4.16) for $d = 2$ and $d = 3$ is presented below in Sect. 4.4 and Sect. 4.5, respectively (see also [195], [148]).

Stability of Three-Dimensional Solutions of the 3-DNLS Equation. We use the same approach as above to investigate the stability of three-dimensional solutions of the *3-DNLS equation* (4.12) (in the non-dissipative case) . First, we rewrite the equation in the Hamiltonian form

$$\partial_t h = \partial_x \left(\delta\mathcal{H}/\delta h \right), \tag{4.36}$$

with the *Hamiltonian* given by [70]

$$\mathcal{H} = \int\limits_{-\infty}^{\infty} \left[\frac{1}{2} |h|^4 + \lambda s h h^* \partial_x \varphi + \frac{1}{2} \kappa \left(\nabla_\perp \partial_x w \right)^2 \right] \mathrm{d}\mathbf{r}, \tag{4.37}$$

where $\partial_x^2 w = h$ and $\varphi = \arg(h)$. Then we investigate whether $\mathcal{H}$ is limited from below under its deformations conserving the projection of the system's momentum $\mathcal{P}_x = (1/2) \int |h|^2 \mathrm{d}\mathbf{r}$, when the *variational equation* in the form (4.20) takes place

$$\delta \left(\mathcal{H} + v\mathcal{P}_x \right) = 0 \tag{4.38}$$

(here, v, as well as in (4.20), is the *Lagrange factor*). Analogous to the above investigation of the properties of the GKP equation, we consider the following scale transformations in the complex vector space $\mathbb{C}$:

$$h(x, \mathbf{r}_\perp) \to \zeta^{-1/2} \eta^{-1} h(x/\zeta, \mathbf{r}_\perp/\eta) \tag{4.39}$$

($\zeta, \eta \in \mathbb{C}$), conserving the projection of the momentum $\mathcal{P}_x$. The Hamiltonian $\mathcal{H}$ as a function of the parameters (ζ, η) is now given by

$$\mathcal{H}(\zeta, \eta) = a\zeta^{-1}\eta^{-2} + b\zeta^{-1} + c\zeta^2\eta^{-2}, \tag{4.40}$$

where

$$a = \frac{1}{2} \int |h|^4 \, d\mathbf{r},$$

$$b = \lambda s \int h h^* \partial_x \varphi d\mathbf{r},$$

$$c = \frac{\kappa}{2} \int (\nabla_\perp \partial_x w)^2 \, d\mathbf{r}.$$

The necessary conditions for the existence of the extremum, namely,

$$\partial_\zeta \mathcal{H} = 0 \quad \text{and} \quad \partial_\eta \mathcal{H} = 0,$$

immediately allow us to obtain the extremum's coordinates

$$\zeta = -\frac{a}{c} \quad \text{and} \quad \eta = \left[-\frac{a}{b} \left(1 + \frac{a^2}{c^2} \right) \right]^{1/2}, \tag{4.41}$$

where $b < 0$ if $\eta \in \mathbb{R} \subset \mathbb{C}$ because $a > 0$ and $c > 0$ by definition, and $b > 0$ if $\eta \in \mathbb{C}$.

The sufficient conditions for the existence of the local minimum of $\mathcal{H}$ at the point (ζ_i, η_i) are given by (4.24), and we therefore obtain for $b < 0$

$$\frac{a}{c} < d = \frac{1}{2\sqrt{2}} \sqrt{13 + \sqrt{185}}. \tag{4.42}$$

Thus it follows from (4.40)–(4.42) that the Hamiltonian $\mathcal{H}$ of (4.36) is limited from below:

$$\mathcal{H} > -\frac{3bd}{1 + 2d^2}, \tag{4.43}$$

where $b < 0$, if condition (4.42) holds. In this case, three-dimensional solutions of the 3-DNLS equation are stable; they are unstable in the opposite case, when $ac^{-1} \geq d$ and $b < 0$. The condition $b < 0$ corresponds to the right-circularly polarized wave propagating in a plasma with $p = (4\pi nT/B^2) > 1$, i.e., when $\lambda = 1$ and $s = -1$ in (4.12) and (4.36), and to the left-circularly polarized wave when $\lambda = -1$ and $s = 1$. It is necessary to note, however, that the change of the signs of $\lambda = 1 \to -1$ and $s = -1 \to 1$ is equivalent to the change $t \to -t$ and $\kappa \to -\kappa$; and for the negative κ the Hamiltonian $\mathcal{H}$ becomes negative in the region "occupied" by the three-dimensional wave weakly limited in the $\mathbf{k}_\perp$-direction, and in this case the condition (4.43) is not fulfilled.

The change of the sign of the integral coefficient b to positive, when $\lambda = 1$ and $s = 1$ or $\lambda = -1$ and $s = -1$ in (4.12) and (4.36), is equivalent to the analytical extension of the solutions from the real (y, z)-plane to the purely imaginary one $(y \to -iy, z \to -iz)$ and, therefore, is equivalent to the change of the sign of the diffraction coefficient κ in the basic equations. In this case, instead of (4.43), the opposite inequality, namely

$$\mathcal{H} < -\frac{3bd}{1+2d^2}, \tag{4.44}$$

where $b > 0$, takes place. From the physical point of view this means that if inequality (4.44) is satisfied, the right-polarized wave with the positive nonlinearity and/or the left-polarized wave with the negative nonlinearity are stable. Note that in the particular case when $\kappa = 0$ in (4.12) and (4.36) (i.e., in the one-dimensional approximation), it is easy to obtain, using the above approach, that instead of inequalities (4.43) and (4.44) the conditions $\mathcal{H} > 0$ and $\mathcal{H} < 0$, respectively, take place. The latter is in complete agreement with the results obtained in Sect. 2.4.3 for the 1-DNLS equation (see also Ref. [33]).

Thus the analysis of the transformation properties of the Hamiltonian of the 3-DNLS equation allows us to determine the ranges of the respective coefficients as well as $\mathcal{H}$ which has the sense of the energy of the system, corresponding to the stable and unstable three-dimensional solutions. The study of the structure and dynamics of three-dimensional Alfvén waves is related to the fact that the 3-DNLS equation (4.12) (and moreover (4.16)) is not a completely integrable system (and therefore this equation cannot be integrated by the IST method). Thus the corresponding problem can be solved only by numerical simulation methods (see Sect. 4.5).

4.2 Asymptotic and Qualitative Analysis of Solutions of GKP Equation and 3-DNLS Equation

In this section, on the basis of the results of Sect. 2.2, we study the structure of (possible) multidimensional solutions of the GKP equation (4.3) with an arbitrary nonlinearity exponent. We employ an approach that takes into account the asymptotics of the solutions along the direction of the wave propagation. The study of the asymptotic behavior of solutions of the 3-DNLS equation along the direction of the wave propagation is a simpler problem because we can explicitly obtain exact solutions in the one-dimensional approximation [32,33]. We also present some considerations on the construction of the eight-dimensional phase space portraits for systems described by the GKP equation based on the results of qualitative analysis of the generalized KdV-class equations.

4.2.1 Basic Equations

Here we first consider the GKP equation and then discuss the 3-DNLS equation. Let us first write the *GKP equation* (4.3) in the following form:

$$\partial_\eta \left(\partial_t u + \alpha u \partial_\eta u - \mu \partial_\eta^2 u + \beta \partial_\eta^3 u + \delta \partial_\eta^4 u + \gamma \partial_\eta^5 u\right) = \kappa \Delta_\perp u, \tag{4.45}$$

where $\Delta_\perp = \partial^2_{\zeta_1} + \partial^2_{\zeta_2}$ and ζ_1 and ζ_2 are the transverse coordinates. When $\mu = \delta = \gamma = 0$, (4.45) is the classic KP equation which represents the completely integrable Hamiltonian system and has in the case $\Delta_\perp = \partial^2_{\zeta_1}$ solutions in the form of a one-dimensional (for $\beta\kappa < 0$) or two-dimensional (for $\beta\kappa > 0$) soliton (see Sect. 3.1)[2]. In Sect. 2.2, using methods of the asymptotic and qualitative analysis, we already studied in detail the asymptotics of the one-dimensional analogue of (4.45), and constructed the sufficiently complete classification of its solutions, including the solutions of soliton and non-soliton types. Now, our purpose is generalization of those results, taking also into account the results of Ref. [83], to the multidimensional cases. To avoid unnecessary cumbersome expressions, we focus on (4.45) in the two-dimensional form assuming that $\Delta_\perp = \partial^2_{\zeta_1}$. Further generalization of the technique used (as well as the results obtained) to the full three-dimensional case $\Delta_\perp = \partial^2_{\zeta_1} + \partial^2_{\zeta_2}$ is rather trivial, as we demonstrate below. Thus here we assume that $\zeta_1 = \zeta$ and, for clarity, $\alpha = 6$; the latter can be easily obtained by the scale transform $u \to (6/\alpha)u$ of (4.45).

Now, let us introduce the new variables, $\bar\eta = \eta + \zeta$ and $\bar\zeta = \eta - \zeta$. Applying first $\bar\eta$ and then $\bar\zeta$ to (4.45), we obtain two one-dimensional equations,

$$\begin{aligned}
&\partial_{\bar\eta}\left(\partial_t u + 6u\partial_{\bar\eta} u - \mu\partial^2_{\bar\eta} u + \beta\partial^3_{\bar\eta} u + \delta\partial^4_{\bar\eta} u + \gamma\partial^5_{\bar\eta} u\right) = \kappa\partial^2_{\bar\eta} u,\\
&\partial_{\bar\zeta}\left(\partial_t u + 6u\partial_{\bar\zeta} u - \mu\partial^2_{\bar\zeta} u + \beta\partial^3_{\bar\zeta} u + \delta\partial^4_{\bar\zeta} u + \gamma\partial^5_{\bar\zeta} u\right) = \kappa\partial^2_{\bar\zeta} u
\end{aligned} \tag{4.46}$$

written in the reference frame with the axes $\bar\eta$ and $-\bar\zeta$ rotated through an angle $\pi/4$ relative to the axes η and ζ. The possibility of representation (4.46) means that the starting equation (4.45) admits two types of one-dimensional solutions, $u(\bar\eta, t)$ and $u(\bar\zeta, t)$, satisfying the first and the second equations of the set (4.46), respectively. It is necessary, however, to bear in mind that the "one-dimensionality" of these solutions nevertheless implicitly assumes the linear dependence of each of the new variables $\bar\eta$ and $\bar\zeta$ on both coordinates η and ζ.

Integrating equations (4.46) over $\bar\eta$ and $-\bar\zeta$, respectively, we obtain two equations of the *generalized KdV equation* type,

$$\begin{aligned}
&\partial_t u + (-\kappa + 6u)\,\partial_{\bar\eta} u - \mu\partial^2_{\bar\eta} u + \beta\partial^3_{\bar\eta} u + \delta\partial^4_{\bar\eta} u + \gamma\partial^5_{\bar\eta} u = 0,\\
&\partial_t u + (-\kappa + 6u)\,\partial_{\bar\zeta} u - \mu\partial^2_{\bar\zeta} u + \beta\partial^3_{\bar\zeta} u + \delta\partial^4_{\bar\zeta} u + \gamma\partial^5_{\bar\zeta} u = 0,
\end{aligned} \tag{4.47}$$

coupled with each other by the way of the change of the coordinates made above. Now, transferring to the coordinates moving along the corresponding

[2] The structure and dynamics of solutions of the model (4.45), which is generally not integrable analytically in the case $\delta = 0$, will be investigated in detail in Sects. 4.4 and 4.5 (see also Refs. [113,148]). In particular, it is demonstrated that when $\mu = 0$, two-dimensional and three-dimensional soliton-type solutions with the monotonous or oscillatory asymptotics can take place, depending on the signs of β, γ, and κ. These solutions in the presence of the viscous-type dissipation in the medium ($\mu > 0$) lose their symmetry and damp with time.

axis with the velocity $-\kappa$, i.e., applying the change $\eta' = \bar{\eta}+\kappa t$ and $\zeta' = \bar{\zeta}+\kappa t$ in (4.47) and omitting "primes" for simplicity, we arrive at

$$\partial_t u + 6u\partial_\eta u - \mu\partial_\eta^2 u + \beta\partial_\eta^3 u + \delta\partial_\eta^4 u + \gamma\partial_\eta^5 u = 0,$$
$$\partial_t u + 6u\partial_\zeta u - \mu\partial_\zeta^2 u + \beta\partial_\zeta^3 u + \delta\partial_\zeta^4 u + \gamma\partial_\zeta^5 u = 0. \tag{4.48}$$

Thus we can now perform the analysis of only one (generalized) equation of the set (4.48), and then, by inverting the change of variables, extend the obtained results to two-dimensional solutions $u(\eta,\zeta,t)$ of (4.45) with $\Delta_\perp = \partial_\zeta^2$.

To consider the *3-DNLS equation*, we first rewrite it in the differential form

$$\partial_\eta \left[\partial_t h + s\partial_\eta \left(|h|^2 h\right) - \mathrm{i}\lambda\partial_\eta^2 h - \nu\partial_\eta^2 h\right] = \kappa\Delta_\perp h, \tag{4.49}$$

where $\Delta_\perp = \partial_{\zeta_1}^2 + \partial_{\zeta_2}^2$. Then assuming for simplicity that $\Delta_\perp = \partial_\zeta^2$ (similarly to the above, generalization to the three-dimensional case is trivial) and introducing, by analogy with the GKP equation, the new variables $\bar{\eta} = \eta + \zeta$ and $\bar{\zeta} = \eta - \zeta$, we again obtain two one-dimensional equations,

$$\partial_{\bar{\eta}} \left[\partial_t h + s\partial_{\bar{\eta}} \left(|h|^2 h\right) - \mathrm{i}\lambda\partial_{\bar{\eta}}^2 h - \nu\partial_{\bar{\eta}}^2 h\right] = \kappa\partial_{\bar{\eta}}^2 h,$$
$$\partial_{\bar{\zeta}} \left[\partial_t h + s\partial_{\bar{\zeta}} \left(|h|^2 h\right) - \mathrm{i}\lambda\partial_{\bar{\zeta}}^2 h - \nu\partial_{\bar{\zeta}}^2 h\right] = \kappa\partial_{\bar{\zeta}}^2 h,$$

written in the reference frame with the axes $\bar{\eta}$ and $-\bar{\zeta}$ rotated through an angle $\pi/4$ relative to the axes η and ζ. Further obvious transformations give us the set

$$\partial_t h + s\partial_{\eta'} \left(|h|^2 h\right) - \mathrm{i}\lambda\partial_{\eta'}^2 h - \nu\partial_{\eta'}^2 h = 0,$$
$$\partial_t h + s\partial_{\zeta'} \left(|h|^2 h\right) - \mathrm{i}\lambda\partial_{\zeta'}^2 h - \nu\partial_{\zeta'}^2 h = 0, \tag{4.50}$$

written in the coordinates $\eta' = \bar{\eta}+\kappa t$ and $\zeta' = \bar{\zeta}+\kappa t$, i.e., in the frame moving along the corresponding axis with the velocity $-\kappa$. Thus, as in the case of the GKP equation, we can perform the analysis for only one equation of the set (4.50) and then, with the inverse change of the system's variables, extend the results to two-dimensional solutions $h(\eta,\zeta,t)$ of (4.49) with $\Delta_\perp = \partial_\zeta^2$.

4.2.2 Generalization to Multidimensional Cases

First, consider generalization of the results obtained in Sect. 2.2 to equations of the GKP class (4.45). Here, we investigate a more general case when equations (4.48) contain the nonlinear term with an arbitrary positive exponent p [112]. Thus, for example, the first equation of the system (4.48) is now given by

$$\partial_t u + 6u^p\partial_\eta u - \mu\partial_\eta^2 u + \beta\partial_\eta^3 u + \delta\partial_\eta^4 u + \gamma\partial_\eta^5 u = 0. \tag{4.51}$$

Recall that in the case $\mu = \delta = \gamma = 0$ equation (4.51) converts to *KdV equation* if $p = 1$, and *MKdV equation* if $p = 2$. Note also that, analogous to the one-dimensional problem, the cases when in (4.51) the nonlinearity exponent is $p = 1, 2$ are interesting from the physical point of view, whereas applications with $p > 2$ are currently unknown. However, we study here the general case $p > 0$ to elucidate the dependence of the solutions on the nonlinearity exponent; we have in mind that equations with arbitrary integer $p > 0$ often reveal similar mathematical characteristics just like the generalized KdV equation considered in Sect. 2.2,.

Taking into account the signs of $\mu > 0$ and $\delta > 0$ (with respect to the physical sense of the corresponding terms, see Sect. 2.2 for details), assuming without loss of generality that $\gamma > 0$ and $\beta = \pm 1$ [112], and substituting $u = Vw$, where V is the velocity of the wave propagating along the axis η and ζ for the first and the second equation of the set (4.48), respectively, we can generalize the results obtained in Ref. [112] for different signs of V and β to (4.46) and, accordingly, to (4.45) with $p \geq 1$ and obtain the following:

1. The exponent p defines the character of the functional dependence $V = f(u)$. For $p > 1$, such a dependence for (4.45), as in the one-dimensional case [112], becomes nonlinear unlike the known linear one for $p = 1$ (for example, in the case of the KP equation). Moreover, solutions of (4.45) can have both positive and negative polarity ($u > 0$ or $u < 0$ for either sign of V) for even p.
2. In the case of the conservative equations of class (4.45) (the cases when $\mu = \delta = 0$) asymptotics of the solutions is defined by:
   (a) When $V > 0$, $\beta = -1$ and $V < 0$, $\beta = -1$ (the upper and lower signs, respectively):
   $$w = A_1 \exp\left[(2\gamma)^{-1/2}\left(C^2 + \sqrt{C^4 \pm 4\gamma}\right)^{1/2}\chi\right]. \tag{4.52}$$
   (b) When $V < 0$, $\beta = 1$:
   $$\begin{aligned} w = A_2 &\exp\left[\left(2C^{-1}\gamma^{1/2}\right)^{-1}\left(2C^{-2}\gamma^{1/2} - 1\right)^{1/2}\chi\right] \\ &\times \cos\left[\left(2C^{-1}\gamma^{1/2}\right)^{-1}\left(2C^{-2}\gamma^{1/2} + 1\right)^{1/2}\chi + \Theta\right], \end{aligned} \tag{4.53}$$
   where A_1, A_2 and Θ are the arbitrary constants, $C = |V|^{-1/4}$, and $\chi = \eta \pm \zeta + (\kappa - V)t$ (the upper(lower) sign here relates to the first(second) equation of the set (4.46), respectively).
   As we can see from expressions (4.52) and (4.53)[3], solitons with monotonous as well as oscillating asymptotics (depending on the signs of V and β) can take place as the solutions $u(\eta, \zeta, t)$ of (4.45) when $\mu = \delta = 0$. Note

[3] Other correlations of the signs of V and β are not realized [112].

that for $\beta = 0$ and any $\gamma > 0$ solutions of (4.46) with $\mu = \delta = 0$ are given by $w = (A_1 + A_2\chi/C)\exp(\chi/\gamma^{1/4}C)$ and, consequently, also describe a soliton with the monotonous asymptotics [148]. Fig. 4.3 shows the result

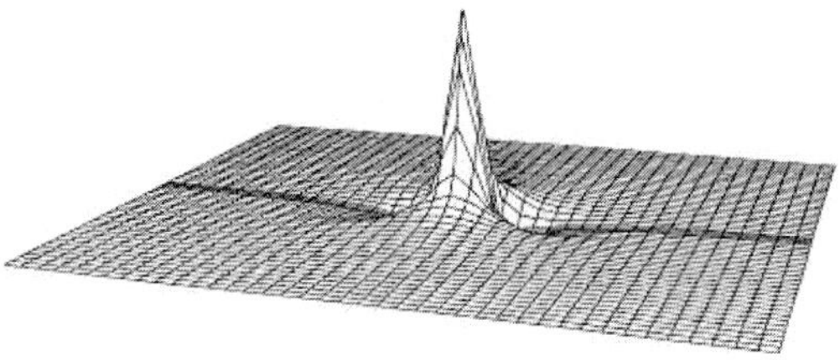

Fig. 4.3. Two-dimensional soliton solution of (4.45) with $\Delta_\perp = \partial_\zeta^2$ for $\mu = \delta = 0$, $p = 1$, $\gamma = 1$, and $\beta = -0.8$ at the time moment $t = 0.2$

of numerical integration of (4.45) for $\mu = \delta = 0$ with the initial condition $u = u_0 \exp(-\eta^2/l_1^2 - \zeta^2/l_2^2)$, that confirms our asymptotic analysis.

3. In the case of the dissipative equations of class (4.45) with an instability (when $\beta = \gamma = 0$) the asymptotics of the solutions is defined by
(a) For $\delta > (4/27)\mu^3 C^8$,

$$w = A_1 \exp\left[(2\delta C)^{-1/3} Q_1^{+}\chi\right] + \exp\left[-(16\delta C)^{-1/3} Q_1^{+}\chi\right]$$
$$\times\left\{A_2 \cos\left[\sqrt{3}\,(16\delta C)^{-1/3} Q_1^{-}\chi + \Theta_1\right]\right.$$
$$\left. + A_3 \sin\left[\sqrt{3}\,(16\delta C)^{-1/3} Q_1^{-}\chi + \Theta_2\right]\right\}. \qquad (4.54)$$

(b) For $\delta = (4/27)\mu^3 C^8$,

$$w = A_1 \exp\left[(\delta C/4)^{-1/3}\chi\right]$$
$$+ A_2 (1 + A_3\chi)\exp\left[-(2\delta C)^{-1/3}\chi\right]. \qquad (4.55)$$

(c) For $\delta < (4/27)\mu^3 C^8$,

$$w = A_1 \exp\left[(\delta C/4)^{-1/3}\,\mathrm{Re}\,(Q^{\pm})\,\chi\right]$$
$$+ A_2 \exp\left[-(2\delta C)^{-1/3}\chi\left(\mathrm{Re}\,(Q^{\pm}) - \sqrt{3}\,|\mathrm{Im}\,(Q^{\pm})|\right)\right]$$
$$+ A_3 \exp\left[-(2\delta C)^{-1/3}\chi\left(\mathrm{Re}\,(Q^{\pm}) + \sqrt{3}\,|\mathrm{Im}\,(Q^{\pm})|\right)\right], \qquad (4.56)$$

where $A1$, $A2$, $A3$, Θ_1, and Θ_2 are arbitrary constants, $Q_1^{\mp} = Q^{+} \pm Q^{-}$,

$$Q^{\pm} = \left[1 \pm \sqrt{1 - 4\mu^3 C^8/27\delta}\right]^{1/3},$$

and $Q^{\pm}$ is real in the cases (a) and (b) and complex in the case (c).
It is easy to see from (4.54)–(4.56) that the solutions $u(\eta, \zeta, t)$ of (4.45) have the oscillating asymptotics in the case (a) and the exponentially decreasing

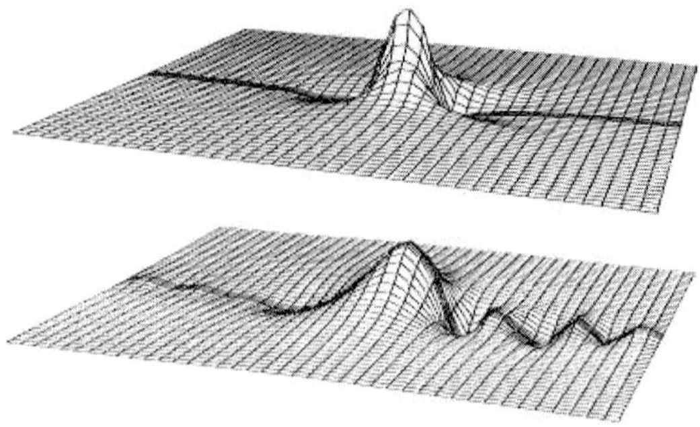

Fig. 4.4. Two–dimensional soliton solution of (4.45) with $\Delta_\perp = \partial_\zeta^2$ for $\beta = \gamma = 0$, $V > 0$, at the time moment $t = 0.3$. *Top* $\mu = 1$, $\delta = 10^{-6}$ (corresponds to the case (c) where $\delta < (4/27)\mu^3 C^8$). *Bottom* $\mu = 1$, $\delta = 1$ (corresponds to the case (a) where $\delta > (4/27)\mu^3 C^8$)

asymptotics in the cases (b) and (c). Fig. 4.4 shows numerical solutions of (4.45) in the cases (c) and (a), respectively, obtained for the initial condition $u = u_0 \exp(-\eta^2/l_1^2 - \zeta^2/l_2^2)$.

4. To properly transform the *phase portraits* of the system and "bind" them (for the two-dimensional equation), we note that in the case $\mu = \delta = 0$ the phase space is eight-dimensional, and in the case $\beta = \gamma = 0$ the phase space is six-dimensional. Here we can employ the results of Ref. [112] coupling the characteristics of every singular point of each equation of the set (4.46) in the eight-dimensional and six-dimensional phase spaces, respectively. Thus the corresponding type of the singularity in either of the four-dimensional or three-dimensional subspaces [112] under the inverse transform of the coordinates, $\eta = (\bar\eta + \bar\zeta)/2$ and $\zeta = (\bar\eta - \bar\zeta)/2$, is not changed, and only parameters of the phase portraits that correspond to solutions of the same class change (leading to the respective changes such parameters as the amplitude, the front steepness, frequency of the oscillations, etc.).

Consider now the *3-DNLS equation* (4.49) with $\Delta_\perp = \partial_\zeta^2$. First we note that (as was already shown in Sect. 4.2.1) (4.49) can be written in the form of the set (4.50), and, as it is known from Refs. [32,33], an exact solution of the one-dimensional DNLS equation is given by

$$h(\eta, t) = (A/2)^{1/2} \left(\mathrm{e}^{-A\eta} + \mathrm{i}\mathrm{e}^{A\eta}\right) \mathrm{e}^{-\mathrm{i}A^2 t} / \cosh^2(2A\eta), \qquad (4.57)$$

where A is the amplitude of the wave (see Sect. 2.4 for more details). Now we can apply the inverse change of the variables, $\eta = (\bar\eta + \bar\zeta)/2$ and $\zeta = (\bar\eta - \bar\zeta)/2$, and, extending solution (4.57) to the two-dimensional case, (4.49) with $\Delta_\perp = \partial_\zeta^2$, write

$$h(\eta, \zeta, t) = (A/2)^{1/2} \left(\mathrm{e}^{-A\chi} + \mathrm{i}\mathrm{e}^{A\chi}\right) \mathrm{e}^{-\mathrm{i}A^2 t} / \cosh^2(2A\chi), \qquad (4.58)$$

where, as for the GKP equation, $\chi = (\eta \pm \zeta + (\kappa - V)t)$, and V is the velocity of the wave propagation relative to the coordinate axis η or ζ for the first or the second equation of the set (4.50), respectively. Fig. 4.5 shows the character of the solution of the first equation of the set (4.50) with $\nu = 0$. The dependence

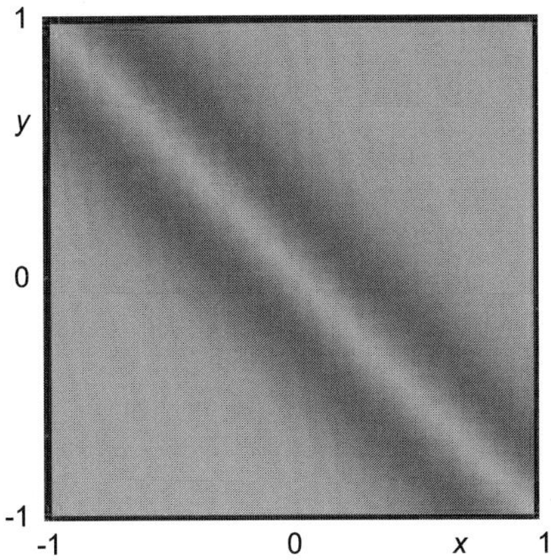

Fig. 4.5. Solution $|h|^2$ of the first equation of the set (4.50) for $A = 1$ and $t = 0$

of the form of the solution on dissipation in the system as well as the dynamic characteristics of the solution for $\nu > 0$ is considered in Sect. 4.5 in detail.

Finally, we do not analyze solutions of the 3-DNLS equation here because, unlike the GKP equation (4.45) and the related set (4.48), the exact solution of each equation of the set (4.50) is well known, and there is no need to construct any special classification of its solutions in the phase space.

4.2.3 Concluding Remarks

To conclude this section, we note that for the GKP equation we considered only the particular cases when $\mu = \delta = 0$ and $\beta = \gamma = 0$ in (4.45) here. For other values of the corresponding coefficients, more complex wave structures resulting from the simultaneous presence of all the effects discussed separately above can be observed. Indeed, results obtained numerically in Sect. 4.4.4 (see also Refs. [114,137]), demonstrate that when $\beta, \mu, \delta \neq 0$ in the presence of the Gaussian random fluctuations of the wave field (for harmonic initial conditions and initial conditions in the form of a solitary pulse) stable wave structures of the soliton-like type can be formed too, with the time evolution. Furthermore, stable soliton structures can be formed also for $\gamma \neq 0$. An analytical study of such cases is highly complicated, however, although the approach considered above can be used for them as well. Note also that the results presented in this section for the GKP equation can be very useful when studying solutions and interpreting the multidimensional phase portraits of more complicated multidimensional model equations (see, for example, Ref. [83]).

4.3 Approaches to Numerical Integration of Equations of GKP-Class and 3-DNLS-Class

Methods of numerical integration of the KP equation and the 1-DNLS equation, considered in Sect. 2.4 and Sect. 3.1, although employed with reasonable success to solve a number of problems, still have some shortcomings. Indeed, the method of stabilizing factor [31], being "static," does not allow us to solve dynamic problems where the principal objects for analysis are the parameters of solitons in motion, including solitons' interactions and collisions, etc. Besides, it was obtained [80] that generalization of this method to investigate the structure of stationary solutions of (4.3) with $\nu = \delta = 0$ [80] leads to divergences when $\gamma < 0$. The "dynamic" method of iterative splitting [64], although having good accuracy characteristics, is cumbersome and therefore highly time-consuming (computation-wise). The hop-scotch method, used in Ref. [58] for investigation of one-dimensional solitons in the KP model when $(\beta/\kappa) < 0$ (see (4.3) with $\nu = \delta = 0$), for example, has a similar problem combined with low accuracy at the boundaries of the integration region (that especially adversely affects the investigation of two-dimensional solutions of the KP equation when $(\beta/\kappa) > 0$). One of the highly effective methods of numerical integration of the DNLS equation based on the Ablowitz–Ladik scheme and considered in Sect. 2.4, was developed only for the one-dimensional approximation, i.e., it does not take into account the integral diffraction term in the right-hand side of the 3-DNLS equation in the form (4.22).

In this section, we consider a few rather simple methods of numerical integration of the GKP- and 3-DNLS-class equations developed in Refs. [77,81,83,98,113,148,195,198]. These methods will be used to study the dynamics of solitons and non-stationary wave packets described by (4.3) and (4.14). The methods are based on the explicit and implicit finite-difference schemes, Sect. 4.3.1 and Sect. 4.3.2, with the $O(\tau^2, h_{x,y}^2)$ and $O(\tau^2, h_{x,y}^4)$ approximations. We consider also the dynamic spectral method, Sect. 4.3.3, including first the Fourier transform of the basic equations in the space variables and then the subsequent solution of the resulting first-order differential equations by the Runge–Kutta method. For every algorithm, we formulate the stability conditions. Unlike the schemes considered in Sect. 3.1, the methods presented here not only enable us to control the evolution of the solutions as well as the dynamics of the solitons and their interactions, they, exhibiting high accuracy characteristics, are less cumbersome than the iterative splitting and the hop-scotch methods used in Refs. [64] and [58] for the KP equation. We consider all the methods on an example of the standard KP equation aiming, first of all, to avoid the inconveniences of those considered earlier as well as to compare all of them. When applied to the 3-DNLS equation, the new methods, unlike the method based on the Ablowitz–Ladik scheme developed in Ref. [33], take into account the multidimensionality of the problem. In conclusion, we discuss the comparative characteristics of the schemes of different types obtained when testing the basic equations on the exact solutions.

4.3.1 Groups of Explicit and Implicit Difference Schemes

Here, we consider the finite-difference methods for the GKP-class equations (on the example of the usual KP equation for simplicity) and the 3-DNLS equation with arbitrary coefficients in the two-dimensional geometry on the basis of the standard models (4.3), with $\nu = \delta = \gamma = 0$, and (4.12), respectively. Generalization of these methods to the three-dimensional equations with addition of the terms accounting for the higher order dispersion correction and the instability (for the KP equation), as well as the effects of the dissipation (for both KP and 3-DNLS equations) is straightforward and generally does not require special explanations (see remarks on this in the end of Sect. 4.3.4). We first consider the problem for the two-dimensional *KP equation*, and then extend the obtained results to the *3-DNLS equation* with $\Delta_\perp = \partial_y^2$.

We write the basic equations in the integral-differential form,

$$\partial_t u + \hat{\mathrm{A}}(t, y)u = f, \tag{4.59}$$

where

$$f = \kappa \int_{-\infty}^{x} \partial_y^2 u \mathrm{d}x,$$

and in the case of the KP equation,

$$\hat{\mathrm{A}}(t, y) = \alpha u \partial_x + \beta \partial_x^3.$$

To integrate (4.59) numerically, both the explicit and implicit methods can be used, with their accuracy depending on the approximation order of the corresponding derivatives and the diffraction term f. Below, we consider the explicit and implicit numerical schemes with different approximation orders separately.

Explicit Schemes. Here, we use for the integration the following *explicit scheme with $O(\tau^2, h_{x,y}^2)$ approximation*:

$$\begin{aligned}\frac{u_{ij}^{n+1} - u_{ij}^{n-1}}{2\tau} &= -\frac{\alpha}{2h_x} u_{ij}^n \left(u_{i+1,j}^n - u_{i-1,j}^n\right) \\ &\quad - \frac{\beta}{2h_x^3}\left(u_{i+2,j}^n - 2u_{i+1,j}^n + 2u_{i-1,j}^n - u_{i-2,j}^n\right) \\ &\quad + f_{ij}^n,\end{aligned} \tag{4.60}$$

(compare with the scheme (1.80) in Sect. 1.3). Here, f_{ij} is the approximation of the integral f by one of the suitable quadratures. In particular, the quadrature prismoidal formula (the *Simpson formula*) gives

$$f_{ij} = \kappa \frac{h_x}{3} \sum_{i=1}^{m} \left(f'_{2i-1,j} + 4f'_{2i,j} + f'_{2i+1,j}\right), \tag{4.61}$$

where

$$f'_{ij} = \frac{1}{h_y^2}\left(u_{i,j+1} - 2u_{ij} + u_{i,j-1}\right),$$

and $m = (N-1)/2$ (i.e., N is odd). According to the estimates of Refs. [83,98], the Simpson formula (4.61) approximates integral f on the solutions of (4.59) with the $O(h_{x,y}^2)$ accuracy for the grid steps $h_x \leq 0.05 \div 0.075 \leq h_y$ (that depends on the particular values of the coefficients α, β, and κ in (4.60) and (4.61). The approximation of the integral by the quadrature *Newton–Cotes formula* with the number of nodes $m > 3$, gives us

$$f_{ij} = \kappa \sum_{k=1}^{n} \sum_{i=1}^{m} A_i^{(m)} f'_{l,j}, \tag{4.62}$$

where $l = i + (m-1)(k-1)$, $A_i^{(m)} = (m-1)h_x C_i^{(m)}$, and $C_i^{(m)}$ are the coefficients of the quadrature formula; this approach requires fewer restrictions on the step of the spatial grid. Approximation (4.62) enables us to compute the integral f with the $O(h_{x,y}^4)$ accuracy for $h_x \leq 0.1 \leq h_y$ [77,83,98].

To estimate the stability of scheme (4.60) we use the Fourier method. Thus we rewrite the difference equation in the form

$$\begin{aligned}\frac{u_{mn}^{p+1} - u_{mn}^{p-1}}{2\tau} &= -\frac{\alpha}{2h_x} u_{mn}^p \Lambda_x u_{mn}^p - \frac{\beta}{2h_x^3} \Lambda_{xxx} u_{mn}^p \\ &\quad + \frac{\kappa}{h_y^2} \int_{-a}^{x} \Lambda_{yy} u_{mn}^p \mathrm{d}x,\end{aligned}$$

where Λ_{k^q} is the difference approximation of a differential operator of the qth order in k; the lower limit of the integration, a, is defined by the size of the region of numerical integration. Solutions to the equation are in the form

$$\begin{aligned} u_{mn}^p &= \lambda_{kl}^p \psi_{mn}^{(k,l)}, \\ m, n &= 1, 2, \ldots, M_{m,n} - 1, \\ \psi_{mn}^{(k,l)} &= 2\sin\left(k\pi m/M_m\right)\sin\left(l\pi n/M_n\right). \end{aligned}$$

Note that

$$\begin{aligned} \Lambda_{\alpha\ldots\alpha}\psi^{(k,l)} &= \Lambda_{\alpha\ldots\alpha}\left(\psi_m^{(k)}\psi_n^{(l)}\right) \\ &= \begin{cases} \psi^{(l)}\Lambda_{\alpha\ldots\alpha}\psi^{(k)}, & \alpha = x, \\ \psi^{(k)}\Lambda_{\alpha\ldots\alpha}\psi^{(l)}, & \alpha = y. \end{cases} \end{aligned} \tag{4.63}$$

For λ_{kl} we then obtain

$$\begin{aligned} \lambda_{kl} = 1 + \frac{\tau}{h_x}\Bigg[&\sin kh_x\left(\frac{4\beta}{h_x^2}\sin^2\frac{kh_x}{2} - \alpha u\right) \\ &- \frac{8\kappa h_x}{h_y^2}\sin^2\frac{lh_y}{2}u^{-1}\int_{-a}^{x} u\mathrm{d}x\Bigg], \end{aligned}$$

where u is the local value of $u(t,x,y)$. The stability takes place when the following inequality is satisfied:

$$\max_{k,l}\left|\lambda_{kl}\right| \leq 1. \tag{4.64}$$

Taking into account that the eigenvalues reach minimum at $k = l = M_{m,n} - 1$, we find

$$\lambda_{M_m-1,M_n-1} = 1 + \frac{\tau}{h_x}\left|\frac{2\beta}{h_x^2} - \alpha u - \frac{4\kappa h_x}{h_y^2} u^{-1}\int_{-a}^{x} u\mathrm{d}x\right|.$$

Thus inequality (4.64) is satisfied when

$$\frac{\tau}{h_x}\left|\alpha v - \frac{2\beta}{h_x^2} + \frac{4\kappa h_x}{h_y^2} v^{-1}\int_{-a}^{x} u\mathrm{d}x\right| \leq 1, \tag{4.65}$$

where

$$v = \max_{k,l}|u|.$$

For a finite a (which is related with the finite product $M_m h_x$), we are able to estimate the integral in the left-hand side of (4.65) on the exact solution (see Sect. 3.1.2) for $n, m = 1, 2$ and $l = 0$:

$$\max_{x,y}\left|\int_{-a}^{x} u\mathrm{d}x\right| < \int_{3\nu^2 t}^{(1/\nu)+3\nu^2 t} u\mathrm{d}x = 2\nu \sim v^{1/2}.$$

Then, for sufficiently small step, $h = \min(h_x, h_y)$, we find from inequality (4.65) that

$$\tau \leq h^3\left|\alpha v h^2 - 2\beta + 4\kappa v^{-1/2} h\right|^{-1}. \tag{4.66}$$

Thus we obtain that the scheme (4.60) is stable under the condition (4.66). This conforms with the corresponding stability condition for the scheme for the KdV equation if we assume here $\kappa = 0$ (see Sect. 1.3).

For better accuracy of numerical integration, (4.59) can be approached by employing the three-layer *explicit scheme with* $O(\tau^2, h_{x,y}^4)$ *approximation*:

$$\begin{aligned}\frac{u_{ij}^{n+1} - u_{ij}^{n-1}}{2\tau} &= \frac{\alpha}{12h_x} u_{ij}^n \left(u_{i+2,j}^n - 8u_{i+1,j}^n + 8u_{i-1,j}^n - u_{i-2,j}^n\right) \\ &+ \frac{\beta}{8h_x^3}\left(u_{i+3,j}^n - 8u_{i+2,j}^n + 13u_{i+1,j}^n - 13u_{i-1,j}^n \right. \\ &\left. + 8u_{i-2,j}^n - u_{i-3,j}^n\right) + f_{ij}^n \end{aligned} \tag{4.67}$$

(compare with (1.81) in Sect. 1.3). Furthermore we have to use expression (4.62) to calculate f_{ij}^n. In this case, to ensure the required accuracy, we choose (unlike (4.61)) another approximation for the second y-derivative given by

$$f'_{lj} = -\frac{1}{12h_y^2}\left(u_{l,j+2} - 16u_{l,j+1} + 30u_{lj} - 16u_{l,j-1} + u_{l,j-2}\right). \tag{4.68}$$

Using the above method, it is possible to demonstrate that for the scheme (4.67) the requirement of stability for sufficiently small h is given by the inequality

$$\tau \le 9h^3 \left|12\alpha v h^2 - 27\beta + 14\kappa v^{-1/2} h\right|^{-1} \tag{4.69}$$

(compare with the corresponding condition for the KdV equation, Sect. 1.3). The limits on the time step set by the inequalities (4.66) and (4.69), taking into account the two-dimensional character of the problem, are indeed important. Besides, it is necessary to store the function values of the two previous layers for every time step. Therefore it is hardly reasonable to use such simple schemes as (4.60) and (4.67) to obtain the asymptotic solutions for $t \to \infty$. These schemes, however, can be useful for simulation of formation and dynamics of solutions at the initial stages of their evolution.

Consider now the possibility to employ the explicit difference schemes for the *3-DNLS equation* in the form (4.12). When integrating this equation numerically we use the approach detailed in Refs. [65–67]. We first perform the change $h = p\bar{h}$ in (4.12) where $p = 1 + \mathrm{i}e$, $\bar{h} = \mathrm{Re}h$, and e is the eccentricity of the polarization ellipse of the Alfvén wave. Thus we obtain

$$\partial_t \bar{h} + 3s\left|p\right|^2 \bar{h}^2 \partial_x \bar{h} - \mathrm{i}\varepsilon \partial_x^2 \bar{h} = \kappa \int_{-\infty}^{x} \Delta_\perp \bar{h} \mathrm{d}x \tag{4.70}$$

which consequently, taking into account the change $\bar{h} = u$, can be represented in the form (4.59) with the operator

$$\hat{\mathrm{A}}(t,u) = 3s\left|p\right|^2 \bar{h}^2 \partial_x - \mathrm{i}\varepsilon \partial_x^2. \tag{4.71}$$

For integration in this case, it is reasonable to choose the group of the three-layer explicit schemes given by

$$\frac{\bar{h}_{ij}^{n+1} - \bar{h}_{ij}^{n-1}}{2\tau} = \alpha\left(\bar{h}_{ij}^n\right)^2 \mathrm{d}_x\left(\bar{h}_{ij}^n\right) + \beta \mathrm{d}_x^2\left(\bar{h}_{ij}^n\right) + f_{ij}^n, \tag{4.72}$$

where $\alpha = -3s|p|^2$, $\beta = \mathrm{i}\varepsilon$, and the integral f_{ij}^n is defined by expressions analogous to (4.61) and (4.62). Depending on the approximation order of the differential operators in the right-hand side of (4.72) by the difference operators, we can obtain the three-layer explicit schemes with different accuracies. For the schemes with the $O(\tau^2, \Delta^2)$ and $O(\tau^2, \Delta^4)$ approxiamtion (here, $\Delta = \max[\Delta_x, \Delta_y]$ is the largest step on the spatial grid), using the Fourier method for the difference approximations of the corresponding order in the KP equation, we obtain for the 3-DNLS equation the restrictions on the time step

$$\tau \le \Delta^2 \left| 8\kappa^2 v^{-1} + \left(2\varepsilon - \alpha v^2 \Delta\right)^2 \right|^{-1/2} \tag{4.73}$$

and

$$\tau \le 3\Delta^2 \left| 49\kappa^2 v^{-1} + \left(7\varepsilon - 4\alpha v^2 \Delta\right)^2 \right|^{-1/2}, \tag{4.74}$$

where

$$v = \max_{i,j,n} \left| \bar{h}_{ij}^n \right| .$$

Thus the three-layer explicit schemes for the 3-DNLS equation (4.70) with the $O(\tau^2, \Delta^2)$ and $O(\tau^2, \Delta^4)$ approximation are stable under conditions (4.73) and (4.74), respectively.

Implicit Schemes. For the KP equation (4.59), consider the following *implicit scheme with $O(\tau^2, h_{x,y}^4)$ approximation*:

$$\begin{aligned}
\frac{u_{ij}^{n+1} - u_{ij}^n}{\tau} &= \frac{\alpha}{24h_x} \left[u_{ij}^n \left(u_{i+2,j}^{n+1} - 8u_{i+1,j}^{n+1} + 8u_{i-1,j}^{n+1} - u_{i-2,j}^{n+1} \right) \right. \\
&\quad \left. + u_{ij}^{n+1} \left(u_{i+2,j}^n - 8u_{i+1,j}^n + 8u_{i-1,j}^n - u_{i-2,j}^n \right) \right] \\
&\quad + \frac{\beta}{16h_x^3} \left(u_{i+3,j}^{n+1} - 8u_{i+2,j}^{n+1} + 13u_{i+1,j}^{n+1} - 13u_{i-1,j}^{n+1} \right. \\
&\quad\quad + 8u_{i-2,j}^{n+1} - u_{i-3,j}^{n+1} + u_{i+3,j}^n - 8u_{i+2,j}^n \\
&\quad\quad \left. + 13u_{i+1,j}^n - 13u_{i-1,j}^n + 8u_{i-2,j}^n - u_{i-3,j}^n \right) \\
&\quad + f_{ij}^n,
\end{aligned} \tag{4.75}$$

where f_{ij}^n is defined by (4.62) and (4.68) and the difference operator in x coincides with the corresponding operators in the scheme (1.84) (see Sect. 1.3). Following the technique used in Sect. 1.3, we represent (4.75) for every fixed j as

$$\begin{aligned}
&-a_{0j}^1 u_{3j}^n + a_{0j}^2 u_{2j}^n - a_{0j}^3 u_{1j}^n + a_{0j}^4 u_{0j}^n = g_{0j}^n, && i = 0; \\
&-a_{1j}^1 u_{4j}^n + a_{1j}^2 u_{3j}^n - a_{1j}^3 u_{2j}^n + a_{1j}^4 u_{1j}^n \\
&\qquad -a_{1j}^5 u_{0j}^n = g_{1j}^n, && i = 1; \\
&-a_{2j}^1 u_{4j}^n + a_{2j}^2 u_{4j}^n - a_{2j}^3 u_{3j}^n + a_{2j}^4 u_{2j}^n \\
&\qquad -a_{2j}^5 u_{1j}^n + a_{2j}^6 u_{0j}^n = g_{2j}^n, && i = 2; \\
&-a_{ij}^1 u_{i+3,j}^n + a_{ij}^2 u_{i+2,j}^n - a_{ij}^3 u_{i+1,j}^n + a_{ij}^4 u_{ij}^n \\
&\quad -a_{ij}^5 u_{i-1,j}^n + a_{ij}^6 u_{i-2,j}^n - a_{ij}^7 u_{i-3,j}^n = g_{ij}^n, && 3 \le i \le N-3; \\
&\qquad a_{N-2,j}^2 u_{Nj}^n - a_{N-2,j}^3 u_{N-1,j}^n \\
&\qquad +a_{N-2,j}^4 u_{N-2,j}^n - a_{N-2,j}^5 u_{N-3,j}^n \\
&+a_{N-2,j}^6 u_{N-4,j}^n - a_{N-2,j}^7 u_{N-5,j}^n = g_{N-2,j}^n, && i = N-2;
\end{aligned} \tag{4.76}$$

$$
\begin{aligned}
-a^3_{N-1,j}u^n_{Nj} + a^4_{N-1,j}u^n_{N-1,j} - a^5_{N-1,j}u^n_{N-2,j} \\
+a^6_{N-1,j}u^n_{N-3,j} - a^7_{N-1,j}u^n_{N-4,j} = g^n_{N-1,j}, \qquad & i = N-1; \\
a^4_{Nj}u^n_{Nj} - a^5_{Nj}u^n_{N-1,j} + a^6_{Nj}u^n_{N-2,j} \\
-a^7_{Nj}u^n_{N-3,j} = g^n_{Nj}, \qquad & i = N.
\end{aligned}
$$

Here for $3 \le i \le N-3$,

$$
\begin{aligned}
g^n_{ij} = -b_{ij}\left(u^{n-1}_{i+3,j} - 8u^{n-1}_{i+2,j} + 13u^{n-1}_{i+1,j} - 13u^{n-1}_{i-1,j} + 8u^{n-1}_{i-2,j} - u^{n-1}_{i-3,j}\right) \\
+u^{n-1}_{ij} + \tau f^{n-1}_{ij}, \qquad j = 1,2,\ldots,M, \quad n = 1,2,\ldots,N_1, \quad (4.77)
\end{aligned}
$$

and

$$
\begin{aligned}
a^1_{ij} &= -\tau\beta/16h^3_x, \quad a^7_{ij} = -a^1_{ij}; \\
a^2_{ij} &= \tau\left[(\beta/h^2_x) - (\alpha u^{n-1}_{ij}/12)\right]/2h_x, \quad a^6_{ij} = -a^2_{ij}; \\
a^3_{ij} &= \tau\left[(\alpha u^{n-1}_{ij}/3) - (13\beta/16h^2_x)\right]/h_x, \quad a^5_{ij} = -a^3_{ij}; \quad (4.78) \\
a^4_{ij} &= 1 + \alpha\tau\left(u^{n-1}_{i+2,j} - 8u^{n-1}_{i+1,j} + 8u^{n-1}_{i-1,j} - u^{n-1}_{i-2,j}\right)/24h_x, \quad b_{ij} = a^7_{ij}.
\end{aligned}
$$

The computations in y are performed for $2 \le j \le M-2$. The values of g^n_{ij} for $i = 0,1,2$ and $j = N-2, N-1, N$ are defined by the same formulas as (1.88) in Sect. 1.3 (here we omit index j for further convenience):

$$
\begin{aligned}
g^n_0 &= -a^1_0u^{n-1}_3 + a^2_0u^{n-1}_2 - a^3_0u^{n-1}_1 + a^4_0u^{n-1}_0, \\
g^n_1 &= -a^1_1u^{n-1}_4 + a^2_1u^{n-1}_3 - a^3_1u^{n-1}_2 + a^4_1u^{n-1}_1 - a^5_1u^{n-1}_0, \\
g^n_2 &= -a^1_2u^{n-1}_5 + a^2_2u^{n-1}_4 - a^3_2u^{n-1}_3 + a^4_2u^{n-1}_2 - a^5_2u^{n-1}_1 + a^6_2u^{n-1}_0, \\
g^n_{N-2} &= a^2_{N-2}u^{n-1}_N - a^3_{N-2}u^{n-1}_{N-1} + a^4_{N-2}u^{n-1}_{N-2} - a^5_{N-2}u^{n-1}_{N-3} \\
&\quad + a^6_{N-2}u^{n-1}_{N-4} - a^7_{N-2}u^{n-1}_{N-5}, \qquad (4.79) \\
g^n_{N-1} &= -a^3_{N-1}u^{n-1}_N + a^4_{N-1}u^{n-1}_{N-1} - a^5_{N-1}u^{n-1}_{N-2} + a^6_{N-1}u^{n-1}_{N-3} \\
&\quad - a^7_{N-1}u^{n-1}_{N-4}, \\
g^n_N &= a^4_Nu^{n-1}_N - a^5_Nu^{n-1}_{N-1} + a^6_Nu^{n-1}_{N-2} - a^7_Nu^{n-1}_{N-3}.
\end{aligned}
$$

The coefficients $a^1_i, \ldots, a^7_i$ for $i = 0,1,2$ and $i = N-2, N-1, N$ are obtained from the boundary conditions of the problem (see Sect. 4.3.2).

The set of equations (4.77) can be solved by employing the monotonous 7-point *sweep method.* The corresponding expressions are given in Sect. 1.3 (see (1.89)–(1.92)) where for the present case it is necessary to change h to g and to take into account that expressions for h^n_i (1.87) and g^n_{ij} (4.77) differ significantly. Thus, as we easily see from (4.79), the algorithm of the monotone sweep is correct if the following conditions for (4.59) and (4.75) are satisfied for $3 \le i \le N-3$:

$$
u^{n-1}_i \ne 12\beta/\alpha h^2_x \qquad \text{and} \qquad u^{n-1}_i \ne 39/16\alpha h^2_x.
$$

Obviously, these inequalities are satisfied for sufficiently small step h_x. Then, following the technique used in Sect. 1.3 we can obtain the conditions similar to (1.93)–(1.95).[4] Testing of the scheme (4.75) demonstrated that the adequate accuracy of the solution is achieved for $h = 0.1$ and $\tau = 0.0025$ (see Sect. 4.3.4). We also note that for $i = 0, 1, 2$ and $i = N-2, N-1, N$, in order to satisfy the conditions (1.91) for the set (4.77) it is sufficient to choose the proper boundary conditions (cf. the next section).

For equation (4.59), consider now another *implicit scheme with* $O(\tau^2, h^2_{x,y})$ *approximation*,

$$\begin{aligned}\frac{u^{n+1}_{ij} - u^n_{ij}}{\tau} &+ \frac{\alpha}{4h_x}\left[u^n_{ij}\left(u^{n+1}_{i+1,j} - u^{n+1}_{i-1,j}\right) + u^{n+1}_{ij}\left(u^n_{i+1,j} - u^n_{i-1,j}\right)\right] \\ &+\frac{\beta}{4h^3_x}\left(u^{n+1}_{i+2,j} - 2u^{n+1}_{i+1,j} + 2u^{n+1}_{i-1,j} - u^{n+1}_{i-2,j} + u^n_{i+2,j}\right. \\ &\qquad \left. -2u^n_{i+1,j} + 2u^n_{i-1,j} - u^n_{i-2,j}\right) = f^n_{ij}, \qquad (4.80)\end{aligned}$$

where f^n_{ij} is defined by (4.61) or (4.62), and (4.68). Analogous to equation (1.96) in Sect. 1.3, we represent (4.80) for each fixed j in the form

$$\begin{aligned}
a^1_{0j}u^n_{2j} - a^2_{0j}u^n_{1j} + a^3_{0j}u^n_{0j} &= g^n_{0j}, && i = 0;\\
a^1_{1j}u^n_{3j} - a^2_{1j}u^n_{2j} + a^3_{1j}u^n_{1j} - a^4_{1j}u^n_{0j} &= g^n_{1j}, && i = 1;\\
a^1_{ij}u^n_{i+2,j} - a^2_{ij}u^n_{i+1,j} + a^3_{ij}u^n_{ij} - a^4_{ij}u^n_{i-1,j} & \\
+a^5_{ij}u^n_{i-2,j} &= g^n_{ij}, && 2 \le i \le N-2; \qquad (4.81)\\
-a^2_{N-1,j}u^n_{Nj} + a^3_{N-1,j}u^n_{N-1,j} - a^4_{N-1,j}u^n_{N-2,j} & \\
+a^5_{N-1,j}u^n_{N-3,j}u^n_{N-4,j} &= g^n_{N-1,j}, && i = N-1;\\
a^3_{Nj}u^n_{Nj} - a^4_{Nj}u^n_{N-1,j} + a^5_{Nj}u^n_{N-2,j} &= g^n_{Nj}, && i = N.
\end{aligned}$$

Here for $2 \le i \le N-2$,

$$g^n_{ij} = \frac{\tau}{4h^3_x}\left(u^{n-1}_{i+2,j} - 2u^{n-1}_{i+1,j} + 2u^{n-1}_{i-1,j} - u^{n-1}_{i-2,j}\right) + u^{n-1}_{ij} + \tau f^{n-1}_{ij},$$

$$a^1_{ij} = -\tau\beta/4h^3_x = -a^5_{ij}$$

$$a^2_{ij} = \tau\left[\alpha u^{n-1}_{ij} - (2\beta/h^2_x)\right]/4h_x = -a^4_{ij}$$

and

$$a^3_{ij} = 1 + \alpha\tau\left(u^{n-1}_{i+1,j} - u^{n-1}_{i-1,j}\right)/4h_x$$

(compare the first equation of this set with (1.99)). The computations in y are performed for $2 \le j \le M-2$. For $i = 0, 1$ and $i = N-1, N$, omitting index j, we have from (4.82)

[4] Note that the existence of these conditions on the correctness of the used algorithm is only related to the nonlinearity of the initial equation. The implicit difference schemes are absolutely stable in the linear case.

$$
\begin{aligned}
g_0^n &= a_0^1 u_2^{n-1} - a_0^2 u_1^{n-1} + a_0^3 u_0^{n-1},\\
g_1^n &= a_1^1 u_3^{n-1} - a_1^2 u_2^{n-1} + a_1^3 u_1^{n-1} - a_1^4 u_0^{n-1},\\
g_{N-1}^n &= -a_{N-1}^2 u_N^{n-1} + a_{N-1}^3 u_{N-1}^{n-1} - a_{N-1}^4 u_{N-2}^{n-1} + a_{N-1}^5 u_{N-3}^{n-1},\\
g_N^n &= a_N^3 u_N^{n-1} - a_N^4 u_{N-1}^{n-1} + a_N^5 u_{N-2}^{n-1}.
\end{aligned}
\tag{4.82}
$$

The coefficients $a_i^1, \ldots, a_i^5$ are defined from the boundary conditions of the initial value problem. The set (4.82) can be solved using the non-monotonous *sweep method* according to the algorithm proposed in Ref. [97]. The scheme (4.82) is correct when the matrix A is not degenerated for all j, i.e., $\det A \neq 0$.

Consider now the family of the implicit schemes for the 3-DNLS equation (4.70),

$$
\frac{\bar{h}_{ij}^{n+1} - \bar{h}_{ij}^{n}}{\tau} = \frac{\alpha}{2}\left[\left(\bar{h}_{ij}^{n}\right)^2 \mathrm{d}_x\left(\bar{h}_{ij}^{n+1}\right) + \left(\bar{h}_{ij}^{n+1}\right)^2 \mathrm{d}_x\left(\bar{h}_{ij}^{n}\right)\right] + \frac{\beta}{2}\mathrm{d}_x^2\left(\bar{h}_{ij}^{n+1} + \bar{h}_{ij}^{n}\right) + f_{ij}^n, \tag{4.83}
$$

where, depending on the order of the space derivatives on the right-hand side of the equation, the integral term f_{ij}^n is defined by (4.61) or (4.62), and (4.68). It is possible to solve the scheme (4.83), as well as the schemes (4.75) and (4.80) using various versions of the sweep method [97]. As shown in Refs. [83,98], the use of the 7-point sweep method for the scheme with the $O(\Delta_x^4, \Delta_y^2)$ approximation of the space derivatives for a rather small step, $\Delta = \Delta_x \leq \Delta_y$, gives the time step restriction (connected with the nonlinearity of the problem) with the value very close to that obtained for 1-DNLS equation (see Sect. 2.4.4).

The analysis of stability of the schemes (4.60), (4.67), and (4.75) approximating the KP equation demonstrates that restrictions on the step τ for the two latter methods are approximately equal and they are less than those for the scheme (4.60). Wider opportunities to choose the time step in (4.80) (the KP equation) save the solution time to a great extent when compared to the other schemes. It is also evident that the implicit schemes (4.75) and (4.80) have one more advantage as compared with the explicit ones (4.60) and (4.67), namely, their two-layer structure ensures lesser requirements to the computer memory resources. Similar reasons can be stated for explicit and implicit schemes for the 3-DNLS equation considered above – the implicit schemes are generally more economical with regard to the computer time and memory requirements compared to the explicit ones with the same order of approximation. It should be noted, however, that for the same steps $h_{x,y}$ ($\Delta_{x,y}$) and τ, it is still more advantageous to use the schemes (4.60), (4.67), and (4.72).

4.3.2 Boundary Conditions and Diffraction Terms

It is well known [24] that the asymptote of the two-dimensional soliton of the KP equation is $u \sim (x^2+y^2)^{-1}$ for $|x, y| \to \infty$ (see Sect. 3.1.2). This relatively slow "algebraic" decrease to zero causes difficulties when approximating the boundary conditions for problem (4.59) and defines the nonlocality of the diffraction term on the right-hand side of the KP equation. The function $|h|^2 = (1+e^2)\bar{h}^2$ (i.e., the envelope of the Alfvén wave packet) in the DNLS equation in general approaches zero exponentially [7,33,37,38], but the function $\bar{h}$ in the model equations (4.70) reveals an oscillating character, which also complicates the problem of the adequate approximation of the boundary conditions and the integral term in (4.59) and (4.70).

For many studies on the dynamics of multidimensional nonlinear waves, important for applications, it is convenient to use the dissipative boundary conditions (for example, to take into account possible effects of radiation from the system [59,63,64]) or to introduce an effective damping near the boundary of the integration region. The latter can be done by using the *Leontovich boundary conditions* of the impedance type or, much easier, by introducing an additional iteration at every time layer [81], or using the expressions [83,98]

$$\begin{aligned} &\tilde{u}_{ij}^n = (1-\tilde{\gamma}_i)\, u_{ij}^{n(num)}, \\ &\tilde{u}_{N+1} = \tilde{u}_{N+2} = \tilde{u}_{N+3} = \tilde{u}_N, \\ &\tilde{\gamma}_i = \tilde{\delta}\exp\left[\tilde{\varepsilon}\,(i-1-N/2)\right]. \end{aligned} \tag{4.84}$$

Here, values of the parameters $\tilde{\delta}$ and $\tilde{\varepsilon}$ in simulations have to be chosen from the condition of the absence of waves reflected from the boundary (i.e., total absorption on the boundaries).

For the KP equation in a number of cases we can choose the "natural" boundary conditions

$$\partial_x^3 u\bigg|_{G_{x,y}} = \varepsilon\partial_x^3\left[\left(x^2+y^2\right)^{-1}\right] = 24\varepsilon\frac{x\left(y^2-x^2\right)}{\left(x^2+y^2\right)^4}, \tag{4.85}$$

where $\varepsilon = \varepsilon(\nu,\xi)$ is chosen in the simulation and ν and ξ are the parameters of the solution (see also Sect. 3.1.2). The simulations show that the boundary conditions (4.85) with the proper choice of the parameter ε have the $O(h_{x,y}^5)$ approximation to the exact solution. Taking into account the conditions (4.84) or (4.85), we can obtain the coefficients $a_i^1, a_i^2, \ldots, a_i^l$ (where l is the number of a point in the template of the difference scheme) and g_i on the boundaries of the simulation region from (4.80) and (4.83) or similar expressions for the schemes modeling the 3-DNLS equation.

When numerically integrating the diffraction term the integral with the infinite limit $\int_{-\infty}^{x}\partial_y^2 u\mathrm{d}x$ is approximated by the integral $\int_{-a}^{x}\partial_y^2 u\mathrm{d}x$ with the finite lower limit. In this case when choosing the boundary conditions (4.85) in the KP model, it is necessary to introduce a "component" approximating $I = \int_{-\infty}^{-a}\partial_y^2 u\mathrm{d}x$ into the right-hand side to ensure stability of the numerical calculation. Thus, if u is a solution of (4.59), the expression for I can be obtained analytically. For example, for $\nu = \mathrm{Re}\nu$ and $\xi = 0$ we have [83,98]

$$I = \frac{8\nu^6\tilde{a}}{(1+\nu^4y^2+\nu^2\tilde{a}^2)^2}\left[\frac{4\nu^4y^2}{1+\nu^4y^2+\nu^2\tilde{a}^2} - 1\right], \tag{4.86}$$

where $\tilde{a} = a - 3\nu^2 t$. If we substitute the integral $\int_{-\infty}^{x} \partial_y^2 u \mathrm{d}x$ by its equivalence,

$$\frac{1}{2}\left[\int_{-\infty}^{x} \partial_y^2 u \mathrm{d}x - \int_{x}^{\infty} \partial_y^2 u \mathrm{d}x\right],$$

then the relation

$$\int_{-\infty}^{x} \partial_y^2 u \mathrm{d}x - \int_{x}^{\infty} \partial_y^2 u \mathrm{d}x = \int_{-a}^{x} \partial_y^2 u \mathrm{d}x - \int_{x}^{a} \partial_y^2 u \mathrm{d}x$$

for $t = 0$ is exact. But for $t \neq 0$ it is again necessary to introduce a "component" approximating $I' = \int_{-\infty}^{-a} \partial_y^2 u \mathrm{d}x - \int_{a}^{\infty} \partial_y^2 u \mathrm{d}x$, and the expression for I' can appear to be more complex than that for I.

For the dissipative boundary conditions [64] and conditions (4.84), taking into account the effective damping near the boundary of the integration region, the problem of introduction the "component" approximating the integral I does not arise. The stability is ensured by the selection of parameters defining the damping in the absence of waves reflected from the boundary. If we consider the oscillating character of the function $\bar{h}$ in the model equations (4.70) as the most preferable when solving the problem of the Alfvén wave dynamics, it is necessary to consider the boundary conditions of the type (4.84). Finally, the standard periodic boundary conditions can also be used for all the above cases.

4.3.3 Dynamic Spectral Method

First we consider the *KP equation* by rewriting it in the differential form

$$\partial_{tx}^2 u + \frac{\alpha}{2}\partial_x^2(u^2) + \beta\partial_x^4 u - \kappa\Delta_\perp u = 0. \tag{4.87}$$

Now, we execute the space Fourier transform $\mathcal{F}$ on the coordinates x, y, and z:

$$U(t,\xi,\zeta,\eta) = \frac{1}{(2\pi)^3}\iiint u(t,x,y,z)\exp\left[-\mathrm{i}(x\xi + y\zeta + z\eta)\right]\mathrm{d}x\mathrm{d}y\mathrm{d}z,$$
$$u(t,x,y,z) = \iiint U(t,\xi,\zeta,\eta)\exp\left[\mathrm{i}(x\xi + y\zeta + z\eta)\right]\mathrm{d}\xi\mathrm{d}\zeta\mathrm{d}\eta. \tag{4.88}$$

Then equation (4.87) is given by

$$\partial_t U + fW + gU = 0, \tag{4.89}$$

where

$$f = \mathrm{i}\alpha\xi/2,$$
$$g = -\mathrm{i}\xi\left[\xi^2\beta + \kappa\xi^{-2}\left(\zeta^2+\eta^2\right)\right],$$

and

$$W = U * U.$$

Furthermore we assume [98,195] that $U = X + \mathrm{i}Y$, $W = Z_1 + \mathrm{i}Z_2$, $f' = -\mathrm{i}f$, $g' = -\mathrm{i}g$, and rewrite (4.89) as the set

$$\begin{aligned}\partial_t X - f'Z_2 - g'Y = 0,\\ \partial_t Y + f'Z_1 + g'X = 0.\end{aligned} \tag{4.90}$$

The functions X and Y at $t = 0$ are defined by the Fourier transform of the initial condition $u(0, x, y, z) = \psi(x, y, z)$ of the initial value (Cauchy) problem for (4.87), and by $W|_{t=0} = \mathcal{F}[\psi^2]$. The convolution W on the next time layers can be obtained by using the convolution theorem [179] according to the scheme

$$\{U\} \to \{\mathcal{F}[U]\} \to \{V\} = \{\mathcal{F}[U]\mathcal{F}[U]\} \to \{W\} = \{\mathcal{F}^{-1}[V]\}.$$

Note that the coefficient g' has a singularity on the plane $\xi = 0$ that must be taken into account in calculations. Here we shall limit ourselves only to the following remark. For $\xi = 0$ and $\zeta, \eta \neq 0$ we have $U = 0$ which can be obtained easily from (4.89). As to the point $(0, 0, 0)$, it is not difficult to demonstrate that for the function u satisfying the KP equation, its Fourier-image is given by $U(t, 0, 0, 0) = \mathcal{F}[u(t, x, y, z)] = 0$ [83,98].

A similar approach is applicable for simulation of the *3-DNLS equation* (4.70). Indeed, writing it in the differential form

$$\partial^2_{tx}\bar{h} + s|p|^2\partial^2_x\left(\bar{h}^3\right) - \mathrm{i}\lambda\partial^3_x\bar{h} - \kappa\Delta_\perp\bar{h} = 0 \tag{4.91}$$

and Fourier transforming it in the space coordinates, see (4.88), we obtain an equation similar to (4.89),

$$\partial_t H + fG + gH = 0, \tag{4.92}$$

where

$$f = \mathrm{i}s|p|^2\xi,$$
$$g = -\mathrm{i}\xi\left[-\mathrm{i}\lambda\xi^2 + \kappa\xi^{-2}\left(\zeta^2+\eta^2\right)\right],$$
$$G = W * H,$$

and

$$W = H * H.$$

Thus, as in the case of the KP equation represented by (4.89), we obtained the equation on the complex functions but with a complex (with the non-zero real part), instead of purely imaginary, factor g. Now it is necessary to solve the complex equation (4.92), instead of the previous real set. The values of H at $t = 0$ are obtained by the Fourier transform of the initial condition $\bar{h}(0,x,y,z) = \psi(x,y,z)$ of the Cauchy problem for (4.91), with $G|_{t=0} = W|_{t=0} * \mathcal{F}[\psi]$ and $W|_{t=0} = \mathcal{F}[\psi^2]$. The convolution values on the next time layers can be obtained by employing the convolution theorem similar to the above procedure for the KP equation.

Regarding equations (4.89), (4.90) and (4.93), we should make the following important remark. The finite numerical integration region causes percolation of the spectral components of the function u (the *Gibbs oscillations*) that is connected with the presence of ruptures of the periodic extrapolation of u on the boundaries of the region. Therefore, in numerical realizations of the Fourier transform, it is necessary to introduce into (4.88) multiplicative *weight functions* to suppress the rupture order and satisfy the possibly larger number of the derivatives of the weighted function u on the boundaries. Thus approximating an integral by finite sums, it is possible to present the direct Fourier transform (4.88) in the form

$$U_{\sigma_1\sigma_2\sigma_3}(t,\xi_p,\zeta_q,\eta_r) = \frac{1}{MNK}\sum_{m=0}^{M-1}\sum_{n=0}^{N-1}\sum_{k=0}^{K-1}\sigma_1(m\Delta x)\sigma_2(n\Delta y)\sigma_3(k\Delta z)$$
$$\times u(t,m\Delta x,n\Delta y,k\Delta z)\exp\left[-\mathrm{i}\left(\xi_p m\Delta x+\zeta_q n\Delta y+\eta_r k\Delta z\right)\right],$$

where

$$\begin{aligned}\sigma_1(m\Delta x) &= \sigma_1\left[(M-m)\,\Delta x\right],\\ \sigma_2(n\Delta y) &= \sigma_2\left[(N-n)\,\Delta y\right],\\ \sigma_3(k\Delta z) &= \sigma_3\left[(K-k)\,\Delta z\right],\end{aligned}$$

$\xi_p = 2\pi p/M\Delta x$, $\zeta_q = 2\pi q/N\Delta y$, $\eta_r = 2\pi r/K\Delta z$, $p = 0,1,2,\ldots,M-1$, $q = 0,1,2,\ldots,N-1$, and $r = 0,1,2,\ldots,K-1$. If as a result of the effect of the weight functions to obtain the minimum distortion of the spectrum in the center of the integration region we are successful in getting the smooth limit of u to zero at the boundaries, the periodic extrapolation of u is infinite (with the accuracy up to the higher order derivatives).

As the weight functions $\sigma_{1,2,3}$, various windows that are widely applied in the spectral analysis can be used. Numerical simulations show that the *weight functions* in the form of the *Blackman–Harris windows* are the most acceptable for the usual KP equation as well as for equations of the type (4.3) [181]:

$$\sigma(j) = c_0 - c_1\cos(2\pi j/N) + c_2\cos(4\pi j/N) - c_3\cos(6\pi j/N),$$

where $j = 0,1,2,\ldots,N-1$. Equations (4.90) can be easily solved by the *Runge–Kutta method* [180]. Let us consider the scheme approximating the set (4.90),

$$\begin{aligned}X^{n+1} - X^n &= \Delta t\left[k_{11} - \frac{1}{2\gamma}(k_{11}-k_{12})\right],\\ Y^{n+1} - Y^n &= -\Delta t\left[k_{21} - \frac{1}{2\gamma}(k_{21}-k_{22})\right],\end{aligned} \qquad (4.93)$$

where

$$k_{11} = f'Z_2^n + g'Y^n,$$
$$k_{12} = k_{11}(1+\gamma\xi),$$
$$k_{21} = f'Z_1^n + g'X^n,$$

and

$$k_{22} = k_{21}(1+\gamma\xi).$$

The scheme (4.93) has the $O(\tau^2)$ approximation independently on the value of γ. Equations of the set (4.93) are solved independently on every time layer.

To elucidate the stability problem of the scheme, let us consider the first equation in the set and rewrite it as [83,98]

$$X^{n+1} - \tau\left[\mathrm{c}(\tau)X^n + \mathrm{d}(\tau)Z_2^n + \mathrm{b}(\tau)Y^n\right] = 0, \tag{4.94}$$

where $\mathrm{b}(\tau) = g'(1+\tau/2)$, $\mathrm{c}(\tau) = 1/\tau$, and $\mathrm{d}(\tau) = f'(1+\tau/2)$. Thus the equation is given by

$$X^{n+1} - \hat{\mathrm{a}}(\tau)X^n = 0, \tag{4.95}$$

whereas the initial equation (4.90) is

$$\partial_t X - \hat{\mathrm{A}}(t)X = 0, \tag{4.96}$$

where $\hat{\mathrm{a}}(\tau)$ and $\hat{\mathrm{A}}(\tau)$ are some operators. The root of the characteristic equation for the difference equation (4.94), $\lambda - \hat{\mathrm{a}}(\tau) = 0$, is $\lambda = \hat{\mathrm{a}}(\tau)$. Since $X^n = X(t_n)$, the coincidence of X^{n+1} with the exact solution, $X(t_n+\tau)$, has an accuracy of τ^{m+1} ($m = 2$ is the approximation order), i.e., an $O(\tau^3)$ accuracy. We obtain from equation (4.96) that

$$X(t_n+\tau) = X(t_n)\exp\left[-\hat{\mathrm{A}}(\tau)\tau_n\right],$$

and it follows from (4.95) that $X^{n+1} = \hat{\mathrm{a}}(\tau)X^n$ and

$$\lambda = \hat{\mathrm{a}}(\tau) = \exp\left[-\hat{\mathrm{A}}(\tau)\tau\right] + O(\tau^3). \tag{4.97}$$

Thus $|\lambda(\tau)| < 1 + C\tau$, where C is a constant. It follows from (4.97) that the power of $\lambda^n(\tau)$ increases if the solution X of (4.96) increases and $\hat{\mathrm{A}}(\tau) < 0$, and vice versa. The analysis for the second equation (4.93) is similar – the convergence of the solution of the difference problem (4.93) to the solution of the problem (4.90) follows from the approximation and its stability.

A similar approach can be elaborated for (4.92) (when it is necessary to solve a difference equation similar to (4.93)). The stability analysis of the scheme, conducted above for the set of equations (4.93), is also completely correct for the difference equation approximating (4.92).

4.3.4 Comparative Characteristics of Different Schemes and Their Use in Numerical Simulation

Testing of the difference schemes (4.60), (4.67), (4.72), (4.75), (4.80), (4.83), and (4.93) presented above was performed in two stages [83,98]. We first investigated the schemes' characteristics related to the integration in x when the coefficient κ in the right-hand side of (4.59) was supposed to be equal to zero. In this case the KP equation transforms into the KdV equation, and the 3-DNLS equation into the 1-DNLS equation. For the KP equation the initial condition was taken in the form of the exact solution of the KdV equation [3]:

$$u(0,x) = (3v/\alpha)\cosh^{-2}\left[\left(v^{1/2}/2\beta\right)(x - x_0)\right] \tag{4.98}$$

with $v = \alpha = 6$ and $\beta = 1$, as well as that of the DNLS equation

$$h(t,x) = (A/2)^{1/2}\left[\exp(-Ax) + \mathrm{i}\exp(Ax)\right]\exp(-\mathrm{i}A^2 t)\cosh^{-2}(2Ax), \tag{4.99}$$

with $t = 0$ (see (4.57)) for all values of the variables y and z. The control of the accuracy on all time layers was fulfilled by comparison of the numerical solutions with the exact (analytical) ones. We calculated the relative mean deviation on each time step τ and the mean-square-root deviation of the numerical solutions from the exact ones to yield

$$\varepsilon = \left|u_\tau^{\text{num}} - u_\tau^{\text{exact}}\right| / u_\tau^{\text{exact}},$$

and

$$s = \left[\frac{1}{MNK}\sum_{m=1}^{M}\sum_{n=1}^{N}\sum_{k=1}^{K}\left|(u_{mnk}^{\text{num}})^2 - (u_{mnk}^{\text{exact}})^2\right|\right]^{1/2}.$$

For example, at the time moment $t = 0.4$ (soliton is near the boundary of the integration region) for the scheme (4.75) we obtained the same results as for the scheme (1.84)[5], namely, $\varepsilon = 6.38775 \times 10^{-3}$ and $s = 1.74663 \times 10^{-4}$, that is quite acceptable and approximately on the order of magnitude better than the corresponding results for the schemes proposed in Refs. [79,93] for solution of the KdV equation. When testing, for example, the scheme (4.83) for the DNLS equation, the results $\varepsilon = 2.64536 \times 10^{-3}$ and $s = 2.32451 \times 10^{-3}$ at the time $t = 20$ are also quite satisfactory; unfortunately, there was no chance to compare them with the results obtained by Dawson and Fontán for the modernized Ablowitz–Ladik method (see Sect. 2.4.4) since similar investigations of the scheme were not performed in Ref. [176].

On the second stage, for the KP equation (4.59) with $\alpha = 6$, $\beta = 1$, and $\kappa = 3$, we "switched on" the y-derivative in the right-hand side of the equation and the testing was performed on the exact solution describing the two-dimensional soliton with $n = 1, 2$, $m = 1, 2$, and $l = 1$ (see Sect. 3.1.2)

[5] It is natural because on this stage of testing the scheme (4.75) coincides exactly with the scheme (1.84).

Table 4.1. Results of simulation tests for schemes (4.60), (4.67), and (4.75) ($h_x = 0.1$, $h_y = 0.3$, $\tau = 0.0025$ and $t = 0.4$)

Parameter	Scheme (4.60)	Scheme (4.67)	Scheme (4.75)
ε	9.1×10^{-2}	8.1×10^{-2}	7.9×10^{-2}
s	1.5×10^{-1}	6.0×10^{-2}	5.7×10^{-2}
$\Delta\Im_1$	8.6%	6.5%	6.2%
$\Delta\mathcal{P}$	7.4%	3.6%	3.7%
$\Delta\mathcal{H}$	16.2%	12.9%	12.3%

which has been set as the initial condition at $t = 0$. The test of the three-dimensional DNLS (3-DNLS) equation was not performed because its exact solutions are not known.

Conditions on the boundary of the integration region for the schemes (4.60), (4.67), (4.75), and (4.80) were calculated using the expressions (4.84) and (4.85). In the last case we introduced an additional component into the diffraction term according to (4.86). For the schemes (4.72) and (4.83) used for simulation of the 1-DNLS equation, the boundary conditions like (4.84) as well as the periodic boundary conditions were used. The conservation of the invariants of the corresponding equations for all schemes was controlled on a level with calculation of the parameters ε and s, t (for the KP equation, see the corresponding expressions in Sect. 3.1.2; for the 1-DNLS equation the invariants are $C_0 = \int |h|^2 \mathrm{d}x$ and $C_1 = \mathcal{H}/2$ where $\mathcal{H} = \int \left[(1/2)|h|^4 - h^* h \partial_x \varphi\right] \mathrm{d}x$ and $\varphi = \arg(h)$). The integrals were calculated on every time layer using the Newton–Cotes formulas with an accuracy of at least $O(h^4)$. We note that the order of the control parameters in the one-dimensional test of the schemes for the KP and 1-DNLS equations is approximately the same. The results of the tests on the exact solution describing the two-dimensional soliton are summarized in Tables 4.1 and 4.2. Figure 4.6 shows general evolution of such a soliton found by integrating the KP equation by the spectral method.

Resuming the results of investigation of the characteristics of the difference schemes (4.60), (4.67), (4.72), (4.75), (4.80), (4.83), and (4.93) we can make the following conclusions. The schemes (4.67), (4.75), and (4.83) with the approximation of the space derivatives have the best accuracy characteristics; the accuracy of the solution with the use of the spectral method is a little bit lower. This is mostly stipulated by the complexity of excluding the Gibbs oscillations which appear owing to the finite the integration interval in the numerical solution of the problem. The schemes (4.60), (4.67), and (4.72),

Table 4.2. Results of simulation tests for schemes (4.80) and (4.92) ($h_x = 0.1$, $h_y = 0.3$, $\tau = 0.0025$, and $t = 0.4$)

Parameter	Scheme (4.80)	Scheme (4.92)
ε	9.2×10^{-2}	8.2×10^{-2}
s	1.0×10^{-1}	6.2×10^{-2}
$\Delta\Im_1$	8.2%	6.6%
$\Delta\mathcal{P}$	7.0%	3.9%
$\Delta\mathcal{H}$	16.0%	13.1%

in their turn, are more time efficient although their three-layer character requires more computer memory resources. The scheme (4.93), despite the use of the Fourier transform on every time layer, computes sufficiently faster because of the weak dependence of the accuracy of solutions on the choice of the step h in $x-y-z$-space as well as the time step τ. We also note that the use of the spectral method in fact implies the periodic boundary conditions in the KP and 3-DNLS problems. Moreover, when the localization of the perturbation approaches (as a result of the evolution) the boundary of the integration region, the weight functions introduce an effective damping near the external boundaries. For the schemes (4.60), (4.67), (4.72), (4.75), (4.80), and (4.83), the problem of the boundary conditions for an arbitrary initial perturbation is non-trivial and depends on the considered physical problem.

The schemes considered above can be easily generalized for equations of the class (4.3) and (4.22). Three-dimensionality of the problem in the finite-difference approach is provided by introduction of the second order derivative on the third space coordinate in the integral f of the difference approximation, analogous to the expressions (4.61) and/or (4.68) for f'. The difference approximation of the derivatives of the fourth and fifth order requires either increasing the number of the grid points of the template of the difference scheme in the x-direction, or increasing the approximation order of the scheme by introducing differential corollaries of the initial equation (see, for example, Ref. [200]). When choosing the spectral method for integration of equations (4.3) and (4.22), the changes are limited by introducing additional terms into expressions (4.89) and (4.92) for g to account for the dissipation and dispersion correction proportional to the fifth derivative in x. We also note that the proposed integral-differential approach for simulations of the multidimensional equations can be extended to the modernized Ablowitz–Ladik method [176] used for numerical integration of the 1-DNLS equation. In this case the integral in the right-hand side of the 3-DNLS equation should

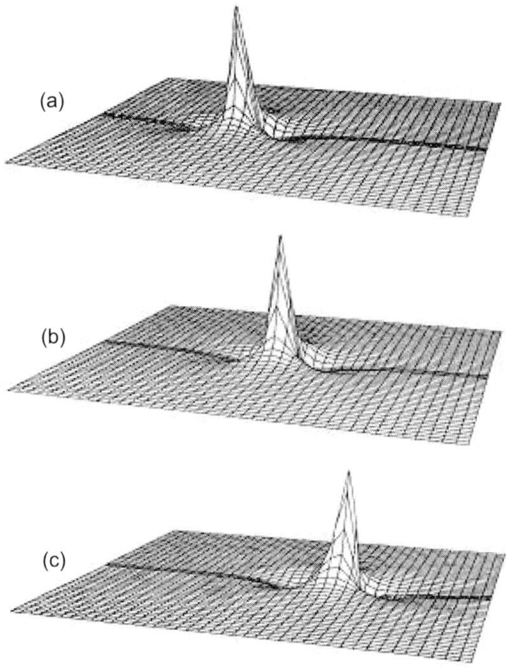

Fig. 4.6. Evolution of an exact soliton solution of (4.1) with $\nu = \delta = \gamma = 0$ obtained by the spectral method. **a** $t = 0$. **b** $t = 0.3$. **c** $t = 0.7$

be approximated by the Newton–Cotes formulas (4.62) and (4.68) with the automatic choice of the number of nodes in the quadrature formula [77].

To conclude this section, we note that the novelty of the schemes (4.60), (4.67), (4.72), (4.75), (4.80), and (4.83) is in the use of the integral-differential representation in their applications to equations of the classes considered above. The finite-difference approximation of the left-hand side and the integral right-hand side are considered together; moreover, such an approach is technically easier to realize than the commonly used differential representations with the mixed derivative $\partial^2_{tx}u$. Also, calculation of an integral in the form $\int_{-\infty}^{x} \partial_y^2 u dx$ in the right-hand side (with remarks of Sect. 4.3.2) enables us to effectively suppress the computation errors connected with the nonlocality of the right-hand side of the initial equation. The dynamic spectral method formulated in subsection 4.3.3, unlike the "static" approach used earlier in Refs. [31,80], allows us to consider the dynamics of the soliton–soliton interactions for the models of the KP class as well as the evolution of non-stationary solutions of the equations of the KP and DNLS classes.

4.4 Dynamics of Two-Dimensional Solitons in Dispersive Media

Here, we consider numerical solutions of the two-dimensional *GKP equation* written in the differential form

$$\partial_x \left(\partial_t u + 6u\partial_x u - \mu\partial_x^2 u - \varepsilon\partial_x^3 u - \lambda\partial_x^5 u\right) = \partial_y^2 u, \qquad (4.100)$$

describing formation and interaction of solitons, and present the evolution of nonstationary wave packets when $\lambda = \pm 1$, ε takes any value, and $\mu \geq 0$ in (4.100). Numerical integration of (4.100) is performed by using both the *dynamic spectral method* (128×128 grid, $\tau = 0.005$, $h_{x,y} = 0.15$) and the *implicit scheme with $O(\tau^2, h_{x,y}^4)$ approximation* and 301×51 grid, $\tau = 0.0025$, $h_x = 0.1$, and $h_y = 0.3$. The total absorption on the boundaries of the computation region (4.84) is imposed. Initial conditions were assumed in the form of the soliton solutions of the KP equation (see Sect. 3.1.2) with $\beta/\kappa > 0$ and various values of ν_n and ζ_n defining the amplitudes, phases, velocities and other soliton parameters. The numerical integration is controlled by conservation of the momentum $\mathcal{P}$ and Hamiltonian $\mathcal{H}$,

$$\begin{aligned}
\mathcal{P} &= \frac{1}{2}\iint u^2 \mathrm{d}x\mathrm{d}y, \\
\mathcal{H} &= \iint \left[-\frac{\varepsilon}{2}\left(\partial_x u\right)^2 + \frac{\lambda}{2}\left(\partial_x^2 u\right)^2 + \frac{1}{2}\left(\int_{-\infty}^{x} \partial_y u \mathrm{d}x\right)^2 - u^3\right] \mathrm{d}x\mathrm{d}y, \qquad (4.101)
\end{aligned}$$

of the soliton solutions of (4.100) with $\mu = 0$.

For the two-dimensional equation (4.100) below, we consider the structure of two-dimensional numerical solutions estimating their stability when $\mu = 0$ (Sect. 4.4.1), the interaction of two-dimensional solitons (Sect. 4.4.2), the influence of the viscous-type dissipation ($\mu > 0$) on their evolution (Sect. 4.4.3), as well as the evolution of two-dimensional solitons in a dispersive medium with stochastic fluctuation of the wave field (Sect. 4.4.4), and the dynamics of solitons in a medium with variable dispersion (Sect. 4.4.5). Everywhere in this section, when referring to the GKP equation, we always mean an equation in the form (4.100), i.e., assume $\Delta_\perp = \partial_y^2$ in the standard three-dimensional GKP equation (4.3) and equation (4.4).

4.4.1 Structure of Two-Dimensional Solutions of GKP-Class Equations

Consider, first, the numerical solutions of (4.100) with $\mu = 0$ for the initial conditions corresponding to the one-soliton solution of the "classic" KP equation, that is, with $n = 1, 2$; $m = 1, 2$; and $l = 1$ in the corresponding expressions (see Sect. 3.1). In this case for $\lambda = 1$ and $\varepsilon \leq 0$, we observe formation of *lump solitons* with the asymptotics very close to that of the KP

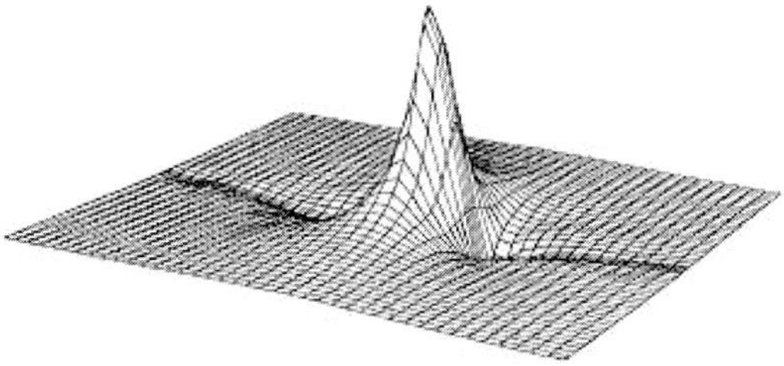

Fig. 4.7. Two-dimensional soliton solution of (4.100) corresponding to $\lambda = 1$, $\varepsilon = -0.6$, and $\mu = 0$ $(t = 0.2)$

rational soliton in the case $\beta/\kappa > 0$. This is shown in Figs. 4.7–4.9. For $\lambda = 1$ and $\varepsilon > 0$, the structure of the soliton solutions changes qualitatively. We can indeed see from a simple analysis of the dispersion relation for the linearized equation (4.100) with $\mu = 0$ and $\partial_y = 0$,

$$\omega \approx c_0 k_x \left(1 + \frac{\varepsilon k_x^2 - k_x^4}{c_0}\right), \tag{4.102}$$

that the inflection point appears on the dispersion curve, and we have the opposite signs of dispersion for the short-wavelength and long-wavelength waves. In this case we observe the formation of solitons of (4.100) from the one-soliton initial condition, oscillating in the x-direction and monotonic in the y-direction (see Figs. 4.10 and 4.11). Velocity V of these solitons satisfies

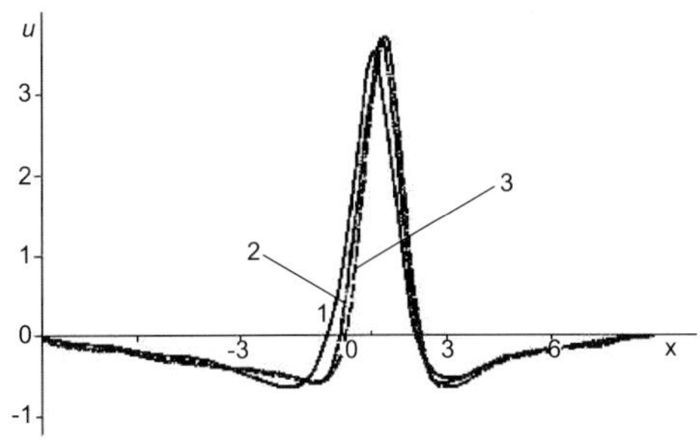

Fig. 4.8. Cross sections of two-dimensional solitons at $y = 0$ for $\mu = 0$ and $\lambda = 1$ $(t = 0.2)$: (1) for $\varepsilon = 0$, (2) for $\varepsilon = -0.3$, (3) for $\varepsilon = -0.6$

the conditions $V_x > 0$ and $V > V^{\mathrm{ph}}_{\mathrm{max}}$, where V^{ph} is the phase velocity of the small-amplitude wave with $k_y = 0$. From the dispersion relation (4.102) we have $V^{\mathrm{ph}} = \varepsilon k_x^2 - k_x^4$. This gives the lower limit for the soliton velocity, namely, $V^{\mathrm{ph}}_{\mathrm{max}} = \varepsilon^2/4$.

Numerical simulations and analysis of the solutions' asymptotics demonstrate [112,113] that the amplitudes and wavelengths of the oscillations depend on ε (see Figs. 4.10 and 4.11). The amplitudes of the oscillations decrease and their wavelengths increase with decreasing ε. When $\varepsilon \to 0$, the soliton asymptotics tends to the algebraic ones described above. On the other hand,

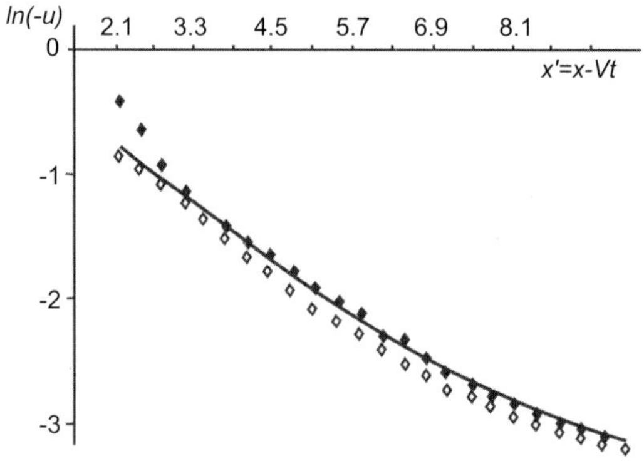

Fig. 4.9. Asymptotics of numerical solutions of (4.100) for $\lambda = 1$, $\varepsilon < 0$: diamonds stand for $\varepsilon = -0.6$ and $\mu = 0$, filled diamonds stand for $\varepsilon = -0.3$ and $\mu = 1$, and solid line represents the soliton of "classic" KP equation

recall that the case $\varepsilon \to -\infty$ corresponds to $\gamma \to 0$ for $\beta =$const< 0 and $\kappa = -c_0/2 < 0$ in the standard form of the two-dimensional *GKP equation*,

$$\partial_x \left(\partial_t u + \alpha u \partial_x u - \nu \partial_x^2 u + \beta \partial_x^3 u + \gamma \partial_x^5 u\right) = \kappa \partial_y^2 u, \tag{4.103}$$

with $\nu = 0$, and in this limit the soliton solution of (4.100) with the algebraic asymptotics transforms into the two-dimensional soliton of the *KP equation*.

In the case $\varepsilon \gg 1$ corresponding to $\gamma \ll 1$ with $\beta =$const> 0 and $\kappa < 0$ in the equation (4.103), the oscillations have very short wavelengths and their average according to the procedure [81,83]

$$\bar{u}\left(x_{i+1/2}\right) = \frac{1}{x_{i+1} - x_i} \int\limits_{x_i}^{x_{i+1}} u dx, \tag{4.104}$$

$$x_{i+1/2} = \frac{1}{2}\left(x_{i+1} + x_i\right),$$

$$i = \begin{cases} m = 1, 2, ..., M-1 \\ n = 1, 2, ..., N-1 \end{cases},$$

where x_m and x_n are the coordinates of the mth maximum and n^{th} minimum of the tail oscillations, respectively, gives us again the algebraic asymptotics similar to that of the KP lump soliton. The asymptotic behavior can be approximated with a good accuracy by the power-law dependence

$$\bar{u}(x) = -ax^{-(2+b)} \tag{4.105}$$

(the coefficients a and b for some particular cases are presented in Table 4.3).

We should stress, however, that the KP equation with $\beta/\kappa < 0$ (this corresponds to $\mu = \lambda = 0$ and $\varepsilon > 0$ in (4.100)) has no two-dimensional lump solutions [16,24], and the soliton solutions of (4.100), decaying in all directions as $|x, y| \to \infty$ in the case $\varepsilon \gg 1$, appear because of $\lambda = 1$, i.e., $\gamma \neq 0$ in (4.103). The described structure of the solitons, appearing for $\mu = 0$

Table 4.3. Coefficients a and b for various λ and ε in (4.105) for the averaged asymptotics

Parameters	Values				
λ	0	1	1	-1	-1
ε	-1	3.16	10	-10	10
a	13.70	54.60	6.05	929.00	2.46
b	0	0.30	-0.60	0.91	-0.97

and $\varepsilon \geq 0$, agrees well with that of the stationary solutions found numerically [80]. For $\lambda = -1$ and either $\varepsilon > 0$ or $\varepsilon \leq 0$, the one-soliton initial condition

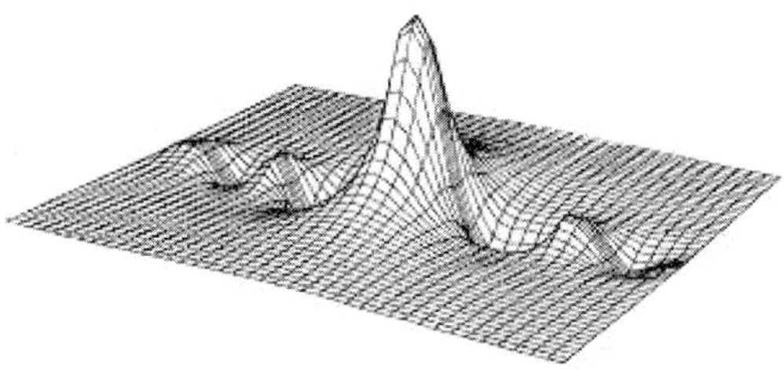

Fig. 4.10. Two-dimensional soliton for $\lambda = 1$ and $\varepsilon = 3.16$ $(t = 0.5)$

leads to the formation (within a wide range of ε values) of a non-stationary spreading wave packet, oscillating in the x-direction and monotonic in the y-direction (see Fig. 4.11, curve 5, and Fig. 4.12). Such "spreading" behavior

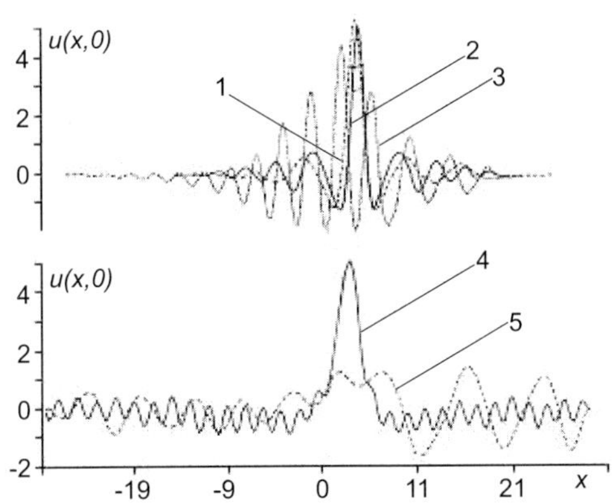

Fig. 4.11. Cross sections of a two-dimensional solution of (4.100) at $y = 0$ for $\mu = 0$ and $\lambda = 1$: (1) $\varepsilon = 2.53$, $t = 0.51$; (2) $\varepsilon = 3.16$, $t = 0.53$; (3) $\varepsilon = 7$, $t = 0.51$, and for $\lambda = -1$: (4) $\varepsilon = -10$, $t = 0.63$; (5) $\varepsilon = 10$, $t = 0.95$

for the examples presented in the above figures has been observed up to $t = 1$. We note that for any ε, the change of the oscillations wavelengths with the increasing $|\varepsilon|$ is qualitatively similar to that in the case $\lambda = 1$, $\varepsilon \to \infty$ ($\varepsilon > 0$), namely: for fast oscillations, when $\varepsilon \to \infty$, the averaging procedure according to (4.104) again gives the algebraic asymptotics (4.105) (see also Table 4.3). For large negative ε in the case $\lambda = -1$ and $\mu = 0$, in agreement with the

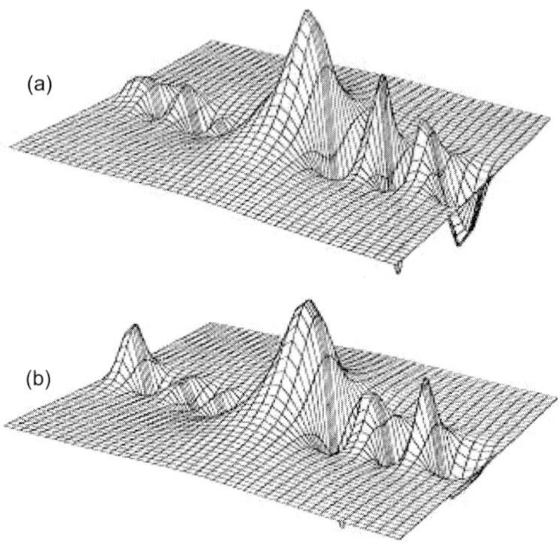

Fig. 4.12. Solutions of (4.100) for $\mu = 0$ and $\lambda = -1$ ($t = 0.2$). **a** $\varepsilon = -0.6$. **b** $\varepsilon = 0.6$

results obtained in Sect. 4.1.3, stable (within the accuracy of the numerical integration) two-dimensional solutions with oscillating in the x-direction tails (Fig. 4.11, curve 4) appear, with averaged algebraic asymptotics of the type (4.105). Approaching such a stationary regime with decreasing ε corresponds to the conservation of the two-dimensional KP-soliton stability, when small (close to zero) *higher order dispersion* correction with $\gamma \le 0$ are introduced into the classic KP equation.

Figure 4.13 shows the change of the amplitude of the solution's main maximum and the integrals $\mathcal{P}$ and $\mathcal{H}$. It illustrates the dynamics of the formation of two-dimensional solitons of (4.100) for $\lambda = 1$ and $|\varepsilon| > 0$. The analysis of the Hamiltonian boundedness on the numerical solutions of (4.100) under the $\mathcal{H}$ deformations on the spatial variables within the class of scale transformations (4.7) (see Sect. 4.1.3) proves the stability of the soliton solutions for $\lambda = 0$ and any ε (within the range of the values of ε investigated numerically) and the instability of the wave packets for $\lambda = -1$ (Fig. 4.14). Note that for $\lambda = 1$, the Hamiltonian $\mathcal{H}$ evolves to $\mathcal{H}_{\min}$. This agrees with the results of Sect. 4.1 on the stationary soliton solutions (compare Figs. 4.13 and 4.14).

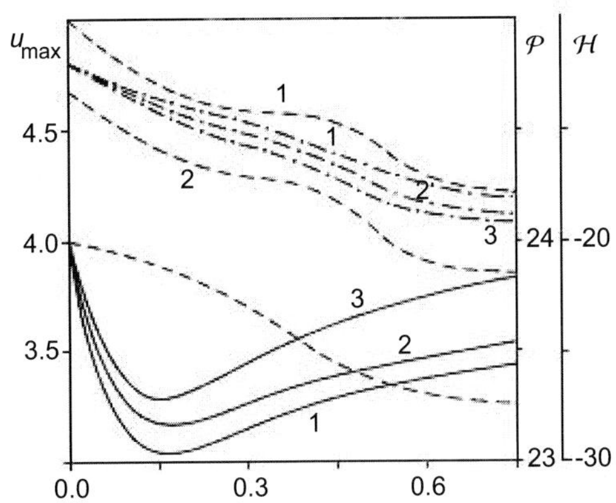

Fig. 4.13. Change of the amplitude of the main maximum $u_{\max}$ (solid lines) of the soliton solution of (4.100), its Hamiltonian $\mathcal{H}$ (dashed lines) and momentum $\mathcal{P}$ (dash-dotted lines) for $\mu = 0$ and $\lambda = 1$: (1) $\varepsilon = -0.6$, (2) $\varepsilon = -0.3$, and (3) $\varepsilon = 0.3$

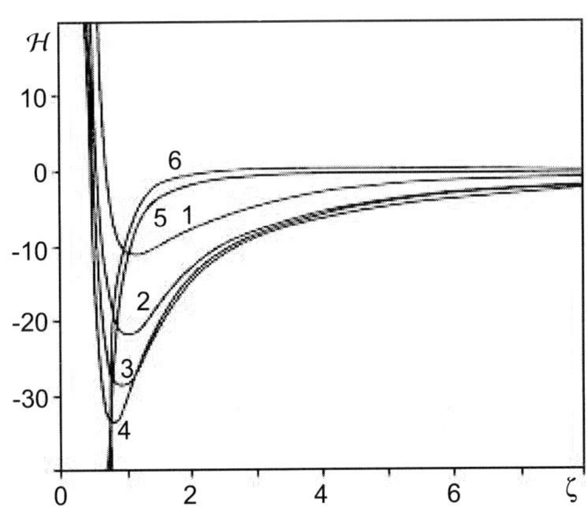

Fig. 4.14. Change of the Hamiltonian $\mathcal{H}(\zeta, \eta)$ in the direction $\eta = [(4b/c)^2\zeta^5]^{1/3}$ under deformations on numerical solutions of (4.100) for $\mu = 0$ and $\lambda = 1$: (1) $\varepsilon = -0.6$, (2) $\varepsilon = -0.3$, and (3) $\varepsilon = 0.3$. (dashed lines) and momentum $\mathcal{P}$ (dash-dotted lines) for $\mu = 0$ and $\lambda = 1$: (1) $\varepsilon = -1.79$, (2) $\varepsilon = -0.3$, (3) $\varepsilon = 0.3$, (4) $\varepsilon = 1.34$, and for $\mu = 0$ and $\lambda = -1$: (5) $\varepsilon = 1.34$, and (6) $\varepsilon = -1.34$

4.4.2 Interactions of Two-Dimensional Solitons

To study interactions of the two-dimensional solitons of the GKP equation (4.100) we use, as before, the initial conditions in the form of the exact solutions (see Sect. 3.1.2) of the KP equation (4.103) with $\nu = \gamma = 0$ for $\beta/\kappa > 0$, but also with $n = 1,2,3,4$; $m = 1,2,3,4$; and $l = 1,2$. They correspond to the two-soliton solutions of the KP equation with positive dispersion and without dissipation ($\mu = 0$). For $\lambda = 0$ and $\varepsilon \leq 0$ for all cases considered, the solitons (formed from the initial pulses) of (4.100) with the algebraic asymptotics interact nonlinearly and exchange their amplitudes and momenta $\mathcal{P}$ (see Figs. 4.15 and 4.16). We see that for $\varepsilon \leq 0$, the form of the solitons and the elastic

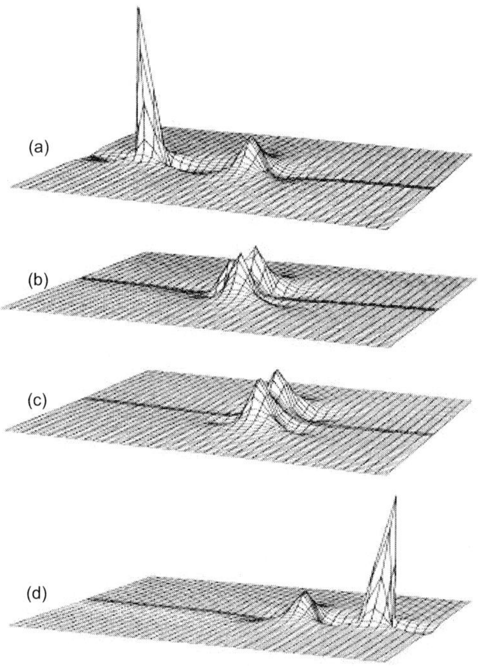

Fig. 4.15. Collision of two two-dimensional solitons propagating along the x-axis with amplitudes $u_1(0) = 16$, $u_2(0) = 4$, and $\Delta x(0) = 4$. Here, we have $\lambda = 1$, $\mu = 0$, and $\varepsilon = -2$. **a** $t = 0$. **b** $t = 0.2$. **c** $t = 0.4$. **d** $t = 1$

(within the accuracy of the numerical integration) character of their collision are qualitatively similar to those of the two-dimensional KP solitons [61,62].

The dynamics of the soliton interactions for $\lambda = 1$ and $\varepsilon > 0$, when the soliton tails are oscillating in the x-direction, is nontrivial and differs significantly from that of the solitons of the two-dimensional KP model with $\beta/\kappa > 0$ and the model described by (4.100) for $\lambda = 1$ and $\varepsilon \leq 0$. The problem was studied in a series of numerical simulations of (4.100) for $\varepsilon = 0.1 - 2.2$ and the two-soliton initial conditions (see Sect. 3.1.2) for various values of the parameters ν_n and ξ_n [81,195]. Figure 4.17 shows the interaction between two pulses having the significantly different amplitudes $u_1(0) = 8$ and $u_2(0) = 1$

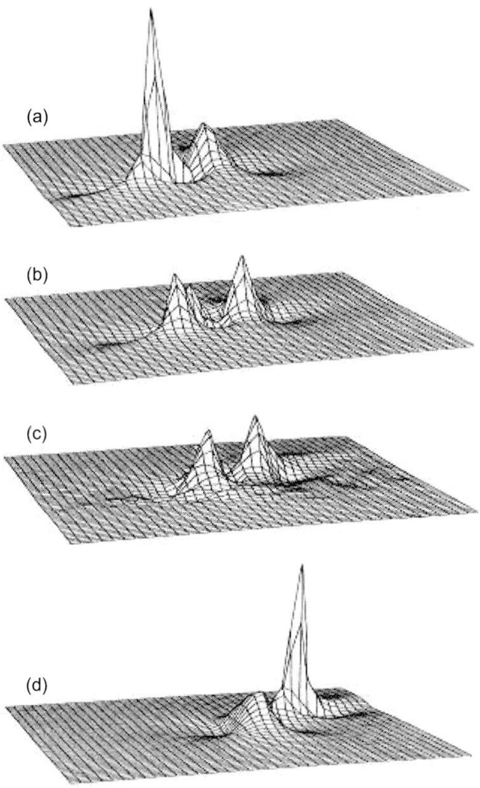

Fig. 4.16. "Oblique" collision of two-dimensional solitons with $u_1(0) = 12$, $u_2(0) = 4$, $\Delta r(0) = 3.3$, $\Delta r_{\min}(0.3) = 0.3$. Here, $\lambda = 1$, $\mu = 0$, and $\varepsilon = 0$. **a** $t = 0$. **b** $t = 0.2$. **c** $t = 0.4$. **d** $t = 0.7$

at $t = 0$ with the initial distance between them being $\Delta x(0) = 4$. We see that

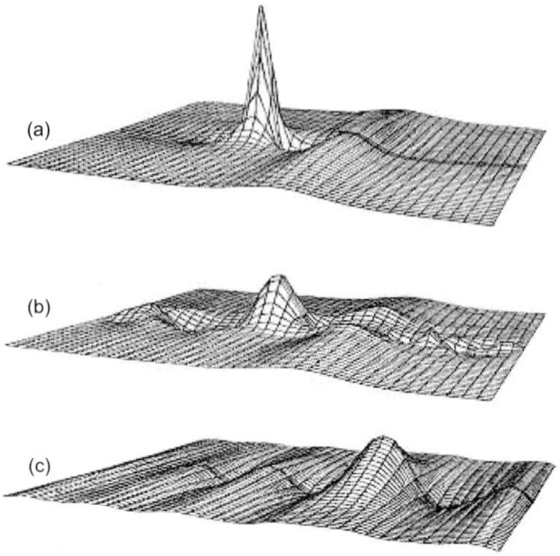

Fig. 4.17. Formation of the two-dimensional soliton with the oscillatory structure from the initial pulses with $u_1(0) = 8$, $u_2(0) = 1$, and $\Delta x(0) = 4$. **a** $t = 0$. **b** $t = 0.3$. **c** $t = 0.8$. The parameters of the GKP equation are $\lambda = 1$, $\mu = 0$, and $\varepsilon = 1.4$

unlike the corresponding solutions of (4.100) with $\varepsilon \leq 0$ and those of the two-dimensional KP equation, the final result of this evolution pattern is the formation of a single pulse with the amplitude $u(t) \approx 3.5$ ($t = 0.8$) and the oscillatory structure corresponding to a two-dimensional soliton of equation (4.100). Note that qualitatively similar results were obtained for interactions of solitons with initially close amplitudes and $\Delta x(0) \leq u_{1,2}(0)$: in the process of the soliton's evolution, the oscillatory structure is also formed and the pulse with the smaller amplitude is "absorbed" by the tail of the bigger one. As a result, a single soliton with $u_2(0) < u < u_1(0)$ with the structure described above is formed (see Fig. 4.18).

For sufficiently large distances $\Delta x(0)$ between the initial pulses (larger than the characteristic sizes of solitons) and similar amplitudes, the final result of the evolution is the formation of two solitons with the oscillatory structure, unlike the cases shown in Figs. 4.17 and 4.18. The result is shown in Fig. 4.19 where we present the evolution of two two-dimensional solitons with the large enough initial distance $\Delta x(0) = 6 > u_{1,2}(0)$ and almost equal initial amplitudes. We see that as soon as the pulses approach each other (up to the distance $\Delta x_{\rm cr} = f[u_1(0), u_2(0)]$), development of the oscillatory structure starts. The characteristic scale of this structure becomes smaller and smaller with increasing time (the wavelengths of the oscillations in the tails and between main maxima decrease and their amplitudes increase up to some limiting values depending on ε). At $t \sim 0.6$, the system can be described as having two bound oscillatory solitons with the constant amplitudes and distances between the pulses, i.e., a two-soliton bound state is formed acquiring (locally) the stationary character and moving as a whole with the time

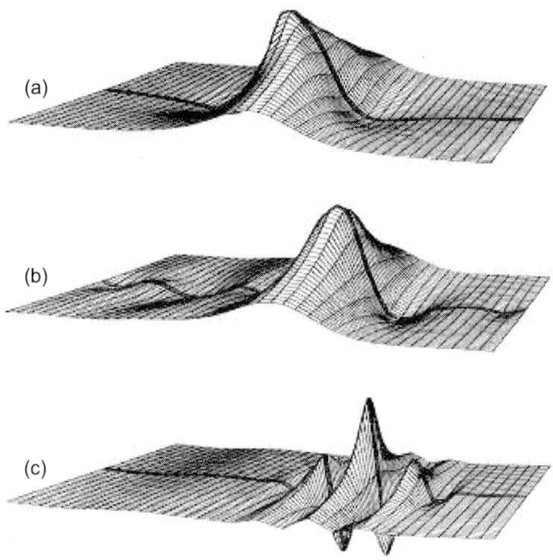

Fig. 4.18. Formation of the two-dimensional soliton with the oscillatory structure from the initial pulses with $u_1(0) = 2$, $u_2(0) = 1.3$, and $\Delta x(0) = 1.4$. **a** $t = 0$. **b** $t = 0.3$. **c** $t = 0.6$. The parameters of the GKP equation are: $\lambda = 1$, $\mu = 0$, and $\varepsilon = 1.9$.

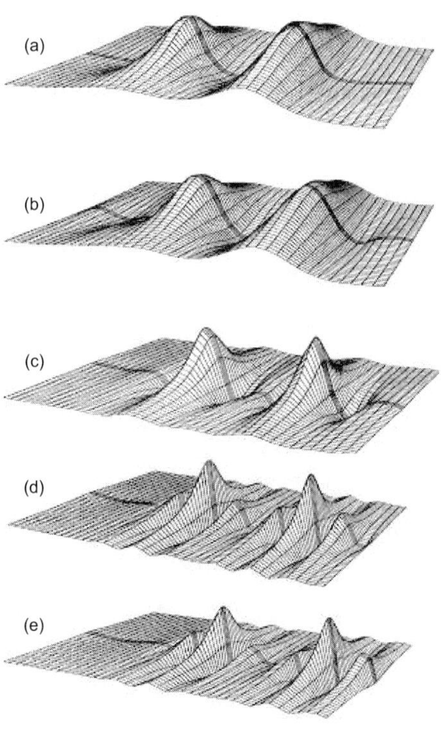

Fig. 4.19. Formation of the two-dimensional bisoliton from the initial pulses with $u_1(0) = 1.35$, $u_2(0) = 1.3$, $\Delta x(0) = 6$. The parameters of GKP equation are the same as in Fig. 4.18. **a** $t = 0$. **b** $t = 0.3$. **c** $t = 0.6$. **d** $t = 0.9$. **e** $t = 1.3$

t increasing further. Thus we observe the formation of a stationary *bisoliton* having two main maxima of equal amplitudes.[6] The bisoliton is symmetric with respect to its center and the distance between the main maxima is $\Delta x(t)$ =const. The structure of the bisoliton is similar to the bound oscillatory solutions found numerically in Ref. [80], and Fig. 4.19 illustrates the dynamics of formation of such a structure.

For all the cases considered (taking into account proper the compensation of "computational radiation" on the system's boundaries), we obtained conservation of the integrals $\mathcal{P}$ and $\mathcal{H}$ with sufficient accuracy ($\sim 5-7\%$ at $t \sim 1.0$). The one- and two-soliton structures are formed as a result of the soliton interaction, and the analysis of the Hamiltonian boundedness on the numerical solutions under the scale transformations (4.7) (see Sect. 4.1) shows that the structures are stable for the investigated range of values $|\varepsilon| \geq 0$ for $\lambda = 1$ (change of $\mathcal{H}$ under these deformations has the form similar to that presented by the curves 1–4 in Fig. 4.14), and the minima of the Hamiltonian, in agreement with the results obtained in Sect. 4.1, are realized on these stationary solutions.

4.4.3 Influence of Dissipation on Evolution of Two-Dimensional Solitons

Now consider the influence of dissipation on the evolution of two-dimensional solitons of (4.100). According to the accepted long-wavelength approach (see Introduction) for the KP-class equations, we restrict ourselves to the study of the role of dissipation of the "viscous" type ($\mu > 0$ in (4.100)) caused by the relaxation processes in a medium. In our numerical simulations we employ, as above, the initial conditions corresponding to the two-dimensional one-soliton solution of the KP equation (see Sect. 3.1), but with $n = 1, 2$; $m = 1, 2$; $l = 1$, and investigate the behavior of solutions of (4.100) with $\lambda = 0, \pm 1$ for various values of the coefficients ε and μ. Figure 4.20 shows the evolution of the initial two-dimensional KP soliton ($\lambda = 0$) in a medium with the viscous dissipation. We can see that even at quite early time moments t, the presence of the dissipation leads to a significant amplitude decrease so that $u(0.1) \cong 0.5u(0)$. The damping rate of the solitons for $\lambda = 1$ are of the same order as for $\lambda = 0$, and the soliton asymptotics for $\varepsilon \leq 0$ and $\mu > 0$ is close to the algebraic one (see Fig. 4.9)); generally, the soliton shape in this case practically does not differ from that shown in Fig. 4.20. Figure 4.21 shows an example of the numerical simulation for $\varepsilon > 0$ and the same initial conditions as in Fig. 4.20. Here we see that the proper account for the fifth derivative leads to the formation of the characteristic oscillatory structure in front of and behind the soliton's maximum in the process of the soliton's evolution. Moreover, for all the cases considered (including $\lambda = 0$) we observe the effect

[6] For the first time the bisoliton formation was observed by Belashov and Karpman in 1990 [62].

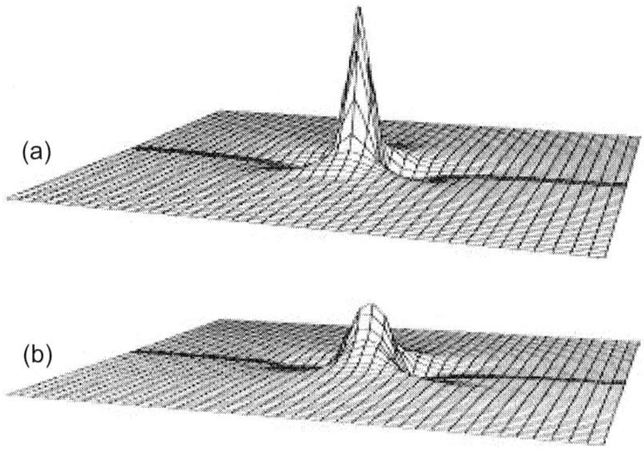

Fig. 4.20. Evolution of the two-dimensional soliton solution of (4.100) with $\lambda = 0$, $\varepsilon = -1$, and $\mu = 1$. **a** $t = 0$. **b** $t = 0.1$

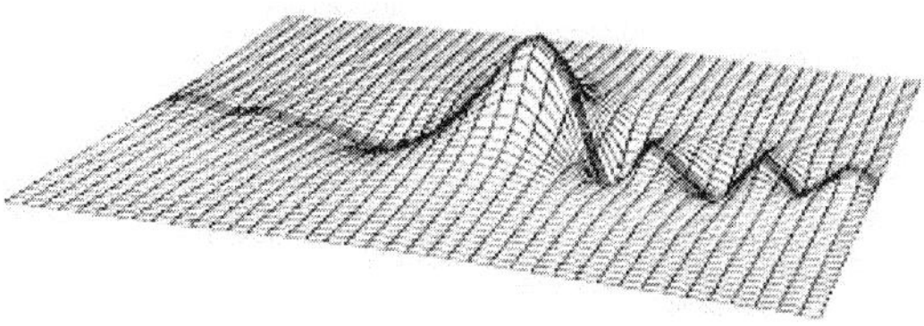

Fig. 4.21. Two–dimensional soliton solution of (4.100) with $\lambda = 1$, $\varepsilon = 0.8$, and $\mu = 1$ at $t = 0.2$

of "stretching" of the oscillatory soliton tail for $\lambda = 1$ and $\varepsilon > 0$ as well as for $\lambda = -1$ and $|\varepsilon| \geq 0$ accompanied by the decrease of the oscillations' frequency and damping of the oscillations behind the main maximum. We also note the associated asymmetrical change of integrals $\mathcal{P}$ and $\mathcal{H}$ in both the front and back "cavities" (where $u < 0$). Indeed, the slope of the "back part" of the soliton becomes less steep than that of the "front part," and the wave field symmetry is broken (this effect is qualitatively similar to that observed in the solutions of the Korteweg–de Vries–Burgers equation [3]). Thus the presence of the viscous-type dissipation in the system leads to (in addition to the trivial general damping of the wave field) the direct influence on the two-dimensional soliton's structure.

4.4.4 Evolution of Two-Dimensional Solitons in Media with Stochastic Fluctuations of the Wave Field

Introductory remarks. In Sect. 2.2.2 we studied evolution of the KdV soliton in media with stochastic fluctuations of the wave field. This problem is highly important since such fluctuations occurred at various scales and caused by numerous reasons,[7] regular as well as stochastic, are practically always present in reality, and their overall influence can be considered with a good approximation as that of stochastic fluctuations of the wave field. Here, we consider the results of the theoretical study of influence of these *stochastic fluctuations of the wave field* on the dynamics of the two-dimensional solitons.

[7] For example, in a plasma of the Earth's ionosphere, by the solar terminator, solar eclipses, man-made explosions, magnetic substorms, and volcano eruptions, etc.

To exclude from our investigation the influence of factors such as dissipation and various instabilities (their type is defined by the particular type of the medium considered), which can obscure the effects caused by the influence of the stochastic field fluctuations on the propagation of nonlinear waves, we consider the classic two-dimensional *KP equation*, accounting only for the term associated with the stochastic fluctuations [151]:

$$\partial_t u - 6u\partial_x u + \partial_x^3 u - \eta(t) = \kappa \int_{-\infty}^{x} \partial_y^2 u \mathrm{d}x, \qquad \kappa = \pm 1 . \tag{4.106}$$

As it is already known from Chap. 3, equation (4.106) with $\eta(t) = 0$ describes evolution of nonlinear waves and solitons in various dispersive media. As in Sect. 2.2, we investigate here the influence of the stochastic fluctuations (described by the external random noise $\eta(t)$) on the two-dimensional KP soliton since for $\eta(t) = 0$ it is a stable structure propagating in the medium without changing its shape. The approach we develop here is sufficiently general, similar to the one-dimensional case.

The inhomogeneous term $\eta(t)$ in (4.106) (similar to (2.93)) describes the external noise when the characteristic width of the soliton, l_s, is much smaller than the coherence length of the noise, l_n. This corresponds to a particular situation of a more general case when the external noise is described by the term $\eta(x, y, t)$. The assumed approximation, being simpler for analytical investigation, allows us to obtain an exact result and provides us useful information for the more general study when other relations between l_s and l_n hold.

In Sect. 2.2 we demonstrated that the *stochastic KdV equation* (see (4.106) with $\kappa = 0$) has a solution in the form of the KdV soliton which, due to the effect of the external noise, is deformed during the propagation. The width and the amplitude of the soliton are proportional to $t^{3/2}$ and $t^{-3/2}$, respectively, when $t \to \infty$. The structure and the dynamics of the KP solitons are still different from those of the KdV solitons, however. Therefore, the study of model (4.106) is of interest not only from the point of view of its applications to processes in particular physical media, but also because of the general theoretical aspect, in the sense of comparing the dynamic behavior of the KP solitons in media with the low-frequency stochastic fluctuations with that of the KdV solitons. This problem was first set up and solved analytically and numerically in Refs. [151,201].

General Approach. Here, we follow the approach used in the one-dimensional case (see Sect. 2.2). First, we note that (4.106) is related to the *KP equation*,

$$\partial_t U - 6U\partial_\xi U + \partial_\xi^3 U = \kappa \int_{-\infty}^{\xi} \partial_y^2 U \mathrm{d}\xi, \tag{4.107}$$

via the *Galilean transform*

$$u(t,x,y) = U(t,\xi,y) + W(t), \tag{4.108}$$

where

$$W(t) = \int_0^t \eta(t)\mathrm{d}t,$$

$$\xi = x + m(t),$$

and

$$m(t) = 6\int_0^t W(t)\mathrm{d}t.$$

Thus, equation (4.106) is integrable and can therefore be integrated using the *IST method* (see Sect. 3.2). Following the consideration of Sect. 2.2 we also assume that the external noise $\eta(t)$ is Gaussian,

$$\langle \eta(t_1)\eta(t_2)\ldots\eta(t_n)\rangle = \begin{array}{ll} 0 & n \text{ is odd,} \\ \sum \Pi\, \langle \eta(t_i)\eta(t_j)\rangle & n \text{ is even,} \end{array} \tag{4.109}$$

and white, $\langle \eta(t)\eta(t')\rangle = 2\varepsilon\delta(t-t')$. In this case formulas (2.73) are also valid [151]. To account for the possibility of both (positive and negative) signs of κ, we consider the problem in the most general way. Let the functional of $U(t,\xi,y)$ be

$$F\left[U(t,\xi,y)\right] = F\left[U(t,\xi,y), \partial_\xi U(t,\xi,y), \ldots\right] = F(t,\xi,y). \tag{4.110}$$

We consider the Fourier transform,

$$\begin{aligned} F(t,\xi,y) &= \frac{1}{2\pi}\int_{-\infty}^{\infty} \mathrm{d}k\tilde{F}\left(t,k,y\right)\mathrm{e}^{\mathrm{i}kx}, \\ \tilde{F}(t,k,y) &= \int_{-\infty}^{\infty} \mathrm{d}xF\left(t,\xi,y\right)\mathrm{e}^{-\mathrm{i}kx}, \end{aligned} \tag{4.111}$$

and obtain [151]

$$\tilde{F}\left(t,k,y\right) = \tilde{F}_0\left(t,k,y\right)\mathrm{e}^{\mathrm{i}km(t)}, \tag{4.112}$$

where

$$\tilde{F}_0\left(t,k,y\right) = \tilde{F}\left(t,k,y\right)\Big|_{m=0} = \int_{-\infty}^{\infty} \mathrm{d}xF\left(t,x,y\right)\mathrm{e}^{-\mathrm{i}kx}. \tag{4.113}$$

The expression for the statistical average is similar to (2.104)–(2.105):

$$\left\langle \tilde{F}\left(t,k,y\right)\right\rangle = \tilde{F}_0\left(t,k,y\right)\tilde{G}(k), \tag{4.114}$$

where for $\tilde{G}(k) = \langle \exp[ikm(t)] \rangle$, and using (4.109) we find

$$\tilde{G}(k) = \mathrm{e}^{-k^2 \langle m^2(t) \rangle /2} \quad \text{and} \quad \langle m^2(t) \rangle = 24\varepsilon t^3 \quad \text{for} \quad t > 0. \tag{4.115}$$

Equation (4.114) shows us that the averaged spectrum (4.112) of the functional $F[U(t,\xi,y)]$ is the product of $\tilde{F}(t,k,y)$, without noise (4.113), and the Gaussian distribution (4.115). Thus, similar to the one-dimensional case, we have

$$\langle F[U(t,\xi,y)] \rangle = \langle F(t,\xi,y) \rangle = \frac{1}{2\pi} \int_{-\infty}^{\infty} \mathrm{d}k \tilde{F}_0(t,k,y)\, \tilde{G}(k) \mathrm{e}^{\mathrm{i}kx}. \tag{4.116}$$

In analogy to the result of the one-dimensional consideration of Sect. 2.2, the convolution theorem allows us to write (4.116) as

$$\langle F[U(t,\xi,y)] \rangle = \int_{-\infty}^{\infty} \mathrm{d}s F[U(t,s,y)]\, G(x-s), \tag{4.117}$$

where $G(x-s)$ is defined by the corresponding expression in Sect. 2.2.2, i.e.,

$$G(s) = \frac{1}{2\pi} \int_{-\infty}^{\infty} \mathrm{d}k \tilde{G}(k) \mathrm{e}^{\mathrm{i}ks} = \frac{\exp\left[-s^2/2 \langle m^2(t) \rangle\right]}{\left[2\pi \langle m^2(t) \rangle\right]^{1/2}}.$$

Expressions (4.116) and (4.117) are now suitable for further investigation of the dynamic behavior of the solitons of (4.106) for different signs of the coefficient κ.

Dynamics of KP Solitons. As an example, we consider here the case when $F[U(t,\xi,y)]$ (4.110) is the functional of the one-soliton solution of (4.102) with $\kappa = 1$. Then, for the one-soliton solution of (4.107) with $\kappa = 1$ (see also Sect. 3.1), we have

$$U(t,\xi,y) = -2\partial_\xi^2 \ln\left[4(\nu+\nu^*)^{-2} + \left|\xi + \varphi + \mathrm{i}\nu y + 3\nu^2 t\right|^2\right]. \tag{4.118}$$

For simplicity of the analysis, let us assume that $\nu = \mathrm{Re}(\nu)$ and $\varphi = 0$ in (4.118). We then obtain from (4.113) that

$$\tilde{F}_0(t,k,y) = 4\pi k \exp\left(3\mathrm{i}k\nu^2 t\right) \sinh(kz), \tag{4.119}$$

where $z = \nu^{-1}\left(1+\nu^4 y^2\right)^{1/2}$. Integrating the right-hand side of (4.116) and taking into account (4.119), we obtain

$$\langle u(0,x,y) \rangle = -4\nu^2 \left(1+\nu^4 y^2 - \nu^2 x^2\right)\left(1+\nu^4 y^2 + \nu^2 x^2\right)^{-2} \tag{4.120}$$

for $t = 0$. This expression is exactly equal to the solution (4.118) with $\nu = \mathrm{Re}(\nu)$ and $\varphi = 0$ at $t = 0$. For large $t > 0$ we obtain the following solution [151]:

$$\begin{aligned}\langle u(t,x,y)\rangle &= \sqrt{\pi}g^{-1}\left[A\exp(A^2)+A^*\exp((A^*)^2)\right] \\ &= 2\sqrt{\pi}g^{-1}\mathrm{e}^{r^2-s^2}\left[r\cos(2rs)-s\sin(2rs)\right], \qquad (4.121)\end{aligned}$$

where $A=(z+\mathrm{i}a)/2\sqrt{g}$, $g=12\varepsilon t^3$, $a=x+3\nu^2 t$, $r=\mathrm{Re}(A)$, and $s=\mathrm{Im}(A)$. We can see from expression (4.121) that due to the effect of the stochastic fluctuations of the wave field, the two-dimensional soliton is deformed in the process of its propagation; moreover, the amplitude of its maximum changes asymptotically as $u\sim t^{-9/2}$, and the soliton acquires the wave-like oscillating structure along the x- and y-directions (see also numerical results and Figs. 4.22 and 4.23 below).[8] Thus, the dynamic behavior of a two-dimensional soliton as well as its structure differ essentially from those of the one-dimensional KdV soliton. However, while the KP soliton is deformed, just like the KdV soliton, the area occupied by the soliton is invariant. In fact, we have from equation (4.121)

$$\iint \langle u(t,x,y)\rangle\, \mathrm{d}x\mathrm{d}y = 0. \qquad (4.122)$$

To conclude, we note that (4.116) and (4.117) can also be used for investigation of the dynamic behavior of the one-dimensional KP solitons in media with the low frequency stochastic fluctuations, i.e., when $\kappa=-1$ in (4.106). For the one-dimensional KP soliton we have [24]

$$u(t,x,y) = -\frac{1}{2}p^2\cosh^{-2}\left[p(c-x)/2\right], \qquad (4.123)$$

where $c=\nu y-4\nu^2 t+\ln(q)+(p+\nu)(p-\nu)t$; p, q, and ν are constants; and instead of equation (4.119) we have

$$\tilde{F}_0(t,k,y) = 2\pi k\mathrm{e}^{\mathrm{i}ck}/\sinh(k\pi/p). \qquad (4.124)$$

Using (4.116) we can further investigate the dynamics of such a soliton in media with external noise, in analogy to the two-dimensional soliton (4.118).

Numerical results. Here, we complete the above analytical study of the dynamics of the KP solitons in media with low-frequency stochastic fluctuations using numerical investigation of the soliton solutions of (4.106) for different levels of the noise intensity [151]. For the numerical integration we employ the already used implicit difference scheme with the $O(\tau^2, h^4_{x,y})$ approximation (see Sect. 4.3 as well as Refs. [83,98]) and the grid 301×51, $\tau=0.0025$, $h_x=0.1$, and $h_y=0.3$. The complete absorption on the boundaries of the computation region is imposed. Initial conditions are taken in the form of the exact soliton solution of KP equation (4.118) for $t=0$. To control the dynamic characteristics of the soliton-noise system, we compute the integral

[8] Note that in Ref. [202], where exactly our setting of the problem and our approach to its solution was used, incorrect results (because of erratic analysis) were obtained regarding the asymptotic behavior of the KP solitons (in particular, $u\sim t^{-3}$). The results of numerical modeling fully agree with our estimations.

(4.122) as well as the integrals which are the dynamic invariants of equation (4.107) with the function $U(t, \xi, y)$ at each time step [24]:

$$\begin{aligned} \mathcal{P}_x &= \frac{1}{2}\iint u^2 \mathrm{d}x\mathrm{d}y, \\ \mathcal{H} &= \iint \left[\frac{1}{2}\left(\partial_x u\right)^2 + \frac{1}{2}\kappa\left(\partial_y w\right)^2 - u^3\right] \mathrm{d}x\mathrm{d}y, \end{aligned} \tag{4.125}$$

where $\partial_x w = u$. The numerical results are presented in Figs. 4.22 and 4.23. Note that these results correspond well to the analytical properties of the soliton solutions obtained above. Indeed, we see that the soliton structure

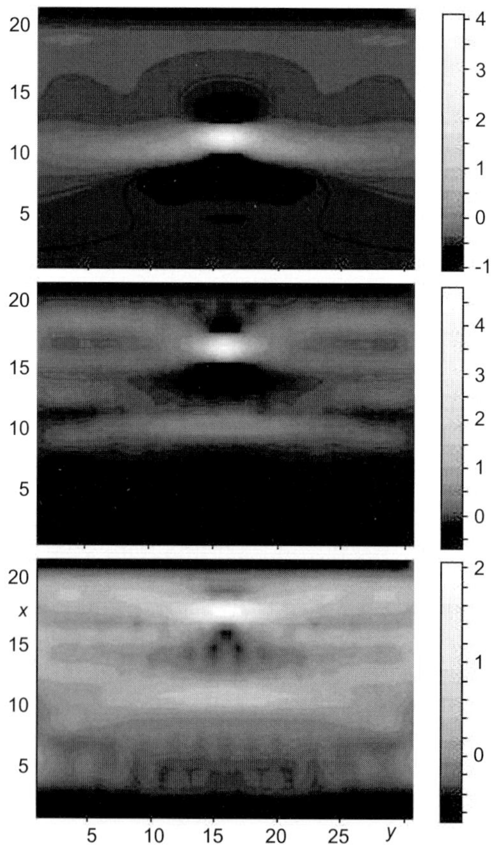

Fig. 4.22. Evolution of a two-dimensional soliton of (4.106) with $\kappa = 1$ at $t = 0.4$, $t = 1.2$, and $t = 2.0$, with the Gaussian noise for the standard deviation $\sigma = 0.01$

along the x- and y-axes acquires the wave-like oscillating character with time. Moreover, the soliton amplitude and the oscillation frequency first increase with the simultaneous (oscillating) decrease of the characteristic lengths of the soliton up to $t \sim 1$, and then the amplitude and frequency start to

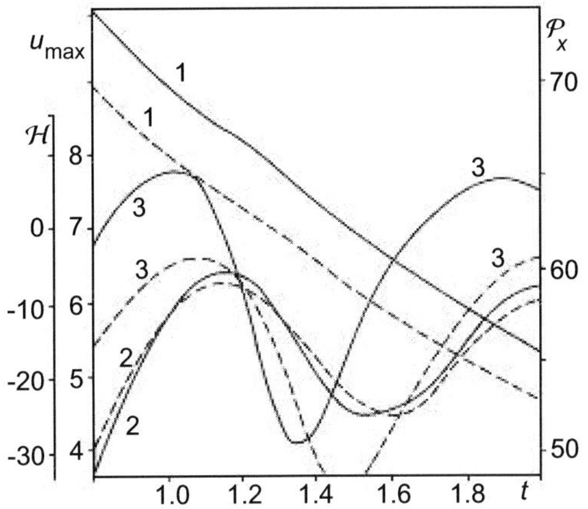

Fig. 4.23. Variations of the fluctuation-averaged values of the amplitude, $u_{\max}$, and the integrals $\mathcal{P}_x$ and $\mathcal{H}$ for the two-dimensional soliton of (4.106) in the process of its propagation in media with the low-frequency stochastic fluctuations for $\sigma = 0.01$ (solid curves) and $\sigma = 0.02$ (dashed curves): (1) $u_{\max}$, (2) $\mathcal{P}_x$, (3) $\mathcal{H}$

decrease according to our analytical prediction (4.120). The variations of $\mathcal{P}_x$ and $\mathcal{H}$ characterize the variations of the momentum and energy of the soliton–noise system comparable with those of the "pure" KP equation when $\mathcal{P}_x$ and $\mathcal{H}$ are constant. These variations are of a quasi-periodic character, i.e., they are related to the the quasi-periodic dependence of l_s on time.

Thus, the numerical results confirm that the structure and the dynamic behavior of the two-dimensional solitons in media with the low-frequency stochastic fluctuations significantly differ from those in the one-dimensional case (i.e., for the KdV solitons). Similar differences are observed for the KP soliton with $\kappa = -1$. In this case, the solitons also acquire the wave-like oscillating structure in the y-direction for large t. To conclude, we note that for a non-white noise, the form of the solutions can in principle also be obtained by the use of the above presented method but it is more complicated in this case. In particular, if the characteristic soliton lengths are comparable to the coherent length of the noise ($l_s \sim l_n$), the *Galilean transform* (4.108) is not correct, and it is necessary to generalize the *IST method* for this case accordingly, for example, as was the case for the KdV equation in Refs. [95,96]. Finally, we note that the results presented above were first reported at the International Conference on Nonlinear Evolution Equation and Dynamical Systems in 1992 (NEEDS'92).

4.4.5 Structure and Evolution of Two-Dimensional Solitons in Media with Variable Dispersion

Here, we consider the dynamics of two-dimensional solitons described by the *KP equation* written in the form

$$\partial_t u + \alpha u \partial_x u + \beta \partial_x^3 u = \kappa \int_{-\infty}^{x} \partial_y^2 u \mathrm{d}x, \tag{4.126}$$

in a medium with the changing in time and/or space dispersion characteristics, namely $\beta = \beta(t, x, y)$. This problem is interesting in view of its obvious applications in the physics of some particular media with dispersion. For example, this situation can be observed in the propagation of *gravity waves* and *gravity-capillary waves* on the surface of *shallow water* [203] when the factor β is given by the expressions $\beta = c_0 H^2/6$ and $\beta = (c_0/6)\left[H^2 - 3\sigma/\rho g\right]$, respectively, where H is the depth, ρ is the density, and σ is the surface tension of the fluid. In these cases, if $H = H(t, x, y)$ (i.e., if there is the change of the bottom depth in space and/or temporal depth "shifts" on the bottom regions), the dispersion parameter β is also a function of the space coordinates and time (note that Sect. 4.6.4 below is specially devoted to this particular application). A similar situation can also take place when studying the evolution of the *FMS waves* in a *magnetized plasma* [203,204] when the factor β in the three-dimensional case is the function of both the Alfvén velocity $v_A = f[B(t, \mathbf{r}), n(t, \mathbf{r})]$ (where n is the plasma density) and the angle $\theta = (\mathbf{k} \wedge \mathbf{B})$ between the wave vector and the magnetic field:

$$\beta = v_A \frac{c^2}{2\omega_{pi}^2} \left(\cot^2 \theta - \frac{m_e}{m_i} \right), \tag{4.127}$$

where ω_{pi} is the ion plasma frequency (see also Sect. 4.6). Obviously, $\beta = \beta(t, \mathbf{r})$ corresponds to the cases of a non-uniform and/or non-stationary plasma and/or magnetic field.

The soliton dynamics in a medium with the variable dispersion of the type $\beta = \beta(t, x)$ was studied in Refs. [203,204] for the one-dimensional model of the KdV solitons. It is well known that solutions of the KdV equation with β =const are divided into two classes depending on the value of the dispersion term: for $|\beta| < u(0, x)l/12$ (where l is the characteristic wavelength of the initial disturbance) they reveal the soliton character, and in the opposite case they are wave packets with the asymptotics proportional to the x-derivative of the Airy function [3]. In both cases, the KdV equation can be integrated analytically by the inverse scattering transform. This approach cannot be used in principle for an arbitrary dependence $\beta = \beta(t, x)$, however, and solving such a problem requires numerical integration. An analogous situation takes place for the non-one-dimensional model described by the KP equation. While the analytical solutions of the KP equation for β =const in the two-dimensional case are well known (see the corresponding expressions in Sect. 3.1), the dispersion term of this equation becomes quasi-linear in the case $\beta = \beta(t, x, y)$ and the model is not exactly integrable (i.e., the inverse scattering transform is not applicable). Below we formulate the problem of numerical integration of the KP equation with $\beta = \beta(t, x, y)$ and consider the main results of the numerical experiments in the investigation of the structure and evolution of two-dimensional solitary waves in a medium with the variable dispersion (the only assumption used is that the medium is described by the *dispersion law* of the type $\omega \approx c_0 k_x \left(1 + k_\perp^2/2k_x^2 + D^2 k_x^2\right)$, where D is the dispersion "length").

To solve the initial problem for the KP equation (4.126) with the variable dispersion, we use the implicit difference scheme (4.75) with the $O(\tau^2, h^4_{x,y})$ approximation. As usual, the initial conditions are chosen in the form of the exact two-dimensional one-soliton solution of the KP equation (see (3.15) in Sect. 3.1) with various values of the parameters ν and ξ, mostly $\nu \neq \mathrm{Re}\nu$ and $\xi \neq 0$ at $t = 0$), the total absorption on the boundaries of the computation region (4.84) is imposed, and the numerical simulations are done for a few types of the model function β. For $t < t_{\mathrm{cr}}$ we assume $\beta = \beta_0 = \mathrm{const}$, and for $t > t_{\mathrm{cr}}$ we either have

$$(1) \qquad \beta(x) = \begin{cases} \beta_0, & x \leq a, \\ \beta_0 + c, & x > a; \end{cases} \tag{4.128}$$

$$(2) \qquad \beta(x,t) = \begin{cases} \beta_0, & x \leq a, \\ \beta_0 + nc, \quad n = (t - t_{\mathrm{cr}})/\tau = 1, 2, ..., & x > a; \end{cases} \tag{4.129}$$

or

$$(3) \qquad \begin{aligned} \beta(t) = \beta_0 \left(1 + k_0 \bar{\beta} \sin \omega t\right), \quad & \bar{\beta} = \left(\beta_{\max} - \beta_{\min}\right)/2, \\ 0 < k_0 < 1, \quad & \pi/2\tau < \omega < 2\pi/\tau. \end{aligned} \tag{4.130}$$

The terms a and c in the above equations are constants. Note that in terms of the problem of the wave propagation on the surface of *shallow water* this means that when reaching t_{cr} we either have: (1) the sudden “breaking up” of the bottom, (2) the gradual “change of the depth” of the bottom, or (3) the “oscillation” of the bottom’s depth with time. More detailed consideration of this application can be found in Sect. 4.6.4.

The first series of numerical experiments is aimed at studying the soliton dynamics with the spasmodic character of the wave dispersion (i.e., $\beta(t, x, y)$ has the form of the step). First, we investigate evolution of the initial pulse for the case when at t_{cr} the step-like change of β is behind the soliton (the “negative” step when $c < 0$ in (4.128) and (4.129)). In this case, the dependence of the spatial structure of the solution on the parameter a in the models (4.128) and (4.129) is studied. The obtained results (see an example in Fig. 4.24) demonstrate that for all cases the evolution of the initial soliton leads to the formation of the wave-like oscillating tail which is not related to the soliton moving away ahead, and is caused only by the local influence of the sudden change of the relief $\beta(t, x)$. Consequently, formation of the oscillatory structure is related mostly to the step-like change of β in space, not the (decreased) role of the dispersion effects behind the soliton.

In the next series of our numerical experiments we consider evolution of the two-dimensional soliton in the case when the sudden change of the dispersion factor takes place directly under the front or in front of the initial pulse (the “negative” step). An example of the results of those simulations is shown in Fig. 4.25. Analysis of the results of the whole series demonstrates that for a character with the relief of the function β, the perturbation caused by the sudden change of the dispersion factor also reveals the local character,

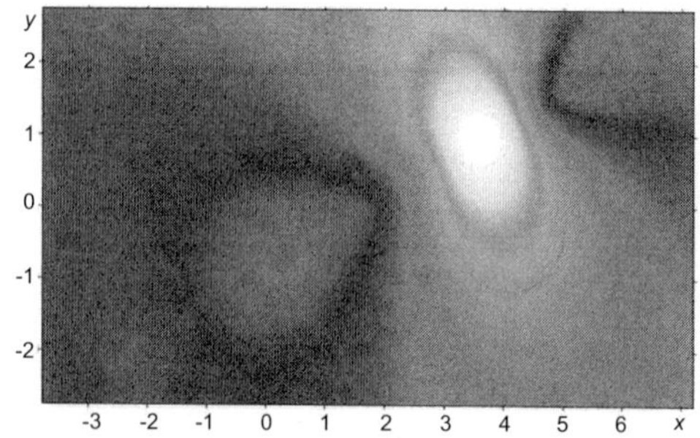

Fig. 4.24. Evolution of a two-dimensional soliton of (4.126) for the dispersion law (4.128) for $a = 5.0$, $c = -0.0038$, and $t = 0.6$

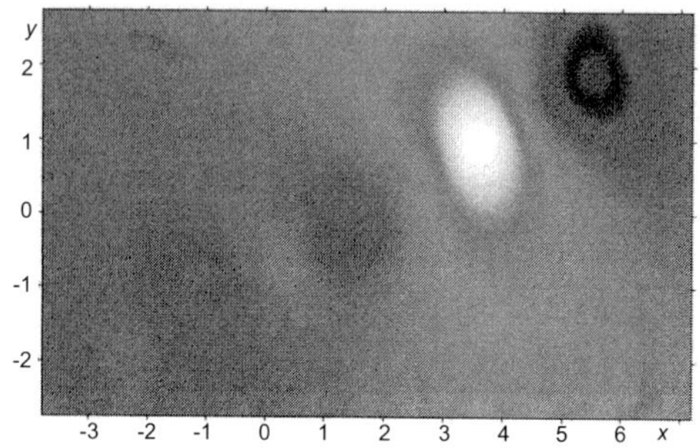

Fig. 4.25. Evolution of a two-dimensional soliton of (4.126) for the dispersion character (4.129) for $a = 4.0$, $c = -0.0038$, and $t = 0.6$

i.e., it does not propagate together with the (moving away) soliton. But, unlike the cases considered in the first series of experiments, the asymptotics of the soliton moving away becomes oscillating (within the time limits of the numerical experiment); besides, on the background of the long-wave tail oscillations we can also note the appearance of the wave fluctuations. These effects can be attributed to the fact that in the regions on the wave surface with the different (local) wave numbers k_x, the factors describing dispersion effects are also different. As a result, the intensity of the phase mixing of the Fourier-harmonics within the $x - y$-region varies with the coordinates and, therefore, it reacts differently to the nonlinear generation of the harmonics with various (in particular, large) wavenumbers k_x.

In the third series of the numerical experiments the dispersion parameter was also changed according to (4.128) and (4.129), but we consider here the cases when the "positive" step ($c > 0$ in (4.128) and (4.129)) is both in front of and behind the initial pulse for a wide range of values of the parameter a. The most interesting example of the results is shown in Fig. 4.26. We can see that when the "positive" step is far in front of the maximum of the function $u(0, x, y)$, the soliton evolution on the initial time steps does not differ qualitatively from that for β =const (see Fig. 4.26a), but for larger time scales, the evolution character becomes defined by the presence of the step. Processes caused by the same effects as noted above for the results of the second series of the simulations, start to develop. As we can see in Fig. 4.26b, a significant change of the soliton structure which can in principle lead to the

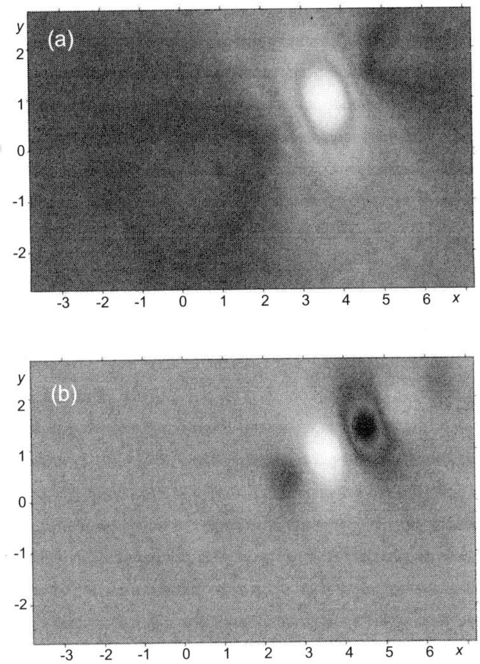

Fig. 4.26. Evolution of a two-dimensional soliton of (4.126) for the dispersion character (4.129) for $a = 5.0$, $c = 0.0038$. **a** $t = 0.6$. **b** $t = 0.8$

wave breaking is observed because of intensive generation of the harmonics with high k_x in the soliton front region. The process takes place even for relatively small steps (i.e. when the parameter c in (4.128) and (4.129) is small). Thus, as it follows from the results of the series, the perturbation of the propagating two-dimensional soliton caused by the sudden change in time and space of the dispersion factor β with $c > 0$ also reveals the local character.

As for the last case of the change of the factor β (the harmonic oscillations of the parameter with time on the whole $x-y$-plane, see (4.130)), a series of numerical experiments for various fixed k_0 =const and the variable frequency ω has demonstrated that for some ω the (locally) stationary standing waves can be formed. For other cases formation of the stationary periodical wave structures is possible, and in the intermediate case, the chaotic regime is usually realized. In the numerical simulations done for various k_0 and fixed ω =const, we found that the stable (within the computation time) solutions can be formed only for $k_0 \leq \beta_0$ in (4.130), and the solutions are unstable for all other cases. Summarizing the above results, we note that numerical simulations of the evolution of the two-dimensional solitons described by the model KP equation with $\beta = \beta(t, x, y)$ allowed us to find various types of stable and unstable solutions, including the solution of the mixed "soliton–non-soliton" type for various dependencies of the change of the dispersion in time and space. The results obtained open new perspectives for investigation of a number of applied problems in the dynamics of the non-one-dimensional nonlinear waves in particular physical media.

4.5 Evolution of Three-Dimensional Nonlinear Waves in Dispersive Media

Here, we numerically investigate the structure and evolution of the three-dimensional solutions of the *GKP equation* written in the form

$$\partial_x \left(\partial_t u + 6u\partial_x u - \mu\partial_x^2 u - \varepsilon\partial_x^3 u - \lambda\partial_x^5 u\right) = \Delta_\perp u, \tag{4.131}$$

with the corresponding factors $\lambda = \pm 1$, $|\varepsilon| \geq 0$, and $\mu \geq 0$ (Sect. 4.5.1 and Sect. 4.5.3). We also study the *3-DNLS equation*

$$\partial_t h + s\partial_x \left(|h|^2 h\right) - \mathrm{i}\lambda\partial_x^2 h - \nu\partial_x^2 h = \kappa \int\limits_{-\infty}^{x} \Delta_\perp h \mathrm{d}x, \tag{4.132}$$

with $\lambda = \pm 1$, $s = \pm 1$, and $|\kappa| > 0$ (Sects. 4.5.2 and 4.5.3) in the axially symmetric geometry, i.e., when $\Delta_\perp = \partial_\rho^2 + \partial_\rho/\rho$, $\rho^2 = y^2 + z^2$. For the numerical integration we use the methods presented above in Sect. 4.3 (they are developed in Refs. [81,83,98,148,195,198]). In Sects. 4.5.1 and 4.5.2 we present results of the numerical study of the structure of the solutions, estimate their stability, and consider the dynamics of the evolution of the three-dimensional axially symmetric pulses in the respective GKP and 3-DNLS models for $\mu = 0$ and $\nu = 0$ (see also Refs. [65-70,148,198]). Finally, in Sect. 4.5.3 we present numerical results of the investigation of the influence of a viscous-type dissipation (the GKP and 3-DNLS models with $\mu > 0$ and $\nu > 0$, respectively) on the evolution of the obtained three-dimensional solutions [195,198].

4.5.1 Structure and Evolution of Three-Dimensional Solutions of GKP-Class Equations

For the numerical investigation of (4.131), we can write it in the mixed integral-differential form (see (4.59) of Sect. 4.3) and integrate using the implicit difference schemes (4.75) and (4.83) on the grid 301×51, $\tau = 0.0025$, $h_x = 0.1$, $h_\rho = 0.3$ with the total absorption on the boundaries of the computation region, taking into account the condition $\partial_\rho u|_{\rho=0} = 0$. We impose the initial condition in the form of the axially symmetric solitary pulse,

$$u(0, x, \rho) = 2\partial_x^2 \ln \left[4\left(\nu + \nu^*\right)^{-2} + |x - \mathrm{i}\nu\rho - \xi|^2\right], \tag{4.133}$$

for various values of the parameters ν and ξ defining amplitudes, phases, velocities, and other characteristics of the initial pulses. To control the changes of the momentum and energy of the solutions of (4.131), we use the expressions

$$\mathcal{P} = \frac{1}{2}\int u^2 \mathrm{d}\mathbf{r},$$
$$\mathcal{H} = \int \left[-\frac{\varepsilon}{2}(\partial_x u)^2 + \frac{\lambda}{2}\left(\partial_x^2 u\right)^2 + \frac{1}{2}(\nabla_\perp \partial_x u)^2 - u^3 \right] \mathrm{d}\mathbf{r}, \qquad (4.134)$$

where $\mathbf{r}^2 = x^2 + \boldsymbol{\rho}^2$.

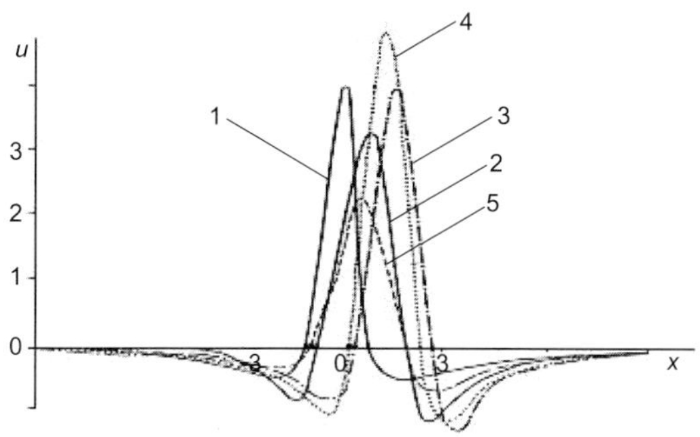

Fig. 4.27. Solutions of (4.131) at $\rho = 0$ for $\lambda = 1$: (1) initial pulse at $t = 0$; (2) $\varepsilon = -1.6$, $\mu = 0$ $(t = 0.25)$; (3) $\varepsilon = -1.6$, $\mu = 0$ $(t = 0.5)$; (4) $\varepsilon = 0$, $\mu = 0$ $(t = 0.6)$; (5) $\varepsilon = -1.6$, $\mu = 1$ $(t = 0.25)$

First, consider numerical solutions for Cauchy problem of (4.131) with $\lambda = 1$. In the case $\mu = 0$ for $\varepsilon \le 0$ and $\varepsilon > 0$, the axially symmetric soliton-like pulses with algebraic (Fig. 4.27, curves 1-4 and Fig. 4.28) and oscillating in the x-direction asymptotics (Fig. 4.29, curves 1 and 2) are formed on the first stage of the nonlinear evolution. We note that the pulse amplitude decreases slightly and small-scale perturbations, arising in the process of formation from the initial pulse (4.133) of a three-dimensional solution satisfying (4.131), are moving out from the integration region in the direction of negative x.

Unlike the two-dimensional solitons considered above in Sect. 4.4, the structures forming in this case are asymmetric along the x-axis: the slope of the tail becomes smaller and for $\varepsilon > 0$ the tail oscillations have the large amplitudes as well as show the more regular character. The dependence on ε of the structure of solutions on the initial stage with $\lambda = 1$ and $\mu = 0$ corresponds to that of the two-dimensional version of (4.131) which we considered above.

Furthermore, the amplitude of pulse increases with time t (see Fig. 4.30, curves 1–3, 6, 7), with the rate $\Gamma = (1/2W)dW/dt \sim 2$ (here, $W = \bar{u^2}/4\pi$ is the wave energy density) which is weakly dependent on ε. In this case, for both $\varepsilon \le 0$ and $\varepsilon > 0$, the characteristic size of the pulse, defined by expressions

$$l_x = x\left(u_{\min 2}\right)\big|_{\rho=0} - x\left(u_{\min 1}\right)\big|_{\rho=0},$$
$$u_{\min 1} = u_{\min}\big|_{x > x(u_{\max})},$$

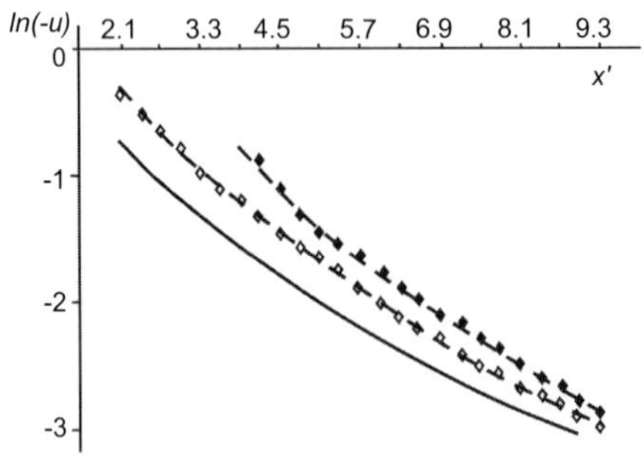

Fig. 4.28. Asymptotics of numerical solutions of (4.131) with $\lambda = 1$, $\varepsilon = -1.58$: the diamonds correspond to $\mu = 0$; the filled diamonds correspond to $\mu = 1.26$; the solid line presents asymptotics of the algebraic KP soliton; $x' = x - Vt$

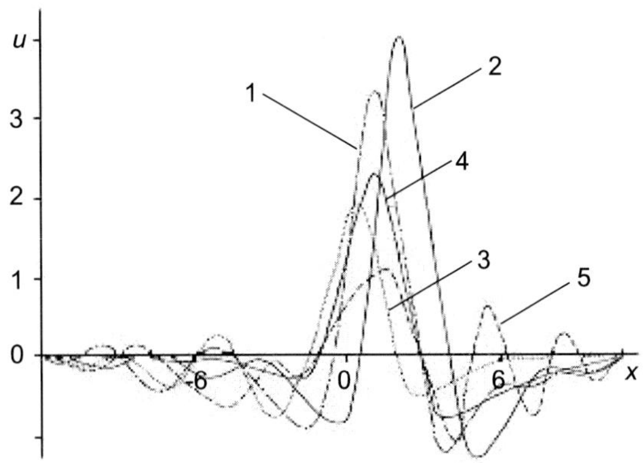

Fig. 4.29. Solutions of (4.131) at $\rho = 0$: (1) $\lambda = 1$, $\varepsilon = 1.6$, and $\mu = 0$ ($t = 0.25$); (2) $\lambda = 1$, $\varepsilon = 1.6$, and $\mu = 0$ ($t = 0.5$); (3) $\lambda = 1$, $\varepsilon = 1$, and $\mu = 1$ ($t = 0.2$); (4) $\lambda = 1$, $\varepsilon = 1.6$, and $\mu = 1.26$ ($t = 0.25$); (5) $\lambda = -1$, $\varepsilon = 1$, $\mu = 1$ ($t = 0.2$)

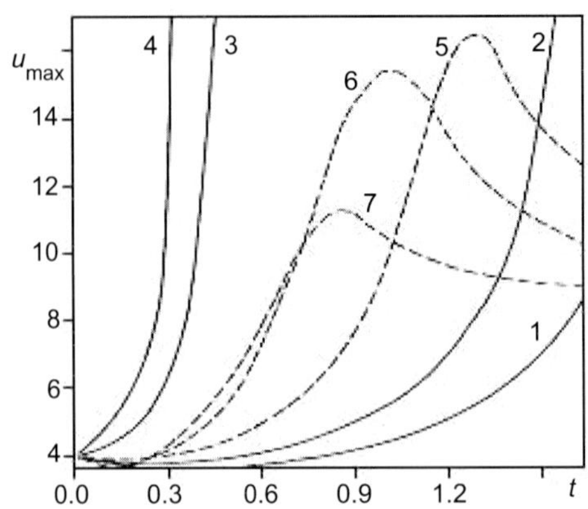

Fig. 4.30. Pulse's maxima versus time for $\lambda = 0$ and $\lambda = 1$ (the solid lines correspond to $\varepsilon < 0$, the dashed lines - to $\varepsilon \geq 0$): (1) $\lambda = 1$, $\varepsilon = -1.34$; (2) $\lambda = 1$, $\varepsilon = -0.89$; (3) $\lambda = 0$, $\varepsilon = -1.34$; (4) $\lambda = 0$, $\varepsilon = -1$; (5) $\lambda = 1$, $\varepsilon = 0$; (6) $\lambda = 1$, $\varepsilon = 1.34$; (7) $\lambda = 1$, $\varepsilon = 2.24$

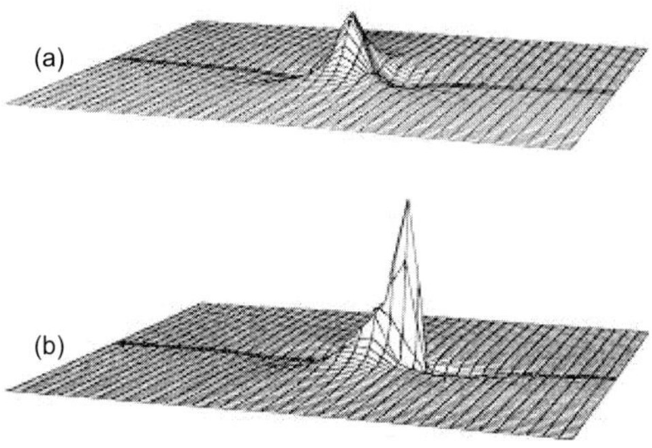

Fig. 4.31. Evolution of the three-dimensional axially-symmetric pulse in the $x-\rho$-plane on the self-focusing stage for $\lambda = 1$ and $\varepsilon = -0.89$. **a** $t = 0$. **b** $t = 1.4$. Here, we have $-13.5 \leq x \leq 13.5$, $-11.25 \leq \rho \leq 11.25$

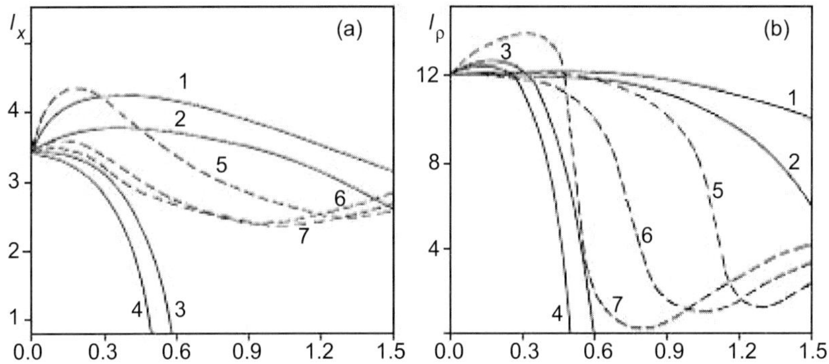

Fig. 4.32. Change with time of the characteristic sizes of the axially symmetric pulses. **a** l_x. **b** l_ρ. The solid and dashed curves correspond to the same cases as in Fig. 4.30

$$u_{\min 2} = u_{\min}\big|_{x<x(u_{\max})},$$

and

$$l_\rho = 2\rho\big|_{u=0.025u_{\max},x=x(u_{\max})},$$

decreases, the pulse "wings" fall behind its central part, and an instability of the *self-focusing* type is developed (see Figs. 4.31 and 4.32, curves 1–3, 6, 7 and Fig. 4.34a). This type of evolution is also characterized by the increase of the momentum $\mathcal{P}$ as well as the decrease (for $\varepsilon \leq 0$ as well as for not very large values $\varepsilon > 0$) of the Hamiltonian $\mathcal{H}$ of the system (Fig. 4.33, curves 1–3, 6, 7) at the expense of the nonlinear term which increases (for this particular time interval of the pulse evolution) faster than the dispersive terms. The difference of the pulse evolution for $\varepsilon > 0$ from its evolution for $\varepsilon \leq 0$ ($\lambda = 1$, $\mu = 0$) on the stage of the wave field focusing in fact only lies in the different character of the pulse asymptotics (for $\varepsilon > 0$ the pulse has an oscillating structure along the x-axis) as well as in the inequality $|\partial\mathcal{H}/\partial t|_{\varepsilon<0} > |\partial\mathcal{H}/\partial t|_{\varepsilon>0}$, owing to change in sign of the integral $(\varepsilon/2)\int(\partial_x u)^2\,\mathrm{d}x\mathrm{d}\boldsymbol{\rho}$ in the Hamiltonian $\mathcal{H}$ with

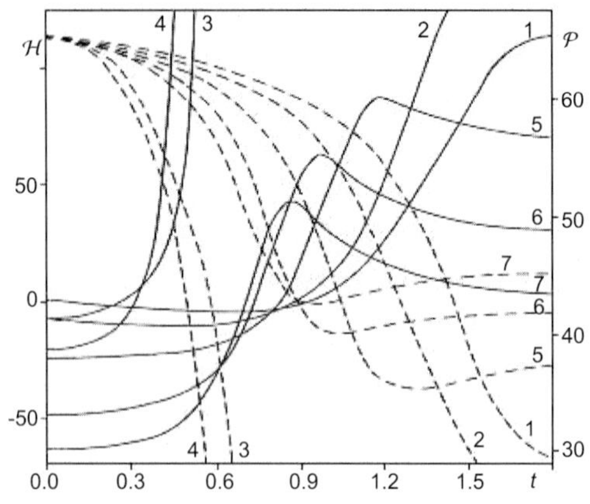

Fig. 4.33. Change with time of the integrals $\mathcal{P}$ and $\mathcal{H}$ of the axially symmetric solution for $\lambda = 0,\ 1$. Solid lines correspond to $\mathcal{H}$ and dashed lines correspond to $\mathcal{P}$. The curve numbers are the same as in Fig. 4.30

the change $\varepsilon \to -\varepsilon$. With further growth of time t, due to the decreasing characteristic sizes l_x and l_ρ of the pulse (see Fig. 4.32), the term proportional to the fifth derivative in (4.131) begins to play the dominant role (that proves to be correct by an analysis of the corresponding change of the integrals in the Hamiltonian $\mathcal{H}$).

As a result, for $\varepsilon \leq 0$ the "flapping" of the pulse "wings" behind the main part of the pulse does not lead to the fast increase of the field density followed usually by formation of the singularity in the region of the main maximum as it takes place for the standard KP equation (see Sect. 3.1) with $(\beta/\kappa) > 0$ [59,63], and forms the circular region of the increased concentration (Fig. 4.34). Further evolution of this structure leads to formation of an additional maximum on the x-axis behind the pulse (see Fig. 4.35a). At this time moment, the compression of the pulse is stopped and the collapse does not take place, as shown in Fig. 4.36.

For $\varepsilon > 0$, an analogous mechanism also plays a certain role but, because of the change of the dispersion's character in the region of small k, the compression and the increase of the amplitude of the pulse are stopped at the earlier stage (see Figs. 4.30 and 4.32, curves 6 and 7), and the evolution proceeds to the defocusing stage. This is analogous to the situation taking place for the usual KP equation with the negative dispersion ($\beta/\kappa < 0$ in the standard form of the KP equation, see Sect. 3.1) [59]. The role of the term proportional to the fifth derivative in (4.131) reveals also the formation of the small-scale oscillations constituing the regular oscillatory structure of the tail (compare Fig. 4.35a and 4.35b).

The final stage of the evolution of the pulse for $\lambda = 1$ and $\mu = 0$ is characterized by the slow decrease of the amplitude of the main maximum (see Figs. 4.30 and 4.36) accompanied by some increase of the amplitude of the second maximum ($\varepsilon \leq 0$) or the amplitudes of the tail oscillations ($\varepsilon > 0$), with the general structure of the solutions being qualitatively conserved.

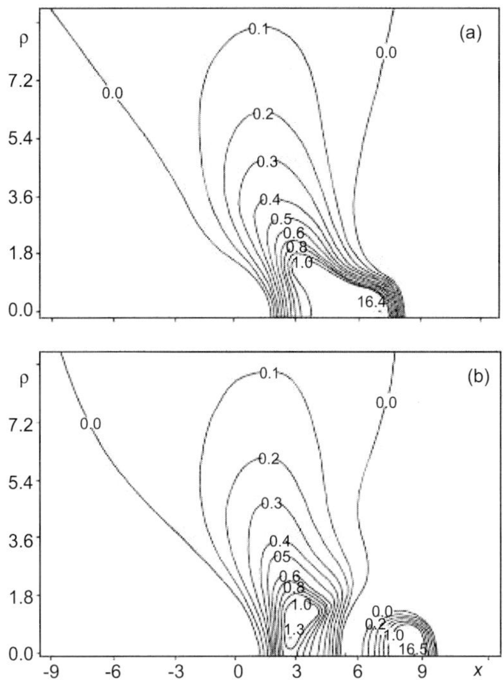

Fig. 4.34. Solution in the $x-\rho$-plane for $\lambda = 1$ and $\varepsilon = -0.89$ corresponding to the stage of the evolution when the pulse amplitude is the largest. **a** $t = 1.65$. **b** $t = 1.8$

The character of the change of the amplitudes, the parameters l_x, l_ρ, as well as the integrals $\mathcal{P}$ and $\mathcal{H}$ allows us to assume the possibility of the solutions approaching a stationary state for $\lambda = 1$ and $|\varepsilon| \geq 0$. However, the computational analysis of further stages of the soliton's evolution is difficult because of the need for the considerable increase of the sizes of the integration region. Therefore, to estimate the stability of the solutions for $\lambda = 1$ and $\mu = 0$, as in the two-dimensional case (Sect. 4.4), an analysis can be done of the boundedness of the Hamiltonian of the equation (4.131) with $\mu = 0$ on the solutions obtained numerically. The result of the analysis (as an example, see Fig. 4.37, curves 1 and 2) confirms that the solutions for large t and $\varepsilon > 0$ approach the stationary ones (corresponding to the minimum of the Hamiltonian $\mathcal{H}$ in Fig. 4.37) and are unstable for $\varepsilon \leq 0$ (for the considered values of ε). Moreover, unlike the standard KP equation with $(\beta/\kappa) > 0$, (4.131) for any $\varepsilon \leq 0$ has no collapsing solutions: the decrease of ε leads to the increase of the "critical" value of the amplitude (see Fig. 4.30), but the time evolution is always finished with defocusing of the wave field. However, for $\varepsilon \to -\infty$ there is the limit transition,

$$\lim_{\varepsilon\to-\infty,\gamma\to 0}\left[-\varepsilon|\gamma|^{1/2}\right] = \text{const} > 0, \qquad \gamma = \lambda|\gamma|, \tag{4.135}$$

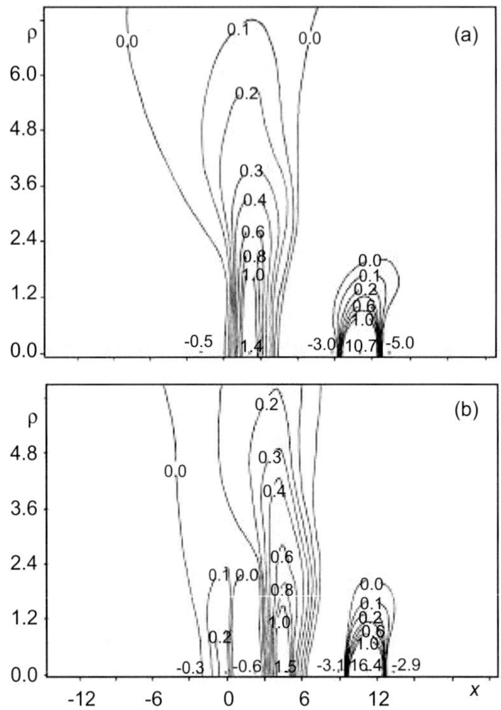

Fig. 4.35. Solutions in the $x-\rho$-plane for $\lambda = 1$ corresponding to the defocusing stage of the evolution. **a** $\varepsilon = -0.89$, $t = 2.1$. **b** $\varepsilon = 0.89$, $t = 1.35$

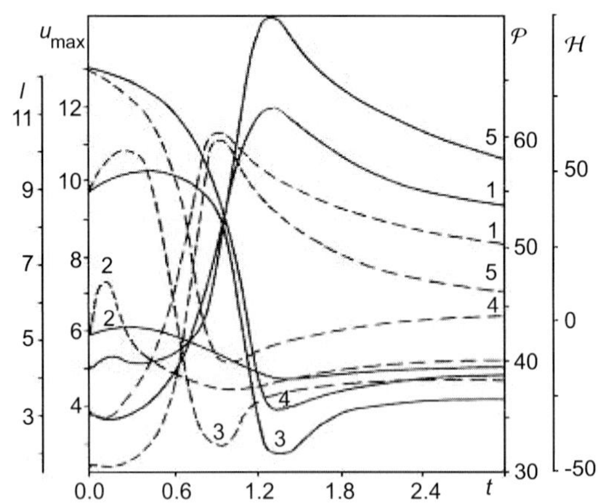

Fig. 4.36. Change with time of the solution's parameters, as well as of the momentum $\mathcal{P}$ and Hamiltonian $\mathcal{H}$ of the system for $\lambda = 1$. The solid lines correspond to $\varepsilon = -0.45$, the dashed lines correspond to $\varepsilon = 1.34$: (1) $u_{\max}$; (2) l_x; (3) l_ρ; (4) $\mathcal{P}$; (5) $\mathcal{H}$

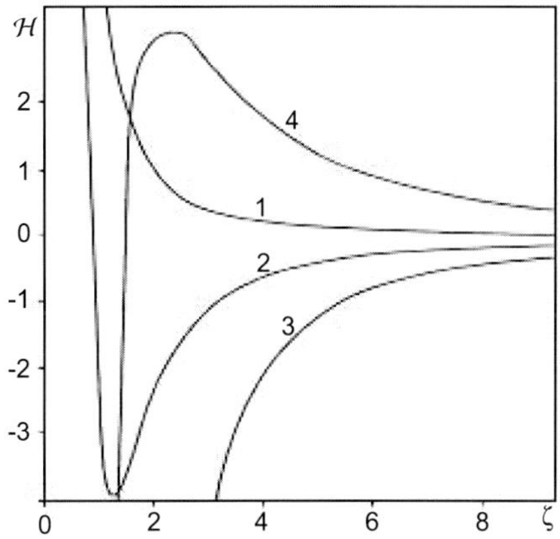

Fig. 4.37. Change of the Hamiltonian $\mathcal{H}$ along the directions $\eta = (2b/c)\zeta^{5/2}$ under deformations on the numerical solutions of (4.131) with $\mu = 0$ and $\lambda = 1$: (1) $\varepsilon = -1.34$; (2) $\varepsilon = 2.24$, and $\lambda = -1$: (3) $\varepsilon = 1.34$; (4) $\varepsilon = -1.34$

leading to the disappearance of the term proportional to the fifth derivative and the transform of (4.131) with $\mu = 0$ into the standard KP equation with the positive dispersion term. The results of the computational analysis for this case ($\mu = \lambda = 0$, $\varepsilon < 0$) leading to the collapse of the solution, are shown in Figs. 4.30, 4.32, and 4.33 (curves 4 and 5).

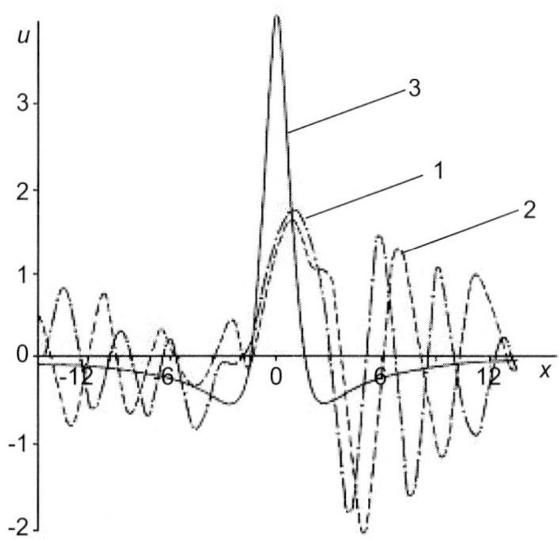

Fig. 4.38. Solutions of (4.131) at $\rho = 0$ and $t = 0.3$ for $\lambda = -1$ and $\mu = 0$: (1) $\varepsilon = -1.34$; (2) $\varepsilon = 1.34$; (3) the initial pulse

Consider now solutions for $\lambda = -1$. Change of the sign of the system's dispersion (i.e., simultaneous change of the signs of ε and λ) relative to the cases considered above leads, as in the two-dimensional version of (4.131), to qualitatively new results. For $\varepsilon < 0$, we observe formation of the spreading with time wave packet with regular oscillations in the x-direction in front of the main maximum, and smaller scale oscillations behind it. The structure of the solutions for $\varepsilon \to -\infty$ changes in a manner similar to the case $\lambda = 1$,

$\varepsilon \to \infty$. With $\varepsilon \to 0$ and its further increase to the positive values, the oscillation frequency decreases and the amplitude of the tail oscillations increases (Fig. 4.38). Thus for significantly large $\varepsilon > 0$, it becomes comparable to the amplitude of the main maximum and to the amplitudes of the oscillations in front of the pulse. An instability of the wave packets for $\mu = 0$ in the case $\lambda = -1$ is confirmed by the analysis of the boundedness of the Hamiltonian $\mathcal{H}$ on the numerical solutions (see Fig. 4.37, curves 3 and 4).

4.5.2 Structure and Evolution of Three-Dimensional Solutions of 3-DNLS-Class Equations

For the numerical investigation, we consider the *3-DNLS equation* in the integral-differential form (4.132) and integrate it in the axially-symmetric geometry $\Delta_\perp = \partial_\rho^2 + (1/\rho)\partial_\rho$, with $\rho^2 = y^2 + z^2$. Here, we use the *implicit scheme with* $O(\tau^2, \Delta_{x,\rho}^4)$ *approximation* (4.83) on the grid 301×51, with $\tau = 0.0025$, $\Delta_x = 0.1$, $\Delta_\rho = 0.3$, and the periodic boundary conditions. We compute the integral on the right-hand side of (4.132) using the *Newton–Cotes formula* (see Sect. 4.3) with an accuracy of at least $O(\Delta_x^4)$ (with the automatic choice of the number of knots in the quadrature expression [77]). The initial conditions are taken in the form of the axially-symmetric solitary pulses of two types:

a. Soliton-like axially symmetric pulse:

$$h(x,\rho,0) = h_0(x) \exp\left[\mathrm{i}\varphi(x) - \rho^2/l_\rho^2\right], \tag{4.136}$$

with

$$h_0(x) = 2\sqrt{2}\delta \sin\nu \left[\cosh\left(4\delta^2 \sin\nu x\right) + \cos\nu\right]^{-1/2}$$

and

$$\varphi(x) = -2s\delta^2 \cos\nu x - \frac{3}{4}s\int_{-\infty}^{x} h_0^2(x)\mathrm{d}x,$$

where $0 < \nu < \pi$.

b Modulated "plane" wave:

$$h(x,\rho,0) = H_0 \exp\left(\frac{2\pi \mathrm{i}x}{\lambda} - \frac{x^2}{l_x^2} - \frac{\rho^2}{l_\rho^2}\right), \tag{4.137}$$

where λ is the wavelength, H_0 is the amplitude, and l_x and l_ρ are the characteristic scales of the Gaussian envelop modulation in the x- and ρ-directions. Note that for $\rho = 0$, the initial conditions (4.136) and (4.137) are equivalent to those (for *DNLS equation*) used for the numerical simulation of the evolution of the one-dimensional Alfvén wave in Refs. [33,176].

To control the numerical computation, we calculate on every time layer the integrals (see Sect. 4.1)

$$C_1 = \mathcal{P}_x = \frac{1}{2} \int_{-\infty}^{\infty} |h|^2 \mathrm{d}\mathbf{r},$$

$$C_2 = \frac{\mathcal{H}}{2} = \frac{1}{2} \int_{-\infty}^{\infty} \left[\frac{1}{2}|h|^4 + \lambda s h h^* \partial_x \varphi + \frac{\kappa}{2} \left(\nabla_\perp \partial_x w\right)^2\right] \mathrm{d}\mathbf{r}, \qquad (4.138)$$

with $\partial_x^2 w = h$ and $\varphi = \arg(h)$. To investigate the structure and evolution of the three-dimensional pulses, we have done a number of simulation runs for both signs of the integral parameter $b = \lambda s \int h h^* \partial_x \varphi \mathrm{d}\mathbf{r}$ and various initial values of the Hamiltonian $\mathcal{H}$ by defining various initial values for the pulse amplitude and the widths l_x and l_ρ. Thus, we have obtained the following results:

1. For $\lambda = 1$, $s = -1$, large $\kappa > 0$, and the initial pulse weakly limited in the transverse ρ-direction when the stability condition (see Sect. 4.1.3) is satisfied, the evolution for large t results in formation of the stable three-dimensional (axially-symmetric) solution (Fig. 4.39).

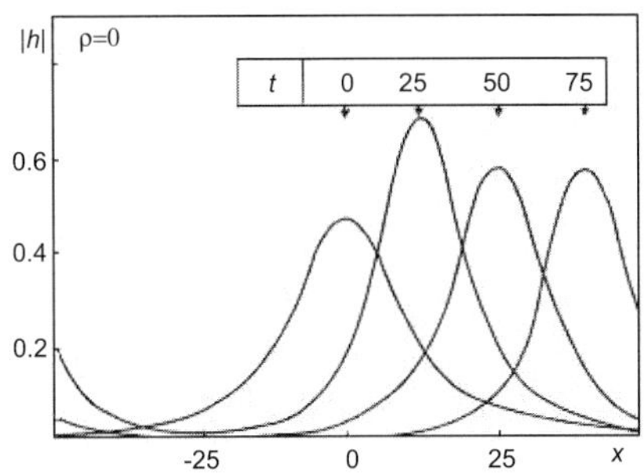

Fig. 4.39. Evolution of the three-dimensional right-circularly polarized nonlinear pulse (4.136) for $\lambda = 1$, $s = -1$, $\kappa = 1$; $\mathcal{H} > -3bd/(1 + 2d^2) > 0$ (see Sect. 4.1.3)

2. For the opposite signs of λ and s (this is equivalent to the change $t \to -t$ and $\kappa \to -\kappa$ in (4.132)), the Hamiltonian $\mathcal{H}$ (4.138) of the 3-DNLS equation becomes negative, and, as it follows from the simulations, the three-dimensional *Alfvén waves* spread with the time evolution (Fig. 4.40).

3. For $\lambda = 1$, $s = -1$, small $\kappa > 0$, and the initial pulse rather strongly limited in the transverse ρ-direction, the condition of the existence of the local minimum of the Hamiltonian $\mathcal{H}$ (see Sect. 4.1.3) is not satisfied, and in the simulations we can observe development of the three-dimensional *collapsing solutions* of the 3-DNLS equation (Figs. 4.41 and 4.42). Note that this effect is typical for all nonlinear systems where the Hamiltonian

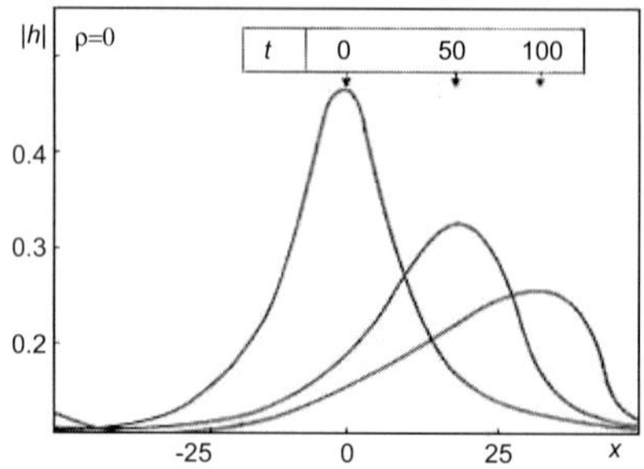

Fig. 4.40. Evolution of the three-dimensional right-circularly polarized nonlinear pulse (4.137) for $\lambda = -1$, $s = 1$, $\kappa = 1$; $\mathcal{H} > 0$ (see Sect. 4.1.3)

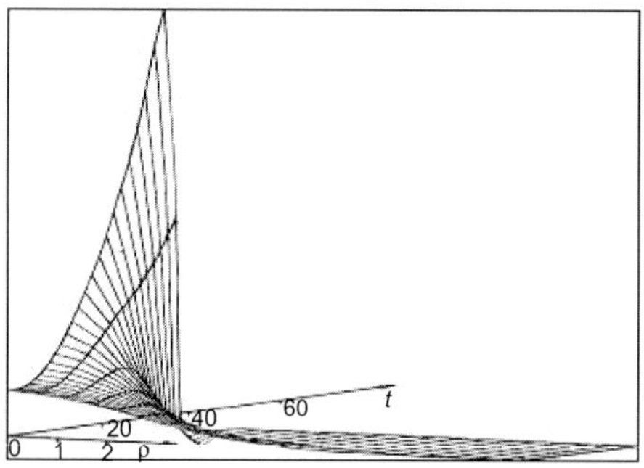

Fig. 4.41. Dynamics of the three-dimensional right-circularly polarized nonlinear pulse (4.137) (the cross-section by the ρ-plane at $h_{\max}$) for $\lambda = 1$, $s = -1$, $\kappa = 0.2$; $0 < \mathcal{H} < -3bd/(1+2d^2)$ (see Sect. 4.1.3)

is unlimited for the fixed first integrals (in this case, for the momentum $\mathcal{P}$), and the quadratic terms in the expression for the Hamiltonian $\mathcal{H}$ (the first and third terms in (4.138)) are positively defined. For example, the same effects have been observed in systems describing the evolution of *FMS waves* [59] and *Langmuir waves* [56] in a plasma. A series of simulation runs done for $b > 0$ when $\lambda = 1$, $s = 1$ and $\lambda = -1$, $s = -1$ in (4.132) with $\nu = 0$ demonstrated that for these conditions in all the cases considered (various initial values of the Hamiltonian $\mathcal{H}$ and the parameters l_x and l_ρ), the initial three-dimensional axially-symmetric pulse is spreading with time. This is quite obvious so far as that such type of conditions for the system's parameters the inequality $\mathcal{H} < -3bd/(1+2d^2)$ (see Sect. 4.1.3) is not satisfied and, therefore, the three-dimensional solutions of 3-DNLS equation are unstable.

If we, however, perform the transform $h \to -sh^*$ in the 3-DNLS equation, i.e., consider the left-circularly polarized waves, the corresponding signs in the expression for the system's Hamiltonian $\mathcal{H}$ (4.138) change to the opposite ones and we observe the mirror-opposite picture for all the cases considered above. Thus the case $\lambda = -1$, $s = -1$, and large $\kappa > 0$, as well as the cases $\lambda = 1$, $s = 1$ and $\lambda = -1$, $s = -1$ for small $\kappa > 0$ correspond to the above cases (1), (2), and (3), respectively, with the opposite signs of

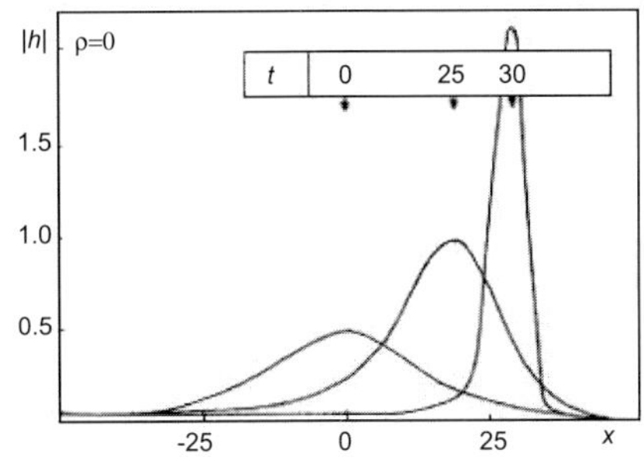

Fig. 4.42. Evolution of the three-dimensional right-circularly polarized nonlinear pulse (4.136) for $\lambda = 1$, $s = -1$, $\kappa = 0.2$; $0 < \mathcal{H} < -3bd/(1 + 2d^2)$

the Hamiltonian $\mathcal{H}$. An example of the dynamics of the three-dimensional left-circularly polarized pulse is shown in Fig. 4.43.

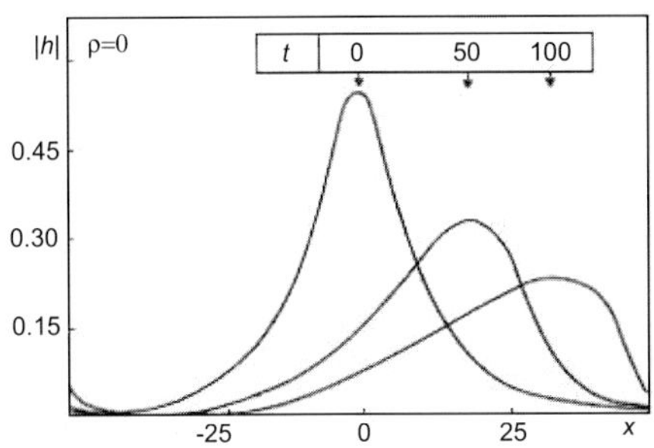

Fig. 4.43. Evolution of the three-dimensional left-circularly polarized nonlinear pulse (4.137) for $\lambda = s = \kappa = 1$; $\mathcal{H} > 0$

To sum up the above, we conclude that the 3-DNLS equation (4.132) with $\nu = 0$ can have the stable three-dimensional solutions in the form of a three-dimensional *Alfvén soliton*, alongside with the three-dimensional collapsing or dispersing with time solutions. The particular form of the solution is defined by the signs of the coefficients λ and s as well as by the chosen initial condition. The results obtained above can be also interpreted in terms of the *self-focusing* phenomenon. Indeed, the (formal) interchange $x \leftrightarrow t$ enables us to convert the Cauchy initial value problem (4.132) and (4.136) or (4.132) and (4.137) to the boundary value problem describing propagation along the x-axis (from the boundary $x = 0$) of a three-dimensional *Alfvén wave beam* localized in the ρ-plane. In this case the above results can be interpreted as follows: (1) the formation of the stationary Alfvén wave beam propagating along the x-axis, (2) the spreading of the Alfvén wave beam, and (3) the self-focusing of the Alfvén wave beam. It is interesting to note that we in fact observe here the dynamics of the Alfvén wave beam propagating in a plasma with $\beta > 0$ at near-to-zero angles with respect to the external magnetic field $\mathbf{B}$, which is qualitatively similar to the dynamics of the *FMS wave beam* propagating in a plasma with $\beta \ll 0$ at an angle close to $\pi/2$ with respect to the external magnetic field [86,196].

4.5.3 Influence of Dissipation on Evolution of Three-Dimensional Nonlinear Waves

The presence of dissipation caused by the relaxation processes of the viscous type in a medium (the particular physical reason of the energy dissipation depends on the type of the medium) changes the character of the evolution of the three-dimensional nonlinear pulse. For the GKP equation for the cases considered above, the role of dissipation ($\mu > 0$) in the formation and evolution of the three-dimensional soliton-like structures and the nonlinear wave packets results in the decrease of the main amplitude maximum according to the exponential law $u(t) = u(0)\exp(\Gamma t)$, where $\Gamma \sim -2.6$. It also leads to the steepening of the wave packet profile in the direction of the wave propagation [148]. In this case, for $\varepsilon > 0$, a stronger suppression of the oscillations in front of the pulse can be observed; the suppression increases with the increase of ε (see Fig. 4.29, curves 3 and 4, Fig. 4.44 and Fig. 4.45). This is also

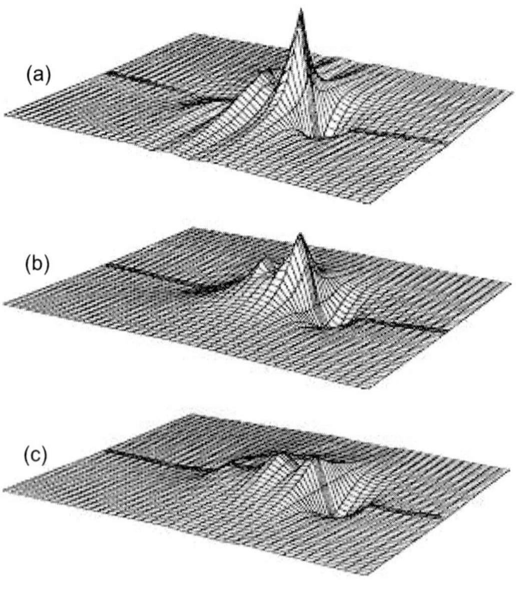

Fig. 4.44. Evolution of the axially symmetric pulse for $\lambda = 1$, $\varepsilon = 3.16$, and $\mu = 1.78$. **a** $t = 0.36$. **b** $t = 0.53$. **c** $t = 0.9$

confirmed by the asymmetric change of the integrals $\mathcal{P}$ and $\mathcal{H}$ in the front and back "caverns" ($u < 0$). As a result of the dissipation, similar to the two-dimensional case considered above (see Sect. 4.4), the oscillating soliton tail stretches out and its overall slope decreases compared to that of the front part of the pulse.

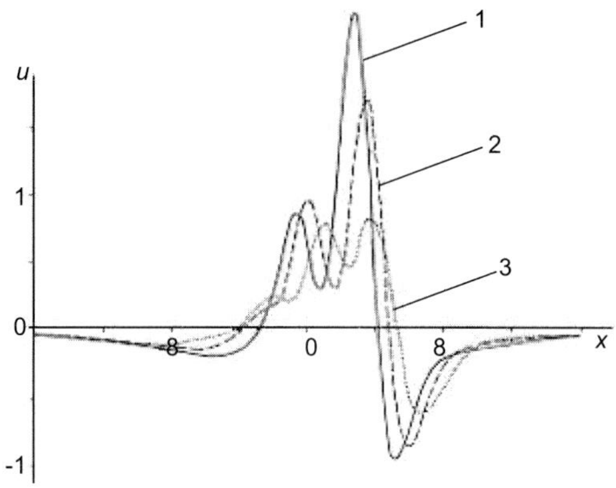

Fig. 4.45. The same results as in Fig. 4.44 but for $\rho = 0$

The results presented in Sect. 4.5.1 for $\mu = 0$ demonstrate that for $\lambda = 1$, the dissipation in the medium is not crucially decisive for the stopping the *wave collapse*. The term $-\mu\partial_x^2 u$, defining the dissipation, plays the role of a competing factor with respect to the nonlinear term in the instability development and, depending on the value $\mu > 0$, in either extent suppresses the instability. Thus for $\varepsilon \leq 0$ and $\mu \sim 1$, as long as the characteristic parameters l_x and l_ρ do not reach the minimum critical values when the term proportional to the fifth derivative begins to play the dominant role, the phase of "flapping" of the soliton wings behind the main maximum is completely absent in the evolution of the pulse. As a result, the second maximum of the solution is absent in this case and the algebraic asymptotics, taking place in the initial stage, is preserved (see Fig. 4.27, curve 5). However, in the case $\varepsilon > 0$ and $\mu > 0$ the structure of the solutions for large time t (excluding the exponential decrease of the amplitude and the steepening of the wave profile) does not qualitatively changes, as shown by a comparison of Fig. 4.35b with Fig. 4.44 and Fig. 4.45.

In the case $\lambda = -1$, the following results are obtained: for $\varepsilon < 0$ and $\mu > 0$, similar to the case $\varepsilon < 0$ and $\mu = 0$ considered in Sect. 4.5.1, the result of the wave evolution is the formation of the spreading with the time wave packet, with the regular oscillations in the direction of its propagation and small-scale oscillations behind its main maximum (see Fig. 4.29, curve 5, and Fig. 4.46). The asymptotic behavior of the solutions for $\varepsilon \to \infty$ and

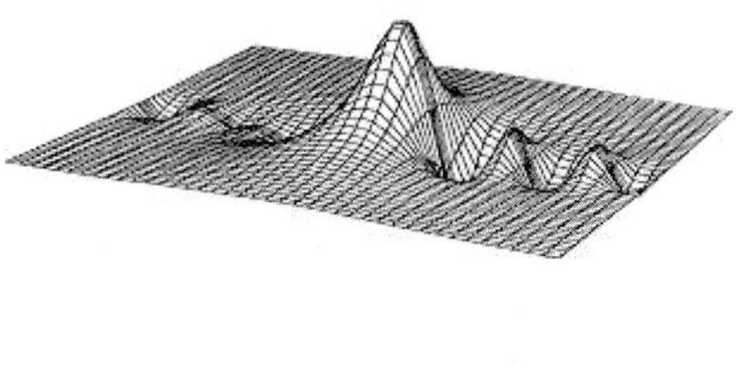

Fig. 4.46. Three-dimensional axially symmetric pulse described by (4.131) with $\lambda = -1$, $\varepsilon = -1.58$, and $\mu = 1.26$ at $t = 0.25$

$\varepsilon \to 0$ is qualitatively analogous to the case $\mu = 0$, shown by a comparison of the curve 5 in Fig. 4.29 and Fig. 4.46. The effect of the dissipation on the structure and evolution of the solutions for $\lambda = -1$ looks similar to that for $\lambda = 1$: first of all, in the steepening of the fronts of the waves in the packet, but unlike the case $\lambda = 1$ for both $\varepsilon < 0$ and $\varepsilon \geq 0$ the tail oscillations are more suppressed. The order of magnitude of the suppression rate for this case is the same as for $\lambda = 1$.

The role played by the dissipation in the 3-DNLS model ($\nu > 0$ in (4.132)) is in many respects analogous to that in the case of the three-dimensional GKP equation. Just as for the GKP model, in the evolution of the Alfvén wave the exponential decrease of its amplitude with time is observed:

$$|h|^2 = (1+e^2)\bar{h}^2(t) = (1+e^2)\bar{h}^2(0)\exp(\Gamma t). \tag{4.139}$$

In this case, the damping rate is of the same order of magnitude as in the GKP model (we have obtained in our numerical simulations the averaged value $\Gamma \approx -3.1$). Moreover, similar to the GKP model, some steepening of the pulse's front takes place, and the back slope of the pulse decreases [68]. The proof of that behavior is, in particular, in the different character of the change of the integrals of motion,

$$\mathcal{P} = (1+e^2)\int\limits_{-\infty}^{\infty} \bar{h}^2 \mathrm{d}\mathbf{r} \tag{4.140}$$

and

$$\mathcal{H} = \int\limits_{-\infty}^{\infty} \left[\frac{1}{2}\left(1+e^2\right)^2 \bar{h}^2 + \lambda s\left(1-e^2\right)\bar{h}^2\partial_x\varphi + \frac{\kappa}{2}\left(\nabla_\perp \partial_x w\right)^2\right] \mathrm{d}\mathbf{r}, \tag{4.141}$$

where $\partial_x^2 w = (1+\mathrm{i}e)\bar{h}$ and $\varphi = \arg[(1+\mathrm{i}e)\bar{h}]$ (see Sect. 4.3), in the regions behind and in front of the main maximum. Indeed, in all cases $\mathcal{P}$ and $|\mathcal{H}|$ decrease faster in front of the pulse.

For various values of the coefficients in the 3-DNLS equation, the character of the evolution is the following:

1. For $\lambda = 1$, $s = -1$, and relatively large $\kappa > 0$, the initial pulse is weakly limited in the direction perpendicular to its propagation, and loses its energy gradually with time ($\mathcal{H} \to 0$ with $t \to \infty$). In this case, the amplitude of the pulse decreases with time (as we noted above, exponentially, with the rate proportional to ν) and, as a result, the solitary wave disperses. Recall here, that in the case of $\nu = 0$, the evolution after initial "sub–focusing" of the pulse leads to the formation of the stable three-dimensional *Alfvén soliton* (see above).

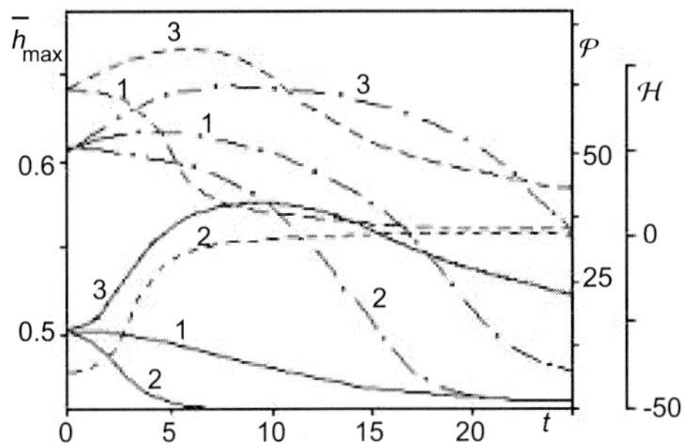

Fig. 4.47. Change of the amplitude of the three-dimensional axially-symmetric Alfvén wave pulse (the solid lines), as well as the momentum (the dash-dotted lines) and the Hamiltonian (the dashed lines) of the 3-DNLS equation with $\nu = 1$: (1) $\lambda = 1$, $s = -1$, and $\kappa = 1.5$; (2) $\lambda = -1$, $s = 1$, and $\kappa = 1.5$; (3) $\lambda = 1$, $s = -1$, and $\kappa = 0.1$

2. For $\lambda = -1$ and $s = 1$, when the *Hamiltonian* $\mathcal{H}$ becomes negative and the Alfvén wave pulse for $\nu = 0$ spreads with its evolution, the presence of the dissipation accelerates this process significantly (for $\nu \sim 1$ we have obtained in the simulations averaged $\Gamma \approx -3.4$). The effect of the steepening of the front of the pulse takes place as well in this case.
3. For $\lambda = 1$, $s = -1$, relatively small $\kappa > 0$, and the initial pulse strongly limited in the transverse ρ-direction, when development of the *wave collapse* is observed in the simulations for $\nu = 0$, the presence of the dissipation can rapidly delay or (for large $\nu > 0$) even stop this process. In this case, the role of the dissipation in the 3-DNLS model is different from that in the model of the three-dimensional GKP equation: it is now the decisive factor in the stopping of the wave collapse.

Figure 4.47 shows the change with time of the amplitude and the integrals $\mathcal{P}$ and $\mathcal{H}$ (averaged throughout the region of numerical integration) for the three cases described above and $\nu = 1$.

4.6 Applications

We already mentioned in the Introduction that equations (0.10) and (0.22) are universal in the sense that they describe the wide class of nonlinear waves in dispersive media in the absence and in the presence of dissipation, respectively, when the dispersion law in the linear approximation is given by

$$\omega \approx c_0 k_x \left(1 + \frac{k_\perp^2}{2k_x^2} - D^2 k_x^2 \right) \tag{4.142}$$

(see the KP equation in its standard form for the waves with $k_\perp \equiv k_y$ in Sect. 3.1) and

$$\omega \approx c_0 k_x \left[1 + \frac{k_\perp^2}{2k_x^2} - \frac{\mathrm{i}\nu k_x}{c_0} + \frac{1}{c_0} \left(-\gamma_1 k_x^2 + \gamma_2 k_x^4 \right) \right] \tag{4.143}$$

(see (0.22) and (4.4)) where c_0 is the phase velocity of the oscillations for $|\mathbf{k}| \to 0$ and D is the dispersion "length." Such situations arise, for example, for the wave perturbations of the acoustic type such as the ion-acoustic waves and magnetosonic waves in a plasma, surface waves in shallow water, and internal gravity waves in the upper atmosphere and ionosphere (applications in the one-dimensional case when $k_\perp = 0$ in the dispersion relations (4.142) and (4.143) are given in Sect. 1.4).

There is a number of works dedicated to applications of these models to nonlinear wave processes in particular physical media, we note here Refs. [3,7,59,80,81,85,138–141,189,196]. In the Introduction and Sects. 2.4 and 4.1 we demonstrated that equations of the DNLS class (0.13), (0.16) and (0.24), (4.14) directly describe the nonlinear evolution of Alfvén waves in a magnetized plasma without and with dissipation in the medium, respectively. Studies of these models were done in Refs. [7,33,37,38,50,65–70].

Here, we consider applications of the results obtained above for multidimensional cases to the investigation of: (a) the propagation of nonlinear ion-acoustic waves in an unmagnetized plasma, including the relativistic effects (Sect. 4.6.1); (b) the dynamics of three-dimensional fast magnetosonic (FMS) waves propagating in a magnetized plasma (Sect. 4.6.2); (c) the dynamics of two-dimensional solitary nonlinear internal gravity waves (IGW), generated in the F-layer of the Earth's ionosphere by fronts of the solar terminator and the solar eclipse as well as by seismic sources, and excitation by them of the traveling perturbations of the plasma electron density (so-called traveling ionospheric disturbances, TID) (Sect. 4.6.3); (d) the evolution of two-dimensional solitons of gravity and gravity-capillary waves on the surface of a shallow water with the bottom relief varying in time and space (Sect. 4.6.4). The main results here are obtained by the analytical and numerical methods detailed above in the previous sections, see also Refs. [50,65–70,81,83,138–141,144,195,196,201,203–206].

4.6.1 Nonlinear Ion-Acoustic Waves in a Plasma

Here we consider applications of the results obtained in the previous sections to description of the structure and dynamics of the multidimensional *ion-acoustic waves* in an *unmagnetized plasma*, including the case of a weakly relativistic plasma. Similar applications in the one-dimensional case were considered above in Sect. 1.4.

Nonrelativistic Approximation. In the absence of the magnetic field and for the negligible ion temperature (see assumptions of Sect. 1.4), the equations of motion and continuity for plasma ions are given by [83]

$$\begin{aligned} \partial_t \mathbf{v} + (\mathbf{v}\cdot\boldsymbol{\nabla})\,\mathbf{v} &= -\frac{e}{m_i}\boldsymbol{\nabla}\varphi, \\ \partial_t n_i + \boldsymbol{\nabla}\cdot(n_i\mathbf{v}) &= 0. \end{aligned} \qquad (4.144)$$

Recall that a comparison with (1.1) shows that in this case the role of the generalized "density" and "sound" plays the ion density n_i and the ion-acoustic velocity $c_s = (T_e/m_i)^{1/2}$; the dispersion "length" is defined by $D^2 = r_D^2/2 = T_e/8\pi n_0 e^2$ where n_0 is the unnperturbed electron density and r_D is the electron Debye length. The electrons in the ion-acoustic wave are Boltzmann distributed,

$$n_e = n_0 \exp\left(\frac{e\varphi}{T_e}\right), \tag{4.145}$$

and the densities of the electrons and ions are related to the electric potential φ via Poisson's equation

$$\Delta\varphi = 4\pi e\left(n_e - n_i\right). \tag{4.146}$$

The dispersion equation for the set (4.144)–(4.146) is written as [189]

$$\omega^2 = \frac{c_s^2\mathbf{k}^2}{1 + r_D^2\mathbf{k}^2}. \tag{4.147}$$

Consider the wave packet propagating in the direction close to the x-axis. We assume that the wave numbers of its harmonics are small satisfying the inequalities,

$$|\mathbf{k}|r_D \ll 1, \qquad k_x^2 \gg \mathbf{k}_\perp^2, \quad \text{and} \quad v' \ll c_s, \tag{4.148}$$

where v' is the x-component of the ion velocity. It is well known that the weakly dispersive (see the first inequality of (4.148)) ion-acoustic wave steepens in the direction of its propagation, therefore, at some time moment the second inequality of (4.148) "switches on." Conditions (4.148) enable us to reduce the dispersion relation (4.147) to the form (4.142). Thus considering the solution in the form of a propagating wave $u = u(t, x - c_s t, \mathbf{r}_\perp)$, limiting ourselves in the nonlinear expansion by the terms quadratic in the wave amplitude, and applying the procedure described above in Sect. 1.1.1 and Sect. 3.1.1), we obtain the nonlinear equation

$$\partial_x\left(\partial_t v + c_s\partial_x v - c_s D^2\partial_x^3 v + v\partial_x v\right) = \pm\frac{c_s}{2}\Delta_\perp v \tag{4.149}$$

(where $\Delta_\perp = \partial_y^2 + \partial_z^2$) that is formally similar to the dimensionless equation (3.3) (compare also with (1.110) of Sect. 1.4). After the homothetic transformation and in the reference frame moving along the x-axis with the velocity c_s, equation (4.149) can be reduced to the standard form of the *KP equation*,

$$\partial_t u + \alpha u\partial_x u + \beta\partial_x^3 u = \kappa\int_{-\infty}^{x}\Delta_\perp u\mathrm{d}x, \tag{4.150}$$

where $\kappa > 0$ is related to the positive dispersion ($\kappa < 0$ corresponds to the negative dispersion, respectively), and the factors at the nonlinear and dispersion terms are (see also Sect. 1.4)

$$\alpha = \frac{3}{2}\frac{c_s}{n_i} \quad \text{and} \quad \beta = \frac{1}{2}c_s r_D^2.$$

Generally speaking, for the ion-acoustic wave the sign on the right-hand side of (4.149) is positive so that the dispersion is negative, $\kappa < 0$ in (4.150). However, for other modes there are cases when the dispersion is positive, i.e. there is the 'minus' sign on the right-hand side of (4.149). The term $c_s\partial_x v$, just as in the one-dimensional case (Sect. 1.4), describes the wave propagation along the x-axis with the "sound" velocity, and other terms responsible for dispersion, nonlinearity, and diffraction describe slow changes of the acoustic field on the background of the wave motion with the velocity c_s. Such acoustic waves is mainly characteristic for isotropic media (e.g., an unmagnetized plasma), but sometimes it can be observed in anisotropic media as well. For example, if the characteristic frequency of the ion-acoustic wave packet significantly exceeds the ion-cyclotron frequency in a magnetized plasma, viz. $\omega \gg \omega_{Bi}$, the plasma anisotropy introduced by the magnetic field can be neglected and therefore (4.149) can be reduced to the standard KP equation (4.150) [189]. In the opposite case, when $\omega \ll \omega_{Bi}$, the anisotropy cannot be neglected.[9] In Refs. [83,195], the isotropic case of (4.150) for the ion-acoustic waves in an unmagnetized plasma was considered, the numerical simulation based on the implicit scheme (4.75) was performed [195], and the spectral method (4.93) and (4.96) was applied for the initial condition $u(0,x,y) = u_0 \exp[-((x/l_x)+(y/l_y))^2/L^2]$ with the periodic boundary conditions [98,195]. Figure 4.48 shows an example of the numerical results obtained for the two-dimensional ($\Delta_\perp = \partial_y^2$) equation (4.150) [195].

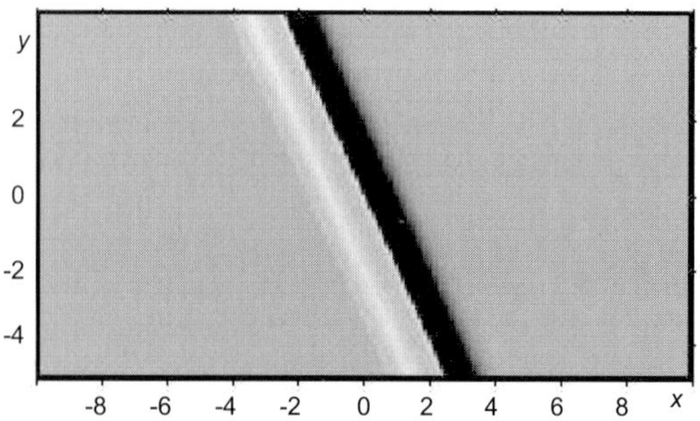

Fig. 4.48. Two-dimensional solution of (4.150) with $\alpha = 6$ and $\beta = 1$

We can see that as a result of the evolution of the two-dimensional acoustic perturbation $u(0,x,y)$ in an isotropic plasma, the one-dimensional soliton of

[9] On the right-hand side of the equation of motion (4.144) appears the additional term $\omega_{Bi}\mathbf{i} \times \mathbf{v}$ ($\mathbf{i}$ is the dimensionless unit vector in the x-direction), and the second term in the dispersion equation changes its sign. In this case, we also obtain an equation of the class (0.3) but with $\mathcal{R}[u] = \mp\Delta_\perp\partial_x u$ [17]. The upper sign in the resulting equation corresponds to the negative dispersion and the lower sign – to the positive dispersion, as in (4.149).

KP equation is formed. The form of the soliton corresponds to that obtained analytically for the negative dispersion in Ref. [16] by employing the *Krylov–Bogolyubov method* and in Ref. [24] by using the *IST method.* It is shown in Refs. [81,83,195] that for sufficiently large t, the soliton velocity and the first three integrals of KP equation $\mathcal{J}_1$, $\mathcal{J}_2$, and $\mathcal{J}_3$ (see Sect. 3.1) are conserved within the limits of the accuracy of the numerical simulation.

Weakly Relativistic Effects. In Sect. 1.4 we have considered the weakly relativistic case for the one-dimensional nonlinear ion-acoustic waves described by KdV equation. Here we use the same approach to study similar waves in the two-dimensional case. As we already demonstrated, the ion-acoustic waves in a plasma can be described by the KP equation (4.150).[10] However, if the velocity of plasma particles approaches the speed of light, the *relativistic effects* start to strongly influence the wave characteristics (such as its phase velocity, amplitude and the characteristic wavelength) in the propagation of the two-dimensional solitary ion-acoustic wave.

For the two-dimensional ion-acoustic solitary waves in a weakly relativistic collisional plasma, the KP equation (4.150) taking into account the relativistic factor u/c can be obtained [207] using the reduced perturbation method [104]. We have

$$\partial_\tau \Phi_1 + \alpha(\vartheta_1)\,\Phi_1 \partial_\xi \Phi_1 + \frac{1}{2}\beta(\vartheta_1)\,\partial_\xi^3 \Phi_1 = -\frac{1}{2}\int_{-\infty}^{\xi} \partial_\eta^2 \Phi_1 \mathrm{d}\xi \tag{4.151}$$

(compare with (1.112)). Here, just as in Sect. 1.4.3, $\Phi_1 = \vartheta_1^{1/2} u_1$ is a small perturbation of the electrostatic potential $\Phi = \varepsilon\Phi_1 + \varepsilon^2\Phi_2 + \ldots$, ε is the small expansion parameter and u_1 is the perturbation of the plasma particle velocity ($u = u_0 + \varepsilon u_1 + \varepsilon^2 u_2 + \ldots$), where

$$\alpha(\vartheta_1) = \beta(\vartheta_1)\left(1 - \frac{\vartheta_2}{\vartheta_1^{3/2}}\right), \qquad \beta(\vartheta_1) = \vartheta_1^{-1/2}, \tag{4.152}$$

and

$$\vartheta_1 = 1 + \frac{3}{2}\left(\frac{u_0}{c}\right)^2, \qquad \vartheta_2 = \frac{3}{2}\frac{u_0}{c^2}.$$

Equation (4.151) is written in the reference frame moving along the x–axis: $\xi = \varepsilon^{1/2}(x - \lambda t)$, $\eta = \varepsilon y$, $\tau = \varepsilon^{3/2} t$, and λ is the phase velocity. Since $\alpha > 0$ (see Sect. 1.4.3), we can obtain the stationary solution as the propagating solitary wave. Following the method of Sect. 1.4 we introduce the new variable $\varsigma = k_x\xi + k_y\eta - \omega\tau$ and substitute it into (4.151). Thus we write the solution in the form of a two-dimensional wave

$$\Phi_1 = \Phi_0 \mathrm{sech}^2\left[\frac{1}{W}\left(\xi + \frac{k_y}{k_x}\eta - \frac{\omega}{k_x}\tau\right)\right], \tag{4.153}$$

[10] Below, we limit the consideration by the two-dimensional case with $\Delta_\perp \equiv \partial_y^2$ in (4.150).

Table 4.4. Comparison of results [81,83] with the extreme cases (three last columns) investigated in Refs. [16,105,106,207]

Parameters	Results [81,83]	$u_0 = 0, \eta \neq 0$	$u_0 = 0, \eta = 0$	$u_0 \neq 0, \eta = 0$
λ	$u_0 + \vartheta_1^{-1/2}$	1	1	$u_0 + \vartheta_1^{-1/2}$
α	$\frac{1-\vartheta_2/\vartheta_1^{3/2}}{\vartheta_1^{1/2}}$	1	1	$\frac{1-\vartheta_2/\vartheta_1^{3/2}}{\vartheta_1^{1/2}}$
β	$\vartheta_1^{-1/2}$	1	1	$\vartheta_1^{-1/2}$
Φ_0	$\frac{3\delta\vartheta_1^{1/2}}{1-\vartheta_2/\vartheta_1^{3/2}}$	3δ	$3s$	$\frac{3s\vartheta_1^{1/2}}{1-\vartheta_2/\vartheta_1^{3/2}}$
W	$\vartheta_1^{-1/4}\left(\frac{2}{\delta}\right)^{1/2}$	$\left(\frac{2}{\delta}\right)^{1/2}$	$\left(\frac{2}{s}\right)^{1/2}$	$\vartheta_1^{-1/4}\left(\frac{2}{s}\right)^{1/2}$

where the amplitude Φ_0 and the characteristic size W are defined by the same expressions as in the one-dimensional case,

$$\Phi_0 = \frac{3\delta}{\alpha(\vartheta_1)}, \qquad W = \left[\frac{2\beta(\vartheta_1)}{\delta}\right]^{1/2}, \tag{4.154}$$

but here $\delta = (\omega/k_x) - (k_y/k_x)^2/2$, and the boundary conditions are $\Phi_1 \to 0$ and $\partial_\xi^n \Phi_1 \to 0$ for $n = 1, 2$ and $|\xi| \to \infty$. The *dispersion law* for these waves is given by [207]

$$\omega = k_x \left[2\beta(\vartheta_1) k_x^2 + \frac{k_y^2}{2k_x^2}\right]$$

(compare with the corresponding expression in Sect. 1.4.3).

We see from (4.152) that the factors at the nonlinear term as well at the dispersion term are defined by the relativistic factor ϑ_1. Equation (4.154) shows the dependence of the amplitude and the characteristic length of the two-dimensional ion-acoustic soliton of the KP equation on the weakly relativistic effects. Comparison of the results following from (4.152)–(4.154) and obtained in Refs. [81,83] with those for the three extreme cases considered in Refs. [16,105,106,207] is given in Table 4.4. Here, just as in the one-dimensional case,

$$s = \left(\frac{\omega}{k}\right) \cong v_0 + \frac{1}{\vartheta_1^{1/2}}\left(1 - \frac{1}{2}k^2\right),$$

v_0 is the velocity of the ion flow (if $v_0 \sim 0$ and the relativistic effects are ignored, we have $s \cong 1 - k^2/2$), but δ is defined by another expression (see above). Table 4.4 demonstrates that the obtained results include all three extreme cases too.

To conclude we note that importance of the problem of the wave propagation in a collisional weakly relativistic plasma was discussed in Sect. 1.4.3; this type of problem appears, for example, in applications to *magnetospheric plasma* [107], *laser plasma* [108] as well as in astrophysics [109].

4.6.2 Nonlinear Effects in Propagation of FMS Waves in a Magnetized Plasma

Recall that in a *magnetized plasma* with $\beta \equiv 4\pi nT/B^2 \ll 1$ for the frequencies $\omega < \omega_{Bi}$ (where $\omega_{Bi} = eB/m_i c$ is the ion-cyclotron frequency), the *FMS waves* can propagate. Their *dispersion law* for $|\mathbf{k}|r_D \ll 1$, $k_x^2 \gg \mathbf{k}_\perp^2$, and $v_x \ll v_A$ is given by (see Sect. 3.2.3 as well as Ref. [196])

$$\omega \approx v_A k_x \left(1 + \frac{\mathbf{k}_\perp^2}{k_x^2} + \chi(\theta)\, D^2 k_x^2\right), \tag{4.155}$$

where $v_A = B^2/4\pi n_i m_i$ is the Alfvén velocity, $\mathbf{k}_\perp$ is the transverse (with respect to the wave propagation) wavenumber, v_x is the x-component of the ion velocity, D is the dispersion length (see below (4.156), and θ is the angle between the wave vector component k_x and the external magnetic field $\mathbf{B}$. Remember also (see Sect. 3.2.3) that the (relatively) weak dispersion means that the principal nonlinear process is the three-wave interaction for small-amplitude waves; the condition of the weak nonlinearity determines a small angle between the interacting waves. For sufficiently high ion temperature ($\beta > m_e/m_i$), the dispersion length is defined by [59]

$$\chi(\theta)\, D^2 = \frac{c^2}{2\omega_{pi}^2}\cot^2\theta - \frac{\rho_i^2}{2}\left(3 - \frac{11}{4}\sin^2\theta\right), \tag{4.156}$$

where $\rho_i = v_{Ti}/\omega_{Bi}$ is the ion Larmor radius. For such type of motions the plasma is quasi-neutral since $\omega \ll \omega_{pi}$ where $\omega_{pi} = (4\pi n_i e^2/m_i)^{1/2}$ is the ion plasma frequency. From (4.156) we can see that the wave dispersion is positive (the phase velocity increases with $|\mathbf{k}|$) except the narrow angle ranges near $\theta = 0$ and $\theta = \pi/2$. For the near-to-transverse propagation with respect to the external magnetic field $\mathbf{B}$ when $|\theta - \pi/2| < (\beta/4)^{1/2}$, the wave dispersion is negative and is defined by the effects associated with the finite ion Larmor radius ρ_i. It is well known that in order to describe the small-amplitude FMS wave propagation within the narrow angular distribution, the KP equation (4.150) can be used [86]. For those angles where the wave dispersion for small $|\mathbf{k}|$ is positive (this also requires a sufficiently high ion temperature), the three-dimensional wave packet of FMS waves in a plasma with $\beta > m_e/m_i$ does not form stable stationary solutions and disperses for angles $|\theta - \pi/2| < (m_e/m_i)^{1/2}$ or collapses outside this cone [59]. In the latter case when a sufficiently intensive *FMS wave beam* is limited in the

$\mathbf{k}_\perp$-direction, the *wave self-focusing* phenomenon can be observed (see Sect. 3.2.3). This problem was solved for the first time in Ref. [86] by averaging the initial equations and their subsequent numerical solution. However, (4.156) is invalid for the angles $\theta < (kc/\omega_{pi})^{1/2}$ where intensive rebuilding of the mechanism of the oscillations' dispersion takes place. For $\beta < m_e/m_i$ the wave dispersion can be determined for any angle θ from the hydrodynamic equations; the FMS wave structure in this case depends on the sign of the dispersion factor

$$\gamma_1 = -v_A \chi(\theta) D^2 = v_A \frac{c^2}{2\omega_{pi}^2} \left(\frac{m_e}{m_i} - \cot^2\theta \right),$$

which is defined by the angle θ, namely: for the near-to-transverse propagation the wave dispersion is negative when $|\theta - \pi/2| \leq (m_e/m_i)^{1/2}$, and it is positive for other angles. Thus the KP equation can also be used [59], and for the sufficiently intensive FMS wave beam limited in the $\mathbf{k}_\perp$-direction, we can expect the self-focusing of the wave beam propagating at those angles θ where the wave dispersion is positive.

As an example, consider a plasma with $\beta < m_e/m_i$ when the *KP equation* (4.149) with $c_s = v_A$ is used [59] for description of the small-amplitude FMS wave propagating within the narrow cone:

$$\partial_t h + \frac{3}{2} v_A \sin\theta h \partial_x h + \gamma_1 \partial_x^3 h + \hat{\nu} h = -\frac{1}{2} v_A \int\limits_{-\infty}^{x} \Delta_\perp h \mathrm{d}x, \qquad (4.157)$$

where $h = B_\sim/|\mathbf{B}|$ is the dimensionless FMS wave amplitude, $B_\sim$ is the magnetic field of the wave, and the factor at the nonlinear term can be determined on the basis of the results of Ref. [3]. In contrast to (4.149), the damping is included into (4.157), and the operator $\hat{\nu}$ determined by the damping rate $\nu(k)$ of the monochromatic wave includes both the collisional and collisionless (Landau) contributions, $\nu = \nu_{\mathrm{col}} + \nu_L$ (see Introduction), where ν_L stands for the collisionless *Landau damping* associated mostly with the plasma electrons. According to the results of Sect. 3.1, equation (4.157) can have, within the range of the angles $|\theta - \pi/2| \leq (m_e/m_i)^{1/2}$, the one-dimensional soliton solutions as well as solutions in the general form of the cnoidal waves. Outside this cone this equation can have solutions in the form of the two-dimensional stationary structures such as a *rational soliton.* The three-dimensional solitons of the KP equation, as we already noted in Sect. 3.1, are unstable in both these cases.

Consider now the one-dimensional shock-like solution of (4.157) with the asymptotics

$$h \to \begin{cases} h_0, & x \to -\infty, \\ 0, & x \to \infty. \end{cases}$$

The condition $h = h_0$ behind the front of the shock corresponds to the density discontinuity $\delta\rho/\rho_0 = h_0 \sin\theta$ and the plasma velocity given by

$v_0 = (4\pi\rho_0)^{-1/2} B h_0 \sin\theta$ [59,81,83]. As a result, the change of the Alfvén velocity is given by $\Delta v_A = 1/2(4\pi\rho_0)^{-1/2} B h_0 \sin\theta$ that, together with the velocity v_0 (in full agreement with (4.157)), gives us $3/4(4\pi\rho_0)^{-1/2} B h_0 \sin\theta$. Since equation (4.157) describes small-amplitude waves, this solution is the weak shock wave. If we assume that the wave damping is also weak (less than the wave dispersion), then, according to Ref. [103], we obtain collisionless *shock waves* with an oscillatory structure behind the front, where the amplitude of oscillations decreases with the distance from the discontinuity. Such an oscillatory structure can be represented as a set of the one-dimensional solitons. In this case, the first one has the linear size $l_s \sim D/(M^2 - 1)^{-1/2}$ where M is the Mach number, and the amplitude is of order h_0; the amplitudes of the subsequent solitons decrease. Provided that the main mechanism of the wave dissipation is due to the electron-ion collisions, the size of the oscillatory tail structure can be estimated as [103]

$$l_d \sim v_A D^2 \frac{\omega_{pe}}{c^2 \nu_{ei}}.$$

It is well known that the laminar oscillatory structure of a shock wave is unstable with respect to decays of the FMS waves involving the Alfvén and the *slow magnetosonic waves* (*SMS waves*), as well as decays inside the FMS branch of oscillations [59,103]. In a *collisionless plasma* with $\beta \ll 1$ and $\beta \neq 0$ the main nonlinear process for $T_e \gg T_i$ is the decay involving SMS waves. For these processes the excited oscillations propagate at large angles to the original wavefront, and because of the finite width of the front the instability can be suppressed. There are no analogous reasons for the instability's stabilization in the nonlinear interaction of FMS waves between themselves, because of their small angles with respect to the front and, therefore, small dispersion of the group velocities. This nonlinear interaction has absolute character, while all other processes are convective. Thus, in the considered case of the weak oblique *shock waves*, the main mechanism determining the structure of the front is the nonlinear interaction of the waves within the FMS branch.

It was shown in Ref. [59] by numerically integrating the KP equation (using the scheme of the iterative splitting) that the growth of the instability of the interacting *FMS waves* leads to the nonlinear deformation of the front structure when the field is pushed out from center of the soliton accompanied by the growth of the magnetic field on the soliton "wings," thus forming one or two collapsing "cavitons".[11] In the process of the growth of the magnetic field in the collapsing cavity there is a moment when the ions start to be reflected from the front (this occurs at the Mach numbers $M = 1.5 - 2.5$). The numerical estimates [59] allow us to conclude that the collapse of the sound oscillations for the collisionless shock waves is the mechanism with

[11] The collapse of sound waves in a dispersive medium is considered in detail in Refs. [59,63,64,103].

which the energy of directional motion converts to other degrees of freedom, such as the low-hybrid noise, fast ions, transverse modulations of the front, etc. The rate of the energy conversion is determined by the parameter ν.

Recent experimental data [208] on the shock waves in the *auroral zones of the Earth's ionosphere* with $\beta < 1$ and $M \sim 1$ agree well with the presented results. Indeed, the data clearly show the presence of the developed magnetohydrodynamic (MHD) turbulence at the front of the shock wave, as well as the energetic ions beams, the ion-cyclotron and low-hybrid oscillations. Thus, for many practically important cases the KP equation model with the dissipative term (4.149) provides adequate description of the dynamics of the nonlinear FMS waves in a magnetized plasma.

For both cases, $\beta > m_e/m_i$ and $\beta < m_e/m_i$, it is necessary to take into account that $\gamma_1 \rightarrow 0$ near the cone where the wave dispersion changes its sign. This does not necessarily mean that the wave dispersion disappears and, consequently, that the description based on KP equation in its standard form becomes invalid. Near the cone of $|\theta - \pi/2| \leq (\beta/4)^{1/2}$, where $\gamma_1 \rightarrow 0$ for $\beta > m_e/m_i$, the results of Ref. [86] need to be more elaborated. For example, the relation (4.155) must be supplemented by the next-order dispersion term which then can play the major role [83,148,196]. A similar situation occurs for $\beta < m_e/m_i$ near the cone of $\theta = \arctan(m_i/m_e)^{1/2}$. Thus, in both cases the dispersion relation takes the form of (4.143). The next order dispersion correction can be obtained from the Taylor's expansion in k of the full dispersion relation and it can be generally written as $\gamma_2 k_x^5$. In the case $\beta < m_e/m_i$, considered below in details, we have [81,196]

$$\gamma_2 = v_A \frac{c^4}{8\omega_{pi}^4} \left[3 \left(\frac{m_e}{m_i} - \cot^2\theta \right)^2 - 4\cot^4\theta \left(1 + \cot^2\theta\right) \right].$$

Thus the character of the wave dispersion becomes more complicated, and it is defined by the correlation of the signs of γ_1 and γ_2 (see Fig. 4.49). For $\gamma_1 > 0$ and $\gamma_2 < 0$ the negative wave dispersion takes place in the region B in Fig. 4.49, whereas for $\gamma_{1,2} > 0$ (the region A) and $\gamma_{1,2} < 0$ (the region C) there is the "mixed" dispersion (when the dispersion sign is different for the small and the large k). Then the propagation of small-amplitude FMS waves within the narrow angle distribution is described by the equation obtained by Belashov and Karpman [113] (see also Refs. [148,196]) which is the generalization of KP equation; for the non-dissipative case the equation is given by

$$\partial_x \left(\partial_t h + \alpha h \partial_x h + \gamma_1 \partial_x^3 h + \gamma_2 \partial_x^5 h \right) = -\left(v_A/2 \right) \Delta_\perp h, \qquad (4.158)$$

where $\Delta_\perp = \partial_y^2 + \partial_z^2$ and $\alpha = (3/2) v_A \sin\theta$. As we noted in Sect. 3.2.3, the nonlinear term, $\alpha h \partial_x h$, is the consequence of the renormalization of the sound velocity and reflects the small probability of other nonlinear processes which can be caused by the vector nonlinearity.

Unlike those for the KP equation, the three-dimensional solutions of (4.158) reveal more complicated structure and dynamics. They depend on

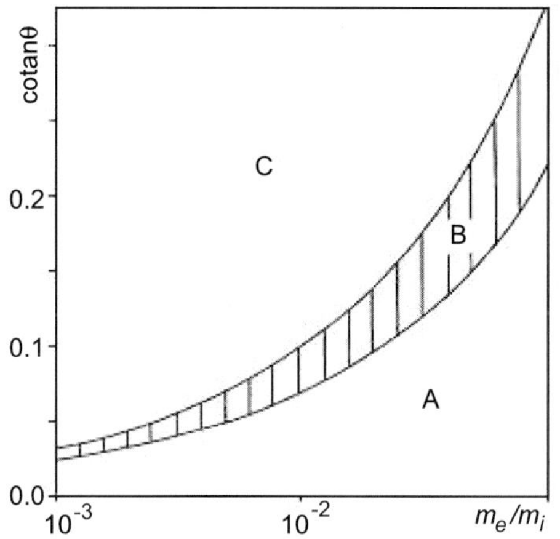

Fig. 4.49. The character of the wave dispersion near the cone $\theta = \arctan(m_i/m_e)^{1/2}$

the correlations of the values and signs of γ_1 and γ_2 [113,148,196,198]. Based on the basis of Sect. 4.1.3 (see also [195]) we obtain the following. In the case $\beta < m_e/m_i$ (in contrast to the the case $\beta > m_e/m_i$ considered in Ref. [86]) the three-dimensional *FMS wave beam* propagating at an angle θ to the external magnetic field does not self-focus and becomes stationary and stable within the cone $\theta < \arctan(m_i/m_e)^{1/2}$ when the inequality [196]

$$\left(m_e/m_i - \cot^2\theta\right)^2 \left[\cot^4\theta\left(1 + \cot^2\theta\right)\right]^{-1} > 4/3$$

is satisfied, i.e., when $\gamma_{1,2} > 0$ in (4.158). This conclusion is confirmed by numerical results [148] for the three-dimensional wave pulses propagating in weakly dispersive media where the presence of the *higher order dispersion* correction in the KP equation stops the *wave collapse* at the initial stage of development of the *self-focusing* instability. This result is important, since before the studies [148,194] no three-dimensional stable wave structure such as soliton was found either analytically or numerically. In order to study the dynamics of the FMS wave beam here we solve, in contrast to [148] where the initial value (Cauchy) problem for the FMS wave was considered, the boundary value problem for such a wave beam having the narrow angular distribution. For that we numerically integrate the corresponding equation since the exact analytical solutions of the generalized KP equation (4.4) in the absence of dissipation are not known.

Thus consider simulation of the dynamics of an *FMS wave beam* in a *magnetized plasma* [81,196]. Let there be a three-dimensional FMS wave beam propagating in the plasma at an angle θ with respect to the external magnetic field near the cone $\theta = \arctan(m_i/m_e)^{1/2}$. For such a beam, we introduce the new variables, $x \to -st$, $y \to -s\kappa^{1/2}y$, $z \to -s\kappa^{1/2}z$, $t \to sx$, $h \to -(6/\alpha)h$, $s = |\gamma_2|^{1/4}$, and $\kappa = v_A/2$, and obtain from (4.158) [81]

$$\partial_t \left(\partial_x h + 6h\partial_t h - \varepsilon\partial_t^3 h - \lambda\partial_t^5 h\right) = \Delta_\perp h, \tag{4.159}$$

which is analogous (with the formal change $t \leftrightarrow x$) to (4.4), with $\mu = 0$ describing the FMS wave beam propagating along the x-axis from the boundary $x = 0$. We assume that $\Delta_\perp = \partial_\rho^2 + (1/\rho)\partial_\rho$ in (4.159) as well as

$$h_0 = h\,(t, 0, \rho) = \cos(mt)\exp(-\rho^2), \tag{4.160}$$

thus selecting the boundary condition to specify a plane localized in (y, z) and to set the time-periodic axially-symmetric FMS wave beam, see Fig. 3.13 in Sect. 3.2.3.

Equation (4.159) with the boundary condition (4.160) was integrated [81] using the implicit scheme (4.75) with the grid 301 × 51 and the total absorption on the boundary (taking into account $\partial_\rho h|_{\rho=0} = 0$), as well as using the dynamic spectral method (see Sect. 4.3) [81,196]. A series of numerical simulations of the propagation of the FMS wave beam for various beam intensities at $x = 0, h_0$ and various angles θ (the cases A, B, and C – see above) gives us the following results. In the region A (corresponding to $\lambda = 1$ and $\varepsilon > 0$), similar to the results obtained above in Sect. 4.5, for any $h(0)$ the spatial evolution of the FMS wave beam leads first to the beam focusing determined initially by the dominant role of the nonlinear processes. On this stage, as for the usual KP equation (Sect. 3.2.3), we observe (see Figs. 4.50 and 4.51, curves 1 and 2) the beam compression in the ρ-direction with its propagation along the x-axis such as $l_\rho(x) \sim l_\rho(0)h_0/h(x)$, where l_ρ is the characteristic size of the wave in the ρ-direction, with the simultaneous fast increase of the wave beam intensity on its axis (cf. expression (3.87) of Sect. 3.2.3). Furthermore, for $x \sim 1$ (depending on ε in (4.159)), the *nonlinear saturation*

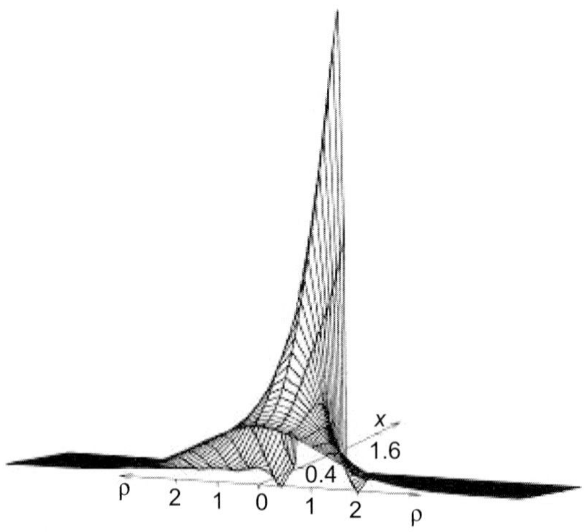

Fig. 4.50. The central part of the wave beam $h(x, \rho)$ with the intensity on the boundary $h_0 = 4$ at the stage of the FMS focusing: $\lambda = 1$, $\varepsilon = 1.34$

mode is realized because of the decreasing l_ρ when the term proportional to the fifth derivative, $\partial_t^5 h$, in (4.159) becomes dominant, thus stopping the *self-focusing*. With further propagation, the defocusing stage starts (see Fig. 4.51, curves 1 and 2). This stage finishes with the formation of a *stationary wave beam* (i.e. $h_{\max}(x)$ =const and $l_\rho(x)$ =const), corresponding to the analytical results obtained of Sect. 4.1.

In regions B and C (see Fig. 4.49) corresponding to $\lambda = -1$ and $|\varepsilon| \geq 0$, the sound wave scatters with its propagation along the x-axis for any beam intensity $h(0)$ on the boundary (see curves 3,4 in Fig. 4.51) similar to the process of the self-influence of the electromagnetic wave in media where the derivatives $\partial^2\omega/\partial k_x^2$ and $\partial^2\omega/\partial k_\perp^2$ have different signs (for example, this takes place for the ion-cyclotron waves, whistlers, etc.) [197]. We can see in Fig. 4.49 that for $\lambda = 0$ when (4.159) transforms into the KP equation with the negative wave dispersion, there are no solutions in the form of the self-focusing FMS wave beam (see Sect. 3.2.3 and Refs. [81,148,194]). Therefore, the effect of self-focusing cannot be observed in the considered model with $\lambda = 0$. The test calculations for (4.159) with $\lambda = 0$ demonstrate that the self-focusing is possible only for $\varepsilon < 0$ (see Fig. 4.51, curve 5), i.e., the FMS wave beam described by this model does not correspond to any real situation, Fig. 4.48.[12]

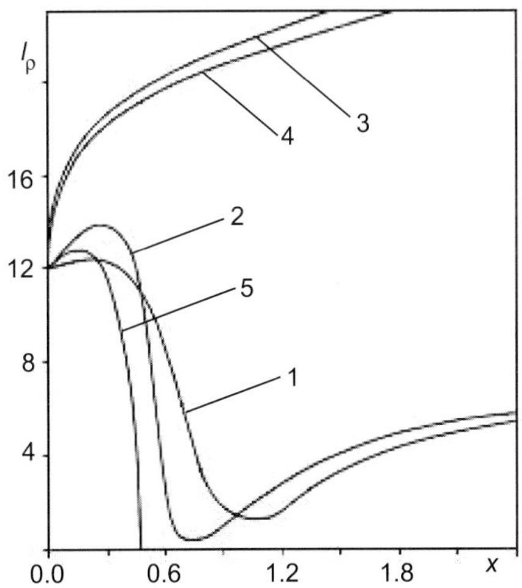

Fig. 4.51. Change of the cross section of a wave beam with its propagation in the x-direction: (1) $\lambda = 1$, $\varepsilon = 1.34$; (2) $\lambda = 1$, $\varepsilon = 2.24$; (3) $\lambda = -1$, $\varepsilon = 1.34$; (4) $\lambda = -1$, $\varepsilon = -1.34$; (5) $\lambda = 0$, $\varepsilon = -1.34$

[12] Thus the self-focusing effect is possible only for the positive wave dispersion sufficiently away from the cone $\theta = \arctan(m_i/m_e)^{1/2}$ where the influence of the higher dispersion corrections is small. We note that the similar self-focusing phenomenon was found for the magneto-elastic solitons in antiferromagnetics [87], thus proving that the area of applications of the KP equation models in physics is sufficiently wide.

Thus results of the study of the *GKP equation* (4.158) demonstrate that the *self-focusing* phenomena are not observed for the FMS wave beam propagating in a plasma at the angles θ to the external magnetic field $\mathbf{B}$ near the cone $\theta = \arctan(m_i/m_e)^{1/2}$, unlike the standard KP equation model, (4.157) with $\hat{\nu} = 0$, even if the wave dispersion for small k is positive. Note that the nonlinear stationary propagation, on a level with the beam scattering, can be observed in this case. We also note that for $|\theta - \pi/2| \gg (m_e/m_i)^{1/2}$ equations (4.155), (4.158), and (4.159) should be supplemented by the terms proportional to the mixed derivatives since we have here $|\mathbf{k}_\perp| \geq k_x$ and the terms proportional to $k_x^i|\mathbf{k}_\perp^j|$, where $i, j = 1, 2, \ldots$ appear in the wave dispersion equation in this case.

If there are stochastic fluctuations of the magnetic field in plasma (which is practically always observed in reality), the corresponding terms should be added into the model. As we demonstrated above in Sect. 4.4 on an example of the two-dimensional KP equation, this can be made by introducing the external noise (which in the most trivial case is the white Gaussian noise) into the left-hand side of the equation. In the present case, the difficulty is that for adequate description of the dynamics of FMS waves in such a fluctuating medium, the three-dimensional GKP equation can not be reduced by any transform to an exact integrable model[13] and therefore can only be integrated numerically.

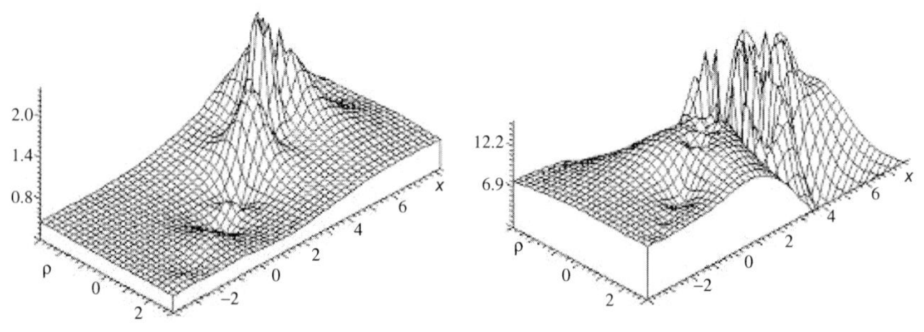

Fig. 4.52. Evolution of the three-dimentional FMS wave in a plasma with the Gaussian noise $\eta = \eta(t)$ with the standard deviation $\sigma = 0.02$; we have $\lambda = 1$ and $\varepsilon = 2.24$

To investigate the influence of *stochastic fluctuations of the magnetic field* on the dynamics of an FMS wave beam we introduce the term $\eta = \eta(t, x, \rho)$ into (4.159). Equation (4.159) with the boundary condition (4.160) is then integrated using the implicit scheme (4.75) with the grid 301×51 with

[13] For the two-dimensional stochastic KP equation this was done with the help of the Galilean transform, thus enabling us to reduce to the classic KP equation model (see Sect. 4.4) which is exactly integrable by the IST method.

the condition of the total absorption on the boundary (taking into account $\partial_\rho h|_{\rho=0} = 0$) [205,206]. A series of numerical simulations in the low-frequency limit when it is possible to assume $\eta = \eta(t)$ as well as in the more general case $\eta = \eta(t, x, \rho)$ allows us to establish that, regardless of the coherence length of the external noise, the initial FMS wave beam is spreading (acquiring the wave structure) with its propagation even in the case when (in the absence of the fluctuations) it should be stabilized, i.e., near the cone $|\theta - \pi/2| \leq (\beta/4)^{1/2}$. Moreover, if the dispersion correction $\gamma_2 \neq 0$ in (4.158)

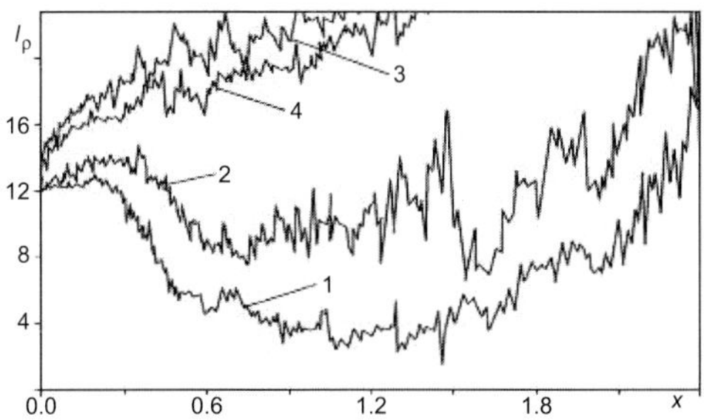

Fig. 4.53. Change of the cross section of the wave beam with its propagation along the x-axis in a plasma with the noise $\eta = \eta(t, x, \rho)$ with $\sigma = 0.04$: (1) $\lambda = 0$, $\varepsilon = -1.34$; (2) $\lambda = 1$, $\varepsilon = 2.24$; (3) $\lambda = -1$, $\varepsilon = 1.34$; (4) $\lambda = -1$, $\varepsilon = -1.34$

(this causes formation of the three-dimensional soliton-like waves with the oscillating asymptotics [114,148,196]), the spatial evolution of the wave beam leads to formation of the turbulent field structure faster since the destruction of the FMS waves by the external noise is intensified with the formation of the tail oscillations arising due to the fine dispersion effects described by the higher order dispersion term. Thus, in all cases, the wave evolution is ended by turbulization of the wave field independent of the values of the dispersion parameters and the initial intensities of both the beam and the noise, as well as any correlation lengths of the noise and the characteristic sizes of the beam. We can see from Figs. 4.52 and 4.53 that evolution of the FMS wave beams leads to their destruction accompanied by the formation of the chaotic turbulent structures which are especially strong in the regions of the wave's maxima.

4.6.3 Solitary Internal Gravity Waves in the F-layer of Earth's Ionosphere

To solve the wide range of problems associated with wave perturbations at the ionospheric heights (in the F-layer), it is necessary to take into account essential factors such as the middle- and large-scale *traveling ionospheric disturbances* (*TID*). TID directly affect variability of the ionospheric parameters as well as those of the Earth's ionosphere waveguide. One of the most convenient approaches to these problems is to study TID dynamics in terms of the *internal gravity waves* (*IGW*) [138–141]. Of special interest are the IGW solitons as traveling in the F-layer stable large-scale wave formations [138], caused by various reasons such as the isolated magnetic substorms, solar terminator [81] and solar eclipse [140,144], seismo-volcanic processes, and high-power artificial explosions [142].

Here we first investigate the dynamics of the solitary nonlinear IGW (as well as TID excited by them at the heights of the ionosphere's F-region) for conditions close to those of the F-layer, by omitting the physical nature of the sources, but assuming that it has the pulse character (more details about excitation of the pulse disturbances by various physical sources are given below as well as in the references listed above). Then we consider applications of the obtained results to the problems of the generation of IGW by the fronts of the solar terminator and the solar eclipse and by the Rayleigh waves excited by the seismic sources.

Two-Dimensional IGW Solitons and Traveling Ionospheric Disturbances of the Electron Density. For the isothermal model of Earth's atmosphere, we take into account $|\mathbf{k}_\perp|^2 \gg k_y^2$ and $|Hk_x| \ll 1$ in the linear approximation, and expanding in k up to the fifth order, write the *dispersion law* as [138]

$$\omega = Vk_x \left[1 + \frac{k_y^2}{2k_x^2} \pm \frac{(\gamma-2)^2}{\gamma^2} H^2 k_x^2 \left(2 + \frac{(\gamma-2)^2}{\gamma^2} \varepsilon H^2 k_x^2 \right) \pm H^2 k_z^2 \right], \tag{4.161}$$

where $V = 2\omega_g H$, $\omega_g = [(\gamma-1)g/\gamma H]^{1/2}$ is the *Brunt–Väisälä frequency*, H is the scale height of the neutral atmosphere, and $\varepsilon = -V/V_{\min}^{\mathrm{ph}}$, where $V_{\min}^{\mathrm{ph}}$ is the minimum phase velocity of the linear oscillations. It is easy to see that the dispersion law (4.161) for $k_z = 0$ is related to the type of (4.143). In this case, taking into account the weak nonlinearity of the function $u = u_z/ac$, $a = \exp(z/2H)$, $c = \sqrt{gH}$, from the hydrodynamic equations for the neutral gas with $\partial_z = 0$ we obtain [138]

$$\partial_t u + ac\frac{2\gamma-1}{\gamma^2} u\partial_\xi u \pm \frac{(\gamma-2)^2}{\gamma^2} VH^2 \times \partial_\xi^3 \left[2u + \frac{(\gamma-2)^2}{\gamma^2} \varepsilon H^2 \partial_\xi^2 u \right] = \frac{V}{2} \int_{-\infty}^{\xi} \partial_y^2 u \mathrm{d}\xi, \tag{4.162}$$

which is written in the reference frame moving along the x-axis with the velocity V ($\xi = x - Vt$). The upper signs in (4.161) and (4.162) correspond to the positive wave dispersion, and the lower signs correspond to the negative one (without loss of generality we further assume that $V < 0$ and, as can be easily seen from (4.161), $\varepsilon < -1$). The obtained equation is the generalization of the KP equation for the velocity of the neutral component at the heights of the F-region, similar to (4.4) with $\mu = 0$ and $\partial_z = 0$. This equation describes the nonlinear *IGW solitons* and nonlinear wave packets, with the structure determined by both the coefficients and the function $u(0, \xi, y)$ corresponding to the initial condition, i.e., it depends on the sort of perturbation and accordingly the type of the source as well.

The structure of the solutions for the initial disturbance of the wave pulse type corresponding to various physical sources as, for example, the terrestrial and anthropogenic factors (as well as the "quasi-one-dimensional" sources of the global character, such as the *solar terminator* and *solar eclipse*), is described in detail in Sect. 4.4 and depends, as for (4.4) with $\mu = 0$ and $\Delta_\perp = \partial_y^2$, on ε (note that in (4.162), unlike (4.4), ε is the factor at the fifth derivative). Indeed, the two-dimensional solitons with the algebraic (for $\varepsilon \ll -1$) or the oscillating (in the direction of propagation, for $\varepsilon \leq -1$) asymptotics correspond to the upper sign in (4.162), whereas the dispersing wave packets and/or the one-dimensional solitons which are stable in the case of the negative dispersion [16] correspond to the lower sign.

Let us consider the case of the upper sign in (4.161) and (4.162) and study the excitation by the IGW solitons of the middle- and large-scale TID for the conditions close to those in the F-layer. Considering the solitary IGW traveling at the near-to-horizontal angles, the continuity equation for the electron density in the F-layer is given by [138,139]

$$\partial_t N = \partial_z \Bigg[\left(\partial_z N + \frac{N}{2H_i} \right) D_0 \exp\left(\frac{z}{H_i} \right)$$
$$- u_z \left(1 - \mathrm{e}^{-\nu t'} \right) N \sin I \cos I \Bigg] - \beta N + Q, \qquad (4.163)$$

where $D_0 \exp(z/H_i) = D_\alpha \sin^2 I$, D_α is the ambipolar diffusion coefficient, H_i is the scale height for ions, I is the magnetic inclination, $\beta = \beta_0 \exp(-Pz/H_i)$ and Q are, respectively, the recombination rate and the ion production rate, the exponent $0 \leq P \leq 2$ characterizes the gas intermixing, $u_z = acu$ is the vertical component of the neutral particles' velocity, and $t' = t - t_0$, t_0 is the moment of the start of the neutral component's perturbation. Now we approximate the profile of the electron density at the height z by $N = N_1 \exp(z/H_i)$, $N_1 = N|_{z=0}$, and obtain that solution of (4.163) is given by [139]

$$N(u,t) = N(u,t_0) \exp\left[G(u,t) \right], \qquad (4.164)$$

where

$$G(u,t) = \int_{t_0}^{t} g(u,t) \mathrm{d}t,$$

$$g(u,t) = C - \left(\frac{1}{H_i} + \frac{1}{2H}\right) f(u,t),$$

$$f(u,t) = acu\,[1 - \exp\left(-\nu t'\right)] \sin I \cos I,$$

$C = 3a_1/H_i^2 - \beta(1-q)$, $a_1 = D_\alpha \sin^2 I$, and $q = Q/\beta N$. Here, the func-

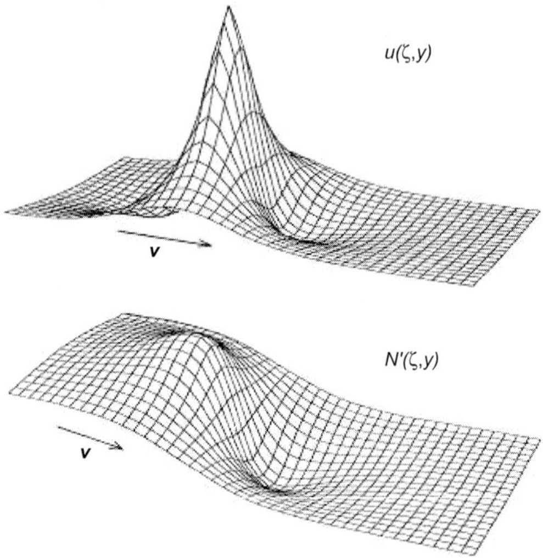

Fig. 4.54. IGW soliton for $\varepsilon = -12$ and associated perturbation of the normalized electron density $N' = [(N(u,t) - N(0,t))/N(0,t)] \times 100\%$

tion u satisfies (4.162). When $\varepsilon \ll -1$ and the solution of (4.162) is the two-dimensional soliton with the algebraic asymptotics, the solution (4.164) for the quasi-pulse source of IGW is shown in Fig. 4.54. If $\varepsilon \leq 1$ then the perturbation of the electron density N as well as the IGW soliton has the oscillating asymptotics shown in Fig. 4.55a.

The solution of (4.162) and (4.164) for the conditions typical for the F-layer gives us the following results. The *solitary IGW* excite in the F-region the *solitary TID* of the electron density, their structure depends on the form of IGW and the ionospheric parameters determined by the photo-chemical and dynamic processes at the height considered. The amplitude of TID increases in the direction of the geomagnetic latitude $\varphi_m = 45°$, the wave front steepens, and at the latitude $\varphi_m = 45°$ the wave becomes similar to the shock wave.

The two-fold increase of the IGW amplitude results in the increase of the TID amplitude: 35% for $\varepsilon \ll -1$; close to 100-105% for $\varepsilon \leq -1$. For all the studied cases we note the phase shift of TID relative to the phase of IGW ($\Delta t \sim 0.5 - 5$min) and the effect of the relaxation of the electron density perturbations which increase with the decreasing ε characterizing essentially the medium's dispersion. Figure 4.55 shows the simulation results for IGW-solitons with a velocity on the order of 200m/s at $z = 0$ and $I = 63.4°$. Thus such ionospheric characteristics as the height of the maximum and the

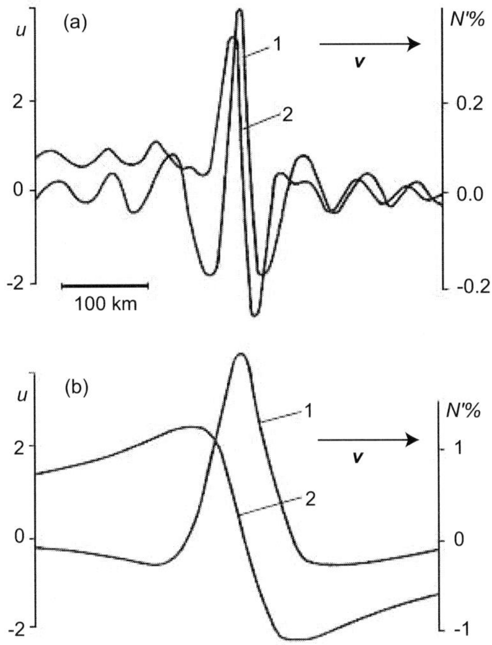

Fig. 4.55. Profiles of the perturbations at $y = 0$ of IGW (the curves 1) and TID (N') (the curves 2). **a** $\varepsilon = -1.2$. **b** $\varepsilon = -12$

critical frequency of the F-layer increase proportionally to the TID amplitude when the two-dimensional nonlinear IGW propagates, as well as experience relaxation similar to the relaxation of the electron density N' [138].

Generation of IGW Solitons by Fronts of Solar Terminator and Solar Eclipse. In addition to the general study of the dynamics of solitary waves in the F-region of the Earth's ionosphere [81,140], the middle-scale and large-scale wave effects associated with motions of the fronts of the *solar terminator* (ST) [81] and the *solar eclipse* (SE) [140] were investigated numerically within the framework of the above developed weakly nonlinear approximation neglecting the dissipation effects. In the reference frame related to the source, the initial condition is given by

$$u_0 = \left[ac\left(1 - \mathrm{e}^{-\nu t'}\right)\sin I \cos I\right]^{-1} \left\{V \frac{H_i}{N_0}\partial_\xi N_0 \left(1 - \mathrm{e}^{-z/H_i}\right)\right.$$
$$\left. + \left[\beta \frac{H_i}{1-P}\left(1 - \mathrm{e}^{(1-P)z/H_i}\right) + \frac{Qz}{N_0}\right]\mathrm{e}^{-z/H_i} + \frac{3}{2}\frac{D_0}{H_i}\mathrm{e}^{z/H_i}\right\},$$

where $t' = \xi/V$ and the parameters are defined by the adopted dynamic model of the F-layer [81], taking into account the characteristic scales of the particular phenomenon studied – the ST or the SE. Then we solve the initial value Cauchy problem for $z = 0, H, 2H$. The solutions obtained are tested by considering generation of the wave precursors by the ST and SE fronts with the periods of order $40-60$min (ST) and $3-10$min (SE) (their scales depend on the parameters of the F-layer and the character of the phenomenon).

Generally, the effect for $z = 0$ is the train of the two-dimensional soliton-like waves (with $k_x \gg k_y$) similar to the multisoliton solutions of the KdV equation (the case $y = 0$). For $z = H, 2H$ the qualitative form of the solution is maintained, although they are less regular and (on the average) the larger-amplitude waves. The characteristics of such soliton-like formations strongly depend on the season and the ionospheric parameters. Simulations for the conditions corresponding to the partial solar eclipse observed on March 18, 1988, and the sunrise and sunset periods on March 1–10, 1990, (an interval of the International Geophysical Calendar) agree well with the results of special targeted experiments on the passive slanted sounding of the ionosphere done in these periods in the Far-Eastern region of Russia [144,209]. Thus we conclude that despite some idealization of the problem, the approach based on the generalized KP equation allows us to predict the effects of the TID dynamics in the F-region of the Earth's ionosphere reasonably well.

Generation of IGW Solitons by Seismic Sources. Effects in the ionospheric plasma stipulated by the *seismic activity* now attracts serious attention of researchers not only by virtue of their pure scientific interest but also because of the urgency in investigation of the capabilities to prevent and reduce consequences of natural catastrophes associated with the seismic hazards. In addition to studying the *seismic-ionospheric phenomena* related to the possibility of prediction the seismic activity, the problem of investigation of the seismic-ionospheric post-effects is important because of many reasons, e.g., for better understanding of the physics of the terro-atmosphere-ionosphere relations, for location of the earthquakes' epicenters, and for selection of oscillations in the spectra of the registered ionospheric data caused by the seismic activity. Despite many works on the seismic-ionospheric effects in the nearest zone of the epicenter (see, e.g., Ref. [142] and numerous references therein), less attention was paid to study effects of the seismic source on the Earth's ionosphere in the farthest zone. This problem was considered for the *surface Rayleigh waves* generated by an earthquake in the three-dimensional model taking into account the weak nonlinearity and dispersion in the ionosphere at the heights of the F-layer [81–83,141,143], as well as the dissipation effects and stochastic fluctuations of the electron density in the F-layer [142] for the ionospheric parameters close to their real values. On the basis of the *GKP equation* model developed above, we present here the main results.

We start with the set of the gas dynamics equations:

$$\partial_t \rho' + \boldsymbol{\nabla} \cdot (\rho_0(z)\mathbf{V}) = 0,$$

$$\rho_0(z)\left[\partial_t \mathbf{V} + (\mathbf{V}\cdot\boldsymbol{\nabla})\,\mathbf{V}\right] = -\boldsymbol{\nabla} p' - \rho' g \mathbf{e}_z$$
$$+\eta(z)\left[\frac{1}{3}\boldsymbol{\nabla}\boldsymbol{\nabla}\cdot\mathbf{V} + \boldsymbol{\nabla}^2\mathbf{V}\right] + \zeta(z)\boldsymbol{\nabla}\left(\boldsymbol{\nabla}\cdot\mathbf{V}\right),$$

and

$$\partial_t p' + (\mathbf{V} \cdot \boldsymbol{\nabla} p_0(z)) = -c^2 \rho_0(z) \boldsymbol{\nabla} \cdot \mathbf{V},$$

where the unperturbed functions are marked by the subscript '0', and the prime denotes perturbations of the corresponding functions if their unperturbed values do not equal zero, and the functions without sub/superscripts correspond to those with the zero unperturbed values. In these equations, ρ is the gas density, p is the gas pressure, and $p_0(z) = \rho_0(z)c^2/\gamma = gH\rho_0(z)$, where $\gamma = C_\mathrm{p}/C_\mathrm{v}$ is the adiabatic exponent, H is the scale height of the neutral atmosphere, $\eta(z) \approx 3c_s^2\rho_0(z)/2\gamma\nu_{nn}(z)$ and $\zeta(z)$ are the dynamic and kinematic viscosities, respectively, and $\nu_{nn}(z)$ is the collision frequency of the neutral particles. We employ the standard assumption of the exponential dependence of the unperturbed density of the atmosphere and ionosphere on the height z, viz., $\rho_0(z) = \rho_0(0)\exp(-z/H)$, and take into account that when the wave propagates on the large distances, the spatial dispersion (the case of the farthest zone) results in the acoustic wave damping accompanied by the shift of the maximum of the spectrum to the lower frequency region [143]. In this case, with the decreasing role of such a source as the acoustic wave pulse caused by the Earth's surface oscillations in the earthquake's epicenter, the role of the *surface Rayleigh waves* increases relatively to the distance [141].

We choose the first boundary condition as that approximating the surface Rayleigh wave at the distance far away from the epicenter [143]:

$$V_z \Big|_{z=0} = \mathrm{d}_t Z(r', t), \tag{4.165}$$

where

$$Z(r', t) = h(t) \exp\left[-(r')^2/L^2\right],$$

$(r')^2 = \xi^2 + y^2$, $\xi = x - v_R t$, and v_R is the velocity of the Rayleigh wave. Thus we consider the problem in the reference frame associated with the Rayleigh wave. The second boundary condition corresponding to $z \to +\infty$ is given by

$$V_z(t, z, \xi, y) \Big|_{z \to +\infty} \longrightarrow 0 \tag{4.166}$$

when $\eta \neq 0$. This formulation provides conservation of the energy flow for $z \to +\infty$.

The Rayleigh wave (4.165) leads to the excitation of IGW propagating upwards with the amplitude increasing exponentially with the height. At the heights of the F-region with the formation of the *solitary IGW* excited by the Rayleigh wave, the nonlinear effects begin to develop [141]. Taking into account the geometry of the problem we assume that $k_x^2 \gg \mathbf{k}_\perp^2$ and $|Hk_x| \ll 1$, i.e., the Boussinesq approximation is valid for the IGW propagating in the F-layer at the small angle to the horizon. Taking into account the dissipation of the viscous type and the weak nonlinearity for the neutral particle velocity $u(t,r',z) = V(t,r',z)|_{x=\xi+vt}$ for $\partial_z = 0$, we thus obtain from the basic equations the *GKP equation* similar to (4.162) [141]:

$$\partial_t u + \frac{2\gamma-1}{\gamma^2} u_z \partial_\xi u - \nu \partial_\xi^2 u + 2\frac{(\gamma-2)^2}{\gamma^2} vH$$
$$\times \partial_\xi^3 \left[u + \frac{(\gamma-2)^2}{2\gamma^2} \varepsilon H^2 \partial_\xi^2 u \right] = \frac{v}{2} \int\limits_{-\infty}^{\xi} \partial_y^2 u \mathrm{d}\xi. \tag{4.167}$$

Note that unlike (4.162), equation (4.167) includes the dissipative term given by

$$\nu = \frac{\rho_0}{2\rho}\left(c_\infty^2 - c_0^2\right)\tau \int\limits_0^\infty \mu\phi(\mu)\mathrm{d}\mu = (2\rho_0)^{-1}\left[\frac{4}{3}\eta + \zeta + \gamma\left(\frac{1}{C_\mathrm{v}} - \frac{1}{C_\mathrm{p}}\right)\right],$$

where c_∞ and c_0 are the velocities of the high-frequency and the low-frequency sound, respectively, see Introduction.

For the solitary IGW propagating in the F-layer at a small angle to the horizon, the continuity equation for the electron density N_e is given by (4.163) and in the Boltzmann approximation $N_e = N_{e0}\exp(z/H_i)$, where $N_{e0} = N_e|_{z=0}$, its solution can be written as (4.164) with the function u satisfying the GKP equation (4.167). Accordingly, the solution for the pulse source (i.e., the Rayleigh wave) for $\nu = 0$ and $\varepsilon \ll -1$ is the classic algebraic *KP soliton* (see Fig. 4.54) and can therefore be obtained analytically by the *IST method.* For $\varepsilon \to -1$ or $\varepsilon \ll -1$ in the case $\nu \neq 0$, the IST method cannot be applied and the problem should be solved numerically. The results of numerical integration of (4.167) and (4.164) with $\nu = 0$ for the ionospheric parameters close to those in the F-layer are shown in Fig. 4.55. The case $\nu \neq 0$ was investigated in detail in Refs. [113,195]. It was shown that the presence of dissipation leads to the exponential decrease of the amplitude with the rate $\Gamma(t) \sim \nu$. Dissipation also leads to the perturbation of the structure and the symmetry of the IGW soliton accompanied by the relaxation in the recovery of the electron density after the wave passes by (see above and also Ref. [83]).

The effects of *stochastic fluctuations of the wave field* $u(t,x,y)$ on the evolution of the ionospheric perturbation excited by the Rayleigh wave can also be accounted for in the basic equations. Thus, accordingly, (4.167) should be complemented by the term like $\chi(t,r',z)$ [142]. In the case of the low-frequency fluctuations when $\chi = \chi(t)$, equation (4.167) for $\varepsilon = 0$ was investigated analytically [151,201] and considered in detail in Sect. 4.4 (where we used the notation η for the function χ). The obtained results can be easily applied to (4.167) with $\chi = \chi(t)$ on the left-hand side. Thus the interpretation of the results [151,201] in terms of the problem (4.164) and (4.167) enables us to conclude that even small stochastic fluctuations of the wave field lead to the damping of the solitary IGW (with its propagation) accompanied by the transform of the wave to an oscillatory structure.

In the case $\chi = \chi(t, r', z)$, however, the analytical study of the process becomes too complicated, and in Ref. [142] numerical integration of (4.164) and (4.167) with the stochastic term was done. The obtained results appear to be qualitatively similar to the case $\chi = \chi(t)$, namely, the decrease of the amplitude of the solitary oscillating IGW is observed, with the subsequent destruction of the wave. The estimates obtained [142] demonstrate that it is practically impossible to pick up the response of the electron density in the F-layer to the earthquake-generated Rayleigh wave already at distances $r \sim 12{-}13L$ (where L is the linear size of the epicenter) from the earthquake's epicenter. To conclude, we note that the analytical and the numerical results obtained above are in good agreement with the results of the radiophysical experiments done in the periods of the seismic activity in the network of stations in the Far-Eastern region of Russia [210].

4.6.4 Two-Dimensional Solitons in Shallow Water

Consider application of the KP equation model to the hydrodynamics, namely to the description of *surface waves* in ideal incompressible shallow (comparatively with the wave length) fluid. In this case, the generalized density and velocity of sound in (1.1)–(1.12) acquire the sense of the fluid depth, H, and velocity, $c = \sqrt{gH}$. The term $gH^2/2$ plays the role of the pressure, this corresponds to the effective adiabatic index $\gamma = 2$ [3]. Then the *Boussinesq equations* (1.7) can be written as

$$\frac{\partial \mathbf{v}}{\partial t} + (\mathbf{v} \cdot \boldsymbol{\nabla})\,\mathbf{v} + \boldsymbol{\nabla} gH + \frac{gh^2}{3}\boldsymbol{\nabla}\Delta H = 0 \tag{4.168}$$

and

$$\frac{\partial H}{\partial t} + \boldsymbol{\nabla}\,(H\mathbf{v}) = 0, \tag{4.169}$$

where h =const is the depth of the fluid; it is easy to add into these equations the terms associated with the capillary effects. Assuming that the curvature of the surface is not too large and the additional pressure to the fluid caused by the surface tension is defined by the Laplace formula $\delta p = \sigma(R_1^{-1} + R_2^{-1})$, where σ is the surface tension coefficient, R_1 and R_2 are the main curvature radii, we can write $\delta p = -\sigma\Delta\eta$ where $\eta(x, y, t)$ is the surface function (the value of η is sufficiently small). Thus we change ρgh in (4.168) to $\rho gH + \delta\rho$ (ρ is the density of the fluid) to obtain

$$\frac{\partial \mathbf{v}}{\partial t} + (\mathbf{v}\boldsymbol{\nabla})\,\mathbf{v} + g\boldsymbol{\nabla} H + \left(\frac{gh^2}{3} - \frac{\sigma}{\rho}\right)\Delta\boldsymbol{\nabla} H = 0. \tag{4.170}$$

Equations (4.169) and (4.170) are the *Boussinesq equations* taking into accounting the *capillary effects* [3]. Change of the factor at the dispersion term in (4.170) leads to the change of the dispersion equation, viz., instead of $\omega = c_0 k(1 - H^2k^2/6 + \ldots)$ we have

$$\omega = c_0 k \left[1 - \left(H^2 - \frac{3\sigma}{\rho g}\right)\frac{k^2}{6} + \ldots\right], \tag{4.171}$$

where $c_0 = \sqrt{gH}$. In this case the dispersion parameter β is defined by

$$\beta = \frac{c_0}{6}\left(H^2 - \frac{3\sigma}{\rho g}\right). \tag{4.172}$$

Furthermore we use the results of Sect. 3.1 and transform (4.168) and (4.169) to the form (3.8), i.e., obtain the *KP equation* for the *gravity-capillary waves* in *shallow water*. Here we note that for sufficiently large $\sigma > \rho g H^2/3$ the dispersion parameter changes its sign that involves the qualitative change of the character of the evolution and the form of the solutions, see Sect. 3.1.

Consider now in more detail the following interesting case. Often there is a case when the factor β is unusually small. Thus, according to (4.172) $\beta = 0$ for $H = (3\sigma/\rho g)^{1/2} \approx 0.48$cm (for pure water). A similar situation occurs for *FMS waves* in a *magnetized plasma* for $\cot^2\theta = m_e/m_i$ (see Sect. 1.1 and Sect. 4.1). However $\beta = 0$ does not mean that there is no dispersion in such a medium. It simply means that in this case the next terms in the Taylor expansion in k of the full dispersion equation must be taken into account (the corresponding additional terms also appear in (4.6) and so on). Then, it is generalized so that the *KP equation* (3.8) can be written as

$$\partial_t u + \alpha u \partial_x u - \beta \partial_x^3 u + \gamma \partial_x^5 u = -\frac{c_0}{2}\int_{-\infty}^{x} \partial_y^2 u dx, \tag{4.173}$$

where the coefficients are

$$\alpha = \frac{3}{2}\frac{c_0}{H}$$

and

$$\gamma = \frac{c_0}{6}\left[H^2\left(\frac{2}{5}H^2 - \frac{\sigma}{\rho g}\right) - \frac{1}{12}\left(\frac{3\sigma}{\rho g} - H^2\right)^2\right].$$

Numerical integration of (4.173) using the *method of stabilizing factor* [31] (see Sect. 3.1.4) enables us to investigate the structure of the two-dimensional solitons in shallow water in the case of the anomalously weak dispersion [80]. It was found that the qualitative form of the solutions depends significantly on the value of $\varepsilon = (\beta/V)(-V/\gamma)^{1/2} \ll 2$, where V is the soliton's velocity in the reference frame moving along the x-axis with the phase velocity c_0.

When $\varepsilon = 0$ the structure of the two-dimensional solitons found numerically does not differ qualitatively from the structure of the algebraic KP-solitons (see Sect. 3.1.2). Such solitons on the surface of the fluid are negative (i.e. appear as the hollow solitons). When $\varepsilon > 0$, for example, in the case of the increasing fluid depth starting from the depth $H = (3\sigma/\rho g)^{1/2}$, the structure of the solitons radically changes: by remaining to decay from their maximum to zero in the transverse direction as before, now their sign varies along the direction of their propagation. For $\varepsilon \to 2$ the number of the oscillations of the tails increases and the solitons the resemble two-dimensional high-frequency trains, i.e., *envelope solitons.* We note that the one-dimensional solitons of the generalized KdV equation ((4.173) with the right-hand side equal to zero) [84] and *IGW solitons* considered above have the similar structure.

Numerical simulation [80] of (4.173) demonstrated that the complex *bisoliton* structures forming the stationary *bound states* are also found. The structure of the field of such bisolitons far from their maxima is qualitatively similar to the structure of the field of a single soliton described above. Therefore, Ref. [80] suggested that more complicated two-dimensional multi-soliton stationary structures can possibly exist. However, their numerical study requires quite high computational power and is therefore very difficult.

Structure and Evolution of Two-Dimensional Solitons of Gravity and Gravity-Capillary Waves with Varying Water Depth. In Sect. 4.4.5, the problem of the evolution of the two-dimensional solitons in media with the dispersion parameter as a function of the coordinates and time, $\beta = \beta(t, x, y)$, was considered in detail. Here, we interpret those results in terms of the propagation of the small-amplitude *surface waves* on a *shallow water.* We consider here the shallow water as some "reservoire," where the relation between its depth and the length of the wave propagating on the water surface is $\lambda \gg H$. The wave amplitude remains small due to the condition $u \ll H$. As we noted in the Introduction and Sect. 4.4, for the *gravity waves* and *gravity-capillary waves* the dispersion parameter β is a function of the water depth H. For these two cases $\beta = c_0 H^2/6$ and $\beta = (c_0/6)(H^2 - 3\sigma/\rho g)$, respectively, where ρ is the density and σ is the coefficient of the surface tension of the fluid. If, in addition, the depth of the reservoire is $H = H(t, x, y)$, i.e., it describes the change in space character of the bottom relief and/or motions in time of the parts of the bottom, then the dispersion parameter also becomes the function of the coordinates and time. Therefore, we can consider the problem for the dispersion parameter defined by these dependencies and interpret the results obtained for the three model cases of Sect. 4.4.5 in terms of the variable in time and/or in space relief of the bottom. Thus the model (4.128),

$$\beta(x) = \begin{cases} \beta_0, & x \leq a, \\ \beta_0 + c, & x > a, \end{cases}$$

describes the sudden break of the bottom at the critical time moment of $t_{\rm cr}$, where the parameters a and c have the sense of the coordinate x and the height of the break, respectively (note that c may be negative – the so-called negative step). In the numerical simulation of the problem we varied, first, the height of the break at a =const and $t_{\rm cr}$ =const and, then, the critical time $t_{\rm cr}$ (at the constant a and c). Finally, we studied the case of the dependence of the structure and the evolution of the surface wave on the localization of the break at the constant c and t_{cr}.

The second model (4.129) is given by

$$\beta(x,t) = \begin{cases} \beta_0, & x \le a, \\ \beta_0 + nc, & x > a, \end{cases}$$

where $n = (t - t_{\mathrm{cr}})/\tau = 1, 2, \ldots$, and is interpreted as the gradual increase/decrease of the depth of the bottom area. In this case, the critical time t_{cr} is chosen at the moment when the depth of the bottom starts to change, the rate of its lifting up (or falling down) is nc/τ (where τ is the time step of the numerical simulation), and the coordinate of the break is a [203]. Variations of a, c, and t_{cr} are analogous to the previous case.

Finally, the last model (4.130),

$$\beta(t) = \beta_0 \left(1 + k_0 \bar{\beta} \sin \omega t\right),$$

where $\bar{\beta} = (\beta_{\max} - \beta_{\min})/2$, $0 < k_0 < 1$, and $\pi/2\tau < \omega < 2\pi/\tau$, can be conveniently interpreted in terms of the harmonic oscillations of the bottom starting at the critical time moment t_{cr}. In this case, in the numerical simulations [203] the parameter k_0 was first varied (at the constant oscillation frequency ω), and then, on the contrary, it was assumed that k_0 =const with the varying frequency ω. As a result, there were more than a thousand numerical runs combined in three series, according to the adopted models of the function $\beta = \beta(t, x, y)$. The general character of the results obtained in these numerical simulations of the evolution of the two-dimensional solitons in media with $\beta = \beta(t, x, y)$ [203,204] was described in Sect. 4.4, and it is clear from the written above that they can be easily interpreted in terms of the hydrodynamics.

5. Appendices

Here, we elaborate and present two technical problems. In the first one, we investigate expansion of a four-dimensional dynamic system linearized in the vicinity of singular points (and the corresponding canonical systems) into two sub-systems. Similarly, we consider expansion of three-dimensional dynamic systems into a two-dimensional system and one equation that used in Sect. 2.2.3 when constructing the *phase portraits* of solutions in four-dimensional and three-dimensional phase spaces, respectively. In the second part, we investigate an algebraic equation of the fourth order appearing when analyzing possible extrema of the Hamiltonian of the GKP equation in Sect. 4.1.

Appendix 1

Each of the sets (2.104) and (2.105) can be represented in the matrix form

$$\dot{\mathbf{x}} = A\mathbf{x}, \tag{5.1}$$

where $\mathbf{x} = (w, x_1, x_2, x_3)$,

$$A = \begin{bmatrix} 0 & 1 & 0 & 0 \\ 0 & 0 & 1 & 0 \\ 0 & 0 & 0 & 1 \\ \alpha_1 & \alpha_2 & \alpha_3 & \alpha_4 \end{bmatrix}, \tag{5.2}$$

and $\alpha_1, \alpha_2, \alpha_3, \alpha_4$ are the factors, respectively, at w, x_1, x_2, x_3 in the last equations of the sets (2.104) and (2.105). The matrix A has its Jordan form J, and the eigenvalues λ of the matrix A define the type of the matrix J. For example, in the neighborhood of the singular point $w_1 = 0$ for $\mu = \delta = 0$ in (2.96) the eigenvalues λ_1 and λ_2 of the matrix (5.2) are real ($\lambda_1 = \lambda_2$), the eigenvalues λ_3 and λ_4 are imaginary ($\lambda_3 = \lambda_4$), and the singular point can be defined as the *stable knot–centre point* in the four-dimensional phase space, by analogy with the classification of the two- and three-dimensional dynamic systems.

Since $\det A \neq 0$ and $\lambda_i \neq 0$ $(i = 1, 2, 3, 4)$, the linear set (5.1) and the corresponding canonical set (which can be obtained from (5.1) by transform $\mathbf{x} = M\mathbf{y}$ where M is the transformation matrix)

$$\dot{\mathbf{y}} = J\mathbf{y}, \quad \text{and} \quad J = \begin{bmatrix} B & 0 \\ 0 & C \end{bmatrix} \tag{5.3}$$

are simple. The matrix J consists of two second order Jordan boxes B and C having in a general case the form

$$\begin{bmatrix} a_j & -b_j \\ b_j & a_j \end{bmatrix},$$

where $a_j, b_j \in \mathbb{R}$ $(j = 1, 2)$ are defined from expressions for the solutions of the secular equations corresponding to the last equation of the sets (2.104) and (2.105) with the coefficients $\alpha_1, \alpha_2, \alpha_3, \alpha_4$, namely, $\lambda_1 = -\lambda_2^* = a_1 - ib_1$ and $\lambda_3 = \lambda_4^* = a_2 - ib_2$. Therefore, the set (5.3) can be represented as two two-dimensional sets:

$$\begin{bmatrix} \dot{y}_1 \\ \dot{y}_2 \end{bmatrix} = B \begin{bmatrix} y_1 \\ y_2 \end{bmatrix}, \quad \text{and} \quad \begin{bmatrix} \dot{y}_3 \\ \dot{y}_4 \end{bmatrix} = C \begin{bmatrix} y_3 \\ y_4 \end{bmatrix}.$$

Therefore, the phase portrait of the linear set (5.1) can be considered in projections of the singular point and spatial phase trajectories onto two planes. The factorization of any set of the type (2.101) linearized in the neighborhood of the corresponding singular points with the coordinates (2.103) into two subsets can be done similarly. It is necessary to take into account that for the special values of the coefficients α_i of the matrix (5.2), when $\lambda_1 = -\lambda_2 = -\lambda_3 = \lambda_4$ or $\lambda_1 = \lambda_2 = \lambda_3 \neq \lambda_4$, the Jordan boxes B and C have another form and, therefore, the factorization considered above is impossible. In our case, however, such a situation does not take place for equations of the type (2.92).

Each of the sets (2.110) and (2.111) can be represented in the matrix form (5.1), where $\mathbf{x} = (w, x_1, x_2)$,

$$A = \begin{bmatrix} 0 & 1 & 0 \\ 0 & 0 & 1 \\ \alpha_1 & \alpha_2 & \alpha_3 \end{bmatrix} \tag{5.4}$$

and $\alpha_1, \alpha_2, \alpha_3$ are the factors, respectively, at $w, x1, x2$ in the last equations of the sets (2.110) and (2.111). Transform the set (5.1) into the canonical form (5.3) with

$$J = \begin{bmatrix} a & -b & 0 \\ b & a & 0 \\ 0 & 0 & \lambda_3 \end{bmatrix},$$

where $a, b, \lambda_3 \in \mathbb{R}$ and $\lambda_1 = \lambda_2^* = a - ib$. Then we can obtain the phase portrait of the set (5.1) from the phase portrait of the canonical set using the transform $\mathbf{x} = M\mathbf{y}$ where M is the transformation matrix. Directions of the trajectories and their angles with respect to the axes in the poles and saddles for the canonical system are known. In the reference frame x_1, x_2, x_3, the axes y_1, y_2, y_3 are straight lines passing through the point $(0, 0, 0)$, their vectors have directions $\mathbf{m}_1, \mathbf{m}_2, \mathbf{m}_3$ and form the matrix M. The eigenvalues $\lambda_1, \lambda_2, \lambda_3$ of the matrix A can be obtained from the equation

$$A\mathbf{m}_i = \lambda_i \mathbf{m}_i, \qquad i = 1, 2, 3,$$

after calculation of the vectors $\mathbf{m}_1, \mathbf{m}_2, \mathbf{m}_3$, that allows us to obtain the directions of the separatrices for the linearized set (5.1).

Appendix 2

Performing the transform $t \to t' + 8a^3b/c^4$ in (4.25), we obtain the reduced equation

$$t'^4 + pt'^2 + qt' + r = 0. \tag{5.5}$$

The cubic resolvent kernel $z^3 + 2pz^2 + (p^2 - 4r)z - q^2 = 0$, using the change $z \to x - 2p/3$, can be reduced to the equation

$$x^3 + p'x + q' = 0, \tag{5.6}$$

where $p' = 2^{10}be^3/c^4$ and $q' = -2^{14}a^2b^2e^4/c^8$, with the discriminant

$$D = 2^{26}b^3c^{-12}e^8\left(2^4 3^{-3}e + a^4bc^{-4}\right). \tag{5.7}$$

In the case $e > 0$ and $|a| \geq 0$ we have $D > 0$, therefore (5.6) as well as the resolvent kernel in the real vector space $\mathbb{R}$ for each quadruple of the values of functions $a, b, c, e \in \mathbb{R}$ have one root. Thus, using the Descartes' rule of signs, we can conclude, that (4.25) for $e > 0$ and $|a| \geq 0$ has one positive root $t \in \mathbb{R}$ (note that $t \leq 0$ does not satisfy (4.25), in this case $\zeta \notin \mathbb{R}$). It follows from the analysis of (4.25) that in the space $\mathbb{R}$ for $S_t \subset \mathbb{R}$ the equalities

$$\inf_{a>0} S_t = \sup_{a<0} S_t = 4\left(be^3\right)^{1/4}/c, \quad \text{and} \quad \inf_{a<0} S_t = 0, \tag{5.8}$$

take place.

Consider now the case $e < 0$ and $|a| \geq 0$. It follows from (4.25) that for $a \leq 0$ this equation foes not have roots $t > 0$ in the space $\mathbb{R}$, therefore, we limit ourselves by an analysis of (4.25) for $a > 0$. When

$$F = a^4b/c^4e < -2^4 3^{-3}, \tag{5.9}$$

we have $D > 0$ from (5.7). It then follows that (5.6) and the resolvent kernel in the space $\mathbb{R}$ for each quadruple of the functions $a, b, c, e \in \mathbb{R}$ have one root, and (4.25), taking into account the rule of signs, has two positive roots $t_{1,2} \in \mathbb{R}$.

Let us estimate boundaries of the set $S_t \subset \mathbb{R}$. With the two changes $t \to t + h$ and $t \to -t + h$ in (4.25), we obtain, respectively, the sets

$$c^4 > 8ia^{4-i}bh^{-i}\left(ah + 2e\right)^{i-1}, \tag{5.10}$$

and

$$(-1)^i c^4 > (-1)^i 8ia^{4-i}bh^{-i}(ah+2e)^{i-1}, \tag{5.11}$$

where $i = 1, 2, 3, 4$. Solving inequalities (5.10) and (5.11) with the condition (5.9), we obtain

$$\begin{aligned}
&\sup S_t = 2^5 3^{-3}[\sqrt{10}\cos(\psi_1/3 + 2\pi/3) - 4]a^{-1}e,\\
&\psi_1 = \text{Arccos}[-11/(2^5 5\sqrt{10})],\\
&\inf S_t = \min\{\max(h_1', h_2'), 2^4 3^{-3}(\sqrt{10} - 8)a^{-1}e\},\\
&h_1' = 8[\sqrt{-2FF'}\cos(\psi_2/3) + F]a^{-1}e,\\
&h_2' = 8[\sqrt{-2FF'}\cos(\psi_2/3 + 4\pi/3) + F]a^{-1}e,\\
&\psi_2 = \text{Arccos}[(2^7F^2 + 32^3F + 3)/(2^3F'\sqrt{-2FF'})],\\
&FF' = 1 - 2F.
\end{aligned} \tag{5.12}$$

If condition (5.9) is not satisfied, we have $D \leq 0$. In this case, a simple analysis shows that (5.6) and the resolvent kernel for each quadruple of the functions $a, b, c, e \in \mathbb{R}$ have one positive and two negative roots. Therefore, equations (5.5) and (4.25) in the real vector space $\mathbb{R}$ have no roots.

References

1. N.J. Zabusky and M.D. Kruskal: Interaction of "solitons" in a collisionless plasma and the recurrence of initial states, Phys. Rev. Lett. **15**, 240 (1965)
2. G.B. Whitham: Linear and nonlinear waves (Wiley-Interscience, New York 1974)
3. V.I. Karpman: Non-linear waves in dispersive media (Pergamon, Oxford 1975)
4. K. Lonngren and A. Scott, Eds.: Solitons in action (Academic, New York 1978)
5. R.K. Bullough and P.J. Caudrey, Eds.: Solitons (Springer, Berlin Heidelberg New York 1980)
6. A.S. Davydov: Solitons in molecular systems (D. Reidel, Dodrdrecht 1985)
7. V. Petviashvili and O. Pokhotelov: Solitary waves in plasmas and in the atmosphere (Gordon and Breach, Philadelphia 1992)
8. F. Kh. Abdullaev: Theory of solitons in inhomogeneous media (Wiley, Chichester 1994)
9. M. Remoissenet: Waves called solitons (Springer, Berlin Heidelberg New York 1994)
10. A. Hasegawa and Yu. Kodama: Solitons in optical communications (Oxford University, Oxford 1995)
11. Yu.S. Kivshar and G.P. Agrawal: Optical solitons: from fibers to photonic crystals (Academic, Amsterdam 2003)
12. D.J. Korteweg and G. de Vries: On the change of form of long waves advancing in a rectangular channel, and on a new type of long stationary waves, Phil. Mag. **39**, 422 (1895)
13. R. Miura: The Kortweg-de Vries equation: a survey of results, SIAM Review **18**, 412 (1976)
14. M. Toda: Theory of nonlinear lattices (Springer, Berlin Heidelberg New York 1981)
15. J.M. Burgers: Application of model system in statistical theory of free turbulence, Proc. Acad. Sci. Amsterdam, **43**, 2 (1940)
16. B.B. Kadomtsev and V.I. Petviashvili: On stability of the solitary waves in weakly dispersive media, Soviet Phys. Doklady **15** 539 (1970) [DAN SSSR **192**, 753 (1970)]
17. V.E. Zakharov and E.A. Kuznetsov: On three-dimensional solitons, Sov. Phys. JETP **38**, 494 (1974) [ZhETF **66**, 594 (1974)]
18. V.P. Maslov and S.Yu. Dobrochotov: Multiphase asymptotics of nonlinear partial equations with a small parameter, Sov. Sci. Rev., 221 (1982)
19. C.S. Gardner, J.M. Green, M.D. Kruskal, and R.M. Miura: Method for solving the Korteweg de Vries equation, Phys. Rev. Lett. **19**, 1095 (1967)
20. V.E. Zakharov and L.D. Faddeev: Korteweg–de Vries equation - a completely integrable system, Funct. Anal. Appl. **5**, 280 (1971) [Funkc. Anal. Prilozh. **5**, 18 (1971)]

21. L.D. Faddeev: 40 years in mathematical physics (World Scientific, Singapore, 1995) [Inverse problem of quantum scattering theory, Sovremennye problemy matematiki **3**, 93 (Moscow, VINITI, 1974)]
22. S.V. Manakov: Method of the inverse scattering problem and two-dimensional evolution equations, Russ. Math. Surv. **31**, 245 (1976) [Usp. Mat. Nauk **31**, 245 (1976)]
23. I.M. Krichever: Methods of algebraical geometry in the theory of nonlinear equations, Russ. Math. Surv. **32**, 185 (1977) [Usp. Mat. Nauk **32**, 183 (1977)]
24. S. Novikov, S.V. Manakov, L.P. Pitaevskii, and V.E. Zakharov: Theory of solitons. The inverse scattering methods (Consultants Bureau, New York 1984)
25. G.L. Lamb: Elements of soliton theory (Wiley-Interscience, New York 1980)
26. M.J. Ablowitz and H. Segur: Solitons and the inverse scattering transform (SIAM, Philadelphia 1981)
27. R.K. Dodd, J.C. Eilbeck, J.D. Gibbon, and H.C. Morris: Solitons and nonlinear wave equations (Academic, London 1982)
28. P.C. Schuur: Asymptotic analysis of soliton problems: an inverse scattering approach (Springer, Berlin Heidelberg New York 1986)
29. V.S. Druma: On analytical solution of two-dimensional Korteweg–de Vries (KdV) equation, JETP Lett. **19**, 219 (1973) [Pis'ma v ZhETF **19**, 753 (1973)]
30. V.E. Zakharov and A.B. Shabat: A scheme for integrating the nonlinear equations of mathematical physics by the method of the inverse scattering problem. I., Funct. Anal. Appl. **8**, 226 (1974) [Funkc. Anal. Prilozh. **8**, 43 (1974)]
31. V.I. Petviashvili: Formation of an extraordinary soliton, Sov. J. PLasma Phys. **2**, 257 (1976) [Fiz. Plazmy **2**, 469 (1976)]
32. K. Mio, T. Ogino, K. Minami, and S. Takeda: Modified nonlinear Schrödinger equation for Alfvén waves propagating along the magnetic field in cold plasmas, J. Phys. Soc. Japan, **41**, 265 (1976)
33. S.P. Dawson and C.F. Fontán: Soliton decay of nonlinear Alfvén waves: numerical studies, Phys. Fluids **31**, 83 (1988)
34. N.F. Cramer: The physics of Alfvén waves (Wiley-VCH, Berlin 2001)
35. D.J. Kaup and A.C. Newell: An exact solution for a derivative nonlinear Schrödinger equation, J. Math. Phys. **19**, 798 (1978)
36. S.V. Vladimirov, V.N. Tsytovich, S.I. Popel, and F.Kh. Khakimov: Modulational interactions in plasmas (Kluwer, Dordrecht Boston London 1995)
37. O.A. Pokhotelov, L. Stenflo, and P. Shukla: Alfvén solitons in the Earth's ionosphere and magnetosphere, J. Geophys. Res. **101A**, 7913 (1996)
38. O.A. Pokhotelov, L. Stenflo, and P. Shukla: Nonlinear structures in the Earth's magnetosphere and atmosphere, Plasma Phys. Rep. **22**, 852 (1996)
39. T. Nagasawa, H. Tsuruta, and Y. Nishida: Excitation of converging ion-acoustic solitons, Phys. Lett. A **79**, 71 (1980)
40. Y. Nishida and T. Nagasawa: Oblique collision of plane ion-acoustic solitons, Phys. Rev. Lett. **45** 1626 (1980).
41. D.J. Kaup: Nonlinear resonances and colliding spherical ion-acoustic solitons, Physica D **2**, 389 (1981)
42. T. Nagasawa and Y. Nishida: Experiments on the ion-acoustic cylindrical solitons, Plasma Phys. **23**, 575 (1981)
43. Y. Nakamura: Experiments on ion-acoustic solitons in plasmas, IEEE Trans. Plasma Sci. **10**, 180 (1982)
44. M. Temerin, K. Cerny, W. Lotko, and F.S. Mozer: Observations of double layers and solitary waves in the auroral plasma, Phys. Rev. Lett. **48**, 1175 (1982)

45. S.G. Alikhanov, N.I. Alinovsky, G.G. Dolgov-Saveliev, et al: Develop of the program on the collisionless shock waves, Plasma Physics and Controlled Nuclear Fusion Research (IAEA, Vienna, 1969) p. 47
46. J.W.M. Paul: Review of experimental studies of collisionless shocks propagating perpendicular to a magnetic field, Collision-Free Shocks in the Laboratory and Space (Frascati, 1969) p. 97
47. A.E. Robson: Experiments on oblique shock waves, Ibid. p. 159
48. V.G. Eselevich, A.G. Eskov, R.H. Kurtmullaev, and A.I. Malutin: Fine structure of shock waves in a plasma and the mechanism of ion-acoustic turbulence saturation, Sov. Phys. JETP **43**, 745 (1971) [ZhETF **60**, 1658 (1971)]
49. Yu.I. Gal'perin, A.S. Leonovich, V.A. Mazur, et al: Motion of a packet of small-scale Alfvén waves in the mid-latitude plasmasphere, Cosmic Res. **24**, 475 (1986)
50. V.Yu. Belashov: Dynamics of the 3D Alfvén waves propagating in magnetized plasma and stability problem, Proc. 1996 Int. Conf. on Plasma Physics (Nagoya, Japan, Sept. 9-13, 1996). Contributed Papers. Vol. 1. (Nagoya, 1997) p. 954
51. L.A. Ostrovskii: Shocks and solitons, Sov. Phys.: Radiophys. and Quantum Electr. **19**, 431 (1976) [Radiofiz. i Elektr. **19**, 661 (1976)]
52. J. Dieter: Experiments on KdV solitons, J. Phys. Soc. Jap. **51**, 1686 (1982)
53 A.M. Obukhov, Ed.: Nelinejnye sistemy gidrodinamicheskogo tipa (Nonlinear systems of hydrodynamic type) (Nauka, Moskva 1974) [in Russian]
54. J.A. Gear and R. Grimshaw: A second-order theory for solitary waves in shallow fluids, Phys. Fluids. **26**, 14 (1983)
55. Yu.V. Katyshev and V.G. Makhankov: Stability of some one-field solitons, Phys. Lett. A **57**, 10 (1976)
56. V.E. Zakharov: Instability and nonlinear oscillations of solitons, Sov. Phys.: JETP Lett. **22**, 172 (1975) [Pis'ma v ZhETF **22**, 364 (1975)]
57. E.W. Laedke and K.H. Spatschek: On the applicability of the variation of action method to some one-field solitons, J. Math. Phys. **20**, 1838 (1979)
58. V.G. Makhankov, E.I. Litvinenko, and A.B. Shvachka: Numerical investigation of the Kadomtsev–Petviashvili soliton stability, Comp. Phys. Comm. **23**, 223 (1981)
59. E.A. Kuznetsov and S.L. Musher: Effect of collapse of sound waves on the structure of collisionless shock waves in a magnetized plasma, Sov. Phys. JETP **64**, 947 (1986) [ZhETF **91**, 1605 (1986)]
60. E.A. Kuznetsov and S.K. Turitsyn: Two- and three-dimensional solitons in weakly dispersive media, Soviet Phys. JETP. **55**, 844 (1982) [ZhETF **82**, 1457 (1982)]
61. N.C. Freeman: Soliton interactions in two dimensions, Adv. Appl. Mech. **20**, 1 (1980)
62. V.Yu. Belashov and V.I. Karpman: Chislennoe issledovanie dinamiki neodnomernykh solitonov v slabo dispergiruyuschikh sredakh (Numerical study of the dynamics of multidimensional solitons in weakly dispersive media), Prepr. IZMIRAN No. 43(928) (Moscow 1990) [in Russian, English Abstract]
63. E.A. Kuznetsov, S.L. Musher, and A.V. Shafarenko: Collapse of acoustic waves in media with positive dispersion, [Sov. Phys.: JETP Lett. **37**, 241 (1983) [Pis'ma v ZhETF **37**, 204 (1983)]
64. S.L. Musher: Kinetika slaboi turbulentnosti i volnovye kollapsy (Kinetic of weak turbulence and wave collapses), DSc Thesis (Novosibirsk, IAiEM SO AN SSSR 1985) [in Russian]
65. V.Yu. Belashov: Nonlinear dynamics of Alfvén waves propagating in magnetized plasma, Proc.XXIV General Assembly of URSI (Kyoto 1993) p. 657

66. V.Yu. Belashov: Nonlinear dynamics of three-dimensional Alfvén waves, Proc. Int. Seminar NEEDS 93 (Italy 1993) p. 242
67. V.Yu. Belashov: Nonlinear dynamics of fast magnetosonic and Alfvén waves propagating in magnetized plasma, Proc. 20th Int. Conf. Fusion 93 (Lisboa 1993) p. 467
68. V.Yu. Belashov: Nonlinear dynamics of three-dimensional Alfvén waves in plasma, Int. Workshop on Theory and Observations of Nonlinear Processes in Near-Earth Environment. Abstracts (Warsaw, Space Research Center 1995) p. 5.2
69. V.Yu. Belashov: Nonlinear dynamics of three-dimensional Alfvén waves in magnetospheric plasma, XX General Assembly of EGS (Kille, Germany 1995). Ann. Geophys. Suppl. **13**, 78 (1995)
70. V.Yu. Belashov: The problem of evolution and stability of 3D Alfvén waves propagating in the magnetosphere-ionosphere plasma along the magnetic field, 19th Annual Apatity Seminar. Abstracts. Preprint PGI N 96-01-99 (Apatity, Kola Sci. Cent. RAS 1996) p. 38
71. B. Alder, S. Fernbach, and M. Rotenberg, Eds.: Methods in computational physics. IX. Plasma physics (Academic, London 1970)
72. A.A. Samarskii: Numerical methods in problems of low-temperature plasma, Adv. Plasma Phys. **5**, 185 (1974)
73. A.A. Samarskii and Yu.P. Popov: Raznostnye skhemy gazovoj dinamiki (Difference schemes of gas dynamics) (Nauka, Moskva 1975) [in Russian]
74. A.A. Samarskii and P.N. Vabishchevich: Computational heat transfer (Wiley, New York 1995)
75. S.K. Godunov, A.V. Zabrodin, M.Ya. Ivanov, et al: Chislennoe reshenie mnogomernykh zadach gazovoj dinamiki (Numerical solution of multidimensional problems of gas dynamics) (Nauka, Moskva 1976) [in Russian]
76. C.A.J. Fletcher: Computational techniques for fluid dynamics (Springer, Berlin Heidelberg New York 1988)
77. V.Yu. Belashov and N.M. Chernova: Effektivnye algoritmy i programmy vychislitel'noj matematiki (Effective algorithms and programs of computational mathematics) (NEISRI FEB RAS, Magadan 1997) [in Russian]
78. Yu.A. Berezin: Chislennoe issledovanie nelinejnykh voln v razrezhennoj plazme (Numerical investigation of nonlinear waves in a rarified plasma) (Nauka, Novosibirsk 1977) [in Russian]
79. Yu.A. Berezin: Modelirovanie nelinejnykh volnovykh processov (Modeling of nonlinear wave processes) (Nauka, Novosibirsk 1982) [in Russian]
80. L.A. Abramyan and Yu.A. Stepanyants: The structure of two-dimensional solitons in media with anomalously small dispersion, Sov. Phys. JETP **61**, 963 (1985) [ZhETF **88**, 1616 (1985)]
81. V.Yu. Belashov: Chislennoe issledovanie dinamiki neodnomernykh nelinejnykh voln v slabo dispergiruyuschikh sredakh (Numerical study of the dynamics of multidimensional nonlinear waves in weakly dispersive media), PhD Thesis (IZMIRAN, Moskva 1991) [in Russian]
82. V.Yu. Belashov: Dinamika neodnomernykh nelinejnykh voln i solitonov v dispergiruyuschikh sredakh (Dynamics of non-one-dimensional nonlinear waves and solitons in dispersive media), DSc Thesis (IZMIRAN, Moskva 1998) [in Russian]
83. V.Yu. Belashov: Uravnenie KP i ego obobscheniya. Teoriya, prilozheniya (The KP equation and its generalizations. Theory, applications) (NEISRI FEB RAS, Magadan 1997) [in Russian]
84. T.J. Kawahara: Oscillatory solitary waves in dispersive media, J. Phys. Soc. Jap. **33**, 260 (1972)

85. E.A. Zabolotskaya and R.V. Khokhlov: Quasi-plane waves in the nonlinear acoustics of confined beams, Sov. Phys. Acoust. **15**, 35 (1969) [Akust. Zhurn. **15**, 35 (1969)]
86. D.Yu. Manin and V.I. Petviashvili: Self-focusing of a magnetosonic wave across the magnetic field, Sov. Phys. JETP Letters **38**, 517 (1983) [Pis'ma v ZhETF **38** 427 (1983)]
87. S.K. Turitsyn and G.E. Fal'kovich: Stability of magnetoelastic solitons and self-focusing of sound in antiferromagnets, Sov. Phys. JETP **62**, 146 (1985) [ZhETF **89**, 258 (1985)]
88. G.B. Whitham: Non-Linear dispersive waves, Proc. R. Soc. A **283**, 238 (1965)
89. N.J. Zabusky: A synergetic approach to problems of nonlinear dispersive wave propagation and interaction, In: Nonlinear Partial Differential Equations, Ed. W. Ames, (Academic, New York 1967) p. 223
90. R.M. Miura, C.S. Gardner, and M.D. Kruskal: Korteweg–de Vries equation and generalizations. II. Existence of conservation laws and constants of motion, J. Math. Phys. **9**, 1204 (1968)
91. M.D. Kruskal, R.M. Miura, C.S. Gardner, and N.J. Zabusky: Korteweg–de Vries equation and generalizations. V. Uniqueness and nonexistence of polynomial conservation laws J. Math. Phys. 11, 952 (1970)
92. E. Ott and R.N. Sudan: Nonlinear theory of ion acoustic waves with Landau damping, Phys. Fluids **12**, 2388 (1969)
93. Yu.A. Berezin and V.I. Karpman: Theory of non-stationary finite amplitude waves in a low density plasma, Sov. Phys. JETP **19**, 1265 (1966) [ZhETF **51**, 1557 (1966)]
94. R. Hirota: Direct methods of finding exact solutions of nonlinear evolution equations, In: Bäcklund transformations, Ed. R.M. Miura (Springer, Berlin Heidelberg New York 1976)
95. H.E. Moses: A generalization of the inverse scattering problem for the one-dimensional Schrödinger equation and application to the Korteweg–de Vries equation, In: Solitons in Action, Eds. K. Lonngren and A. Scott, (Academic, New York 1978) pp. 21-32
96. V.Yu. Belashov: Evoluciya solitonov KdV na "etape nestacionarnosti" (Evolution of KdV solitons on the "non-stationary stage") (Preprint NEISRI FEB RAS, Magadan 1984) [in Russian]
97. A.A. Samarskii and E.S. Nikolaev: Metody resheniya setochnykh uravnenii (The methods of solution of difference equations) (Nauka, Moskva 1978) [in Russian]
98. V.Yu. Belashov: The methods for numerical integration of nonlinear evolutional KP-class equations, Proceed. XX Int. Conf. on Phenomena in Ionized Gases. Contributed papers, Vol. 6 (Pisa 1991) p. 1241
99. S.V. Vladimirov and M.Y. Yu: Nonlinear ion-acoustic waves in a collisional plasma, Phys. Rev. E **48**, 2136 (1993)
100. F.F. Chen: Introduction to plasma physics, (Plenum, New York, 1974)
101. A.B. Mikhailovskii: Theory of plasma instabilities, Vol. 1 (Consultants Bureau, New York 1974)
102. T.D. Rognlien and S.A. Self SA: Ion-acoustic instability of a two-temperature, collisional, fully ionized plasma, Phys. Rev. Lett. **27**, 792 (1971)
103. R.Z. Sagdeev and A.A. Galeev: Nonlinear plasma theory (Benjamin, New York 1969)
104. T. Taniuti and C.C. Wei: Reductive perturbation method in nonlinear wave propagation, J. Phys. Soc. Jap. **24**, 941 (1968)
105. H. Washimi and T. Taniuti: Propagation of ion-acoustic solitary waves of small amplitude, Phys. Rev. Lett. **17**, 966 (1966)

106. G.C. Das and S.N. Paul: Ion-acoustic solitary waves in relativistic plasmas, Phys. Fluids **28**, 823 (1985)
107. J.I. Vette: Summary of particle populations in the magnetosphere, In: Particles and fields in the magnetosphere, Ed. B.M. McCormac, (D. Reidel, Dordrecht 1970) p. 305
108. P.K. Shukla, M.Y. Yu, and N.L. Tsintsadze: Intense solitary laser pulse propagation in a plasma, Phys. Fluids **27**, 327 (1984)
109. J. Arons: Some problems of pulsar physics, Space Sci. Rev. **24**, 417 (1979)
110. R.C. Davidson: Methods in nonlinear plasma theory, (Academic, New York 1972)
111. S.I. Braginskii, Reviews of plasma physics, Vol. 1 (Consultants Bureau, New York 1965) p. 205
112. V.Yu. Belashov and S.G. Tyunina: Qualitative analysis and asymptotics of solutions of generalized KdV-class equations, Radiophys. Quant. Electr. **40**, 210 (1997) [Izv. VUZ: Radiofizika **40**, 328 (1997)]
113. V.I. Karpman and V.Yu. Belashov: Dynamics of two-dimensional solitons in weakly dispersive media, Phys. Lett. A **154**, 131 (1991)
114. V.I. Karpman and V.Yu. Belashov: O strukture dvumernykh oscilliruyuschikh solitonov w slabo dispergiruyuschikh sredakh (On structure of two-dimensional oscillating solitons in weakly dispersive media) (Prepr. IZMIRAN No. 25(972), Moscow 1991) [in Russian, English Abstract]
115. R.M. Miura: Korteweg–de Vries equation and generalizations. I. A remarkable explicit nonlinear transformation, J. Math. Phys. **9**, 1202 (1968)
116. R. Hirota: A new form of Bäcklund transformation and its relation to the inverse scattering problem, Progr. Theor. Phys. **52** 1498 (1974)
117. R. Hirota and J. Satsuma: A variety of nonlinear network equations generated from the Bäcklund transformation for the Toda lattice, Progr. Theor. Phys. Suppl. **59**, 64 (1976)
118. P. Rosenau and J.M. Hyman: Compactons: solitons with finite wavelength, Phys. Rev. Lett. **70**, 564 (1993)
119. C.N. Kumar and P.K. Panigrahi: Compacton-like solutions for modified KdV and other nonlinear equations, arXiv: solv-int/9904020, v1, 23 Apr 1999
120. B. Dey and A. Khare: Stability of compacton solutions, Phys. Rev. E **58**, R2741 (1998)
121. S.V. Vladimirov, M.Y. Yu, and L. Stenflo: Surface wave solitons in an electronic medium, Phys. Lett. A **174**, 313 (1993)
122. S.V. Vladimirov, M.Y. Yu, and L. Stenflo: On solitary surface waves in cold plasmas, Comm. Plasma Phys. Contr. Fusion **15**, 299 (1993)
123. S.V. Vladimirov and M.Y. Yu: Solitary ionizing surface waves on low-temperature plasmas, IEEE Trans. Plasma Sci. **21**, 250 (1993)
124. S.V. Vladimirov, M.Y. Yu, and V.N. Tsytovich: Recent advances in the theory of nonlinear surface waves: Phys. Rep. **241**, 1 (1994)
125. S.V. Vladimirov and M.Y. Yu: Boundary effects on the nonlinear interactions of surface waves, Phys. Fluids B **5**, 2887 (1993)
126. M.Y. Yu: Solitary electron plasma waves, Phys. Lett. A **59**, 361 (1976)
127. M. Chaker, M. Moisan, and Z. Zakrzewski: Microwave and RF surface wave sustained discharges as plasma sources for plasma chemistry and plasma processing, Plasma Chem. Plasma Process. **6**, 79 (1986)
128. M. Moisan and Z. Zakrzevski: Plasmas sustained by surface waves at microwave and RF frequencies: experimental investigation and applications, In: Radiative Processes in Discharge Plasma, Eds. J. M. Prouf and L. H. Luessen (Plenum, New York 1986) p. 381

129. R. Claude, M. Moisan, M. R. Wertheimer, and Z. Zakrzewski: Comparison of microwave and lower frequency discharges for plasma polymerization, Appl. Phys. Lett. **50**, 1797 (1987)
130. G.S. Selwyn, J.E. Heidenreich, and K.L. Haller: Particle trapping phenomena in radio frequency plasmas, Appl. Phys. Lett. **57**, 1876 (1990)
131. M. Moisan, A. Shivarova, and A. W. Trivelpiece: Experimental investigations of the propagation of surface waves along a plasma column, Plasma Phys. **24**, 1331 (1982)
132. O.M. Gradov and L. Stenflo: Linear theory of a cold bounded plasma, Phys. Rep. **94**, 111 (1983)
133. S.V. Vladimirov: Solitary surface waves in plasma slabs, Sov. J. Plasma Phys. 12 (1986) 552 [Fiz. Plazmy **12**, 961 (1986)]
134. D. Henderson: In: Trends in Interfacial Electrochemistry, Ed. A.F. Silva (Reidel, Dordrecht 1986) p. 473
135. J. Topper and T. Kawahara: Approximate equations for long nonlinear waves on a viscous fluid, J. Phys. Soc. Japan **44**, 663 (1978)
136. B.I. Cohen, J.A. Krommes, W.M. Tang, and M.N. Rosenbluth: Non-linear saturation of the dissipative trapped-ion mode by mode coupling, Nucl. Fusion **16**, 971 (1976)
137. T. Kawahara: Formation of saturated solitons in a nonlinear dispersive system with instability and dissipation, Phys. Rev. Lett. **51**, 381 (1983)
138. V.Yu. Belashov: Travelling ionospheric disturbances in the F region of the ionosphere and their influence on the fluctuations of the VLF radiosignals, Proc. 1988 Int. Symp. on EMC. Vol. 1 (Wroclaw 1988) p. 181
139. V.Yu. Belashov: Solitary electron density waves induced by the IGW's solitons in the ionosphere, Proc. 1989 Int. Symp. on EMC. Vol. 1 (Nagoya 1989) p. 228
140. V.Yu. Belashov: Dynamics of nonlinear internal gravity waves at ionosphere F-region heights, Geomagn. and Aeron. **30**, 536 (1990) [Geomagn. i Aeronom. **30**, 637 (1990)]
141. V.Yu. Belashov: On the earthquake-induced IGW in the ionosphere F layer, Proc. 1988 Int. Symp. on EMC. Vol. 1 (Wroclaw 1988) p. 227
142. V.Yu. Belashov: Seismogenic perturbations at heights of ionosphere F layer, Intern. Workshop on Seismo Electromagnetics (IWSE-97, Tokyo, NASDA 1997) p. 225
143. V.Yu. Belashov: Theoretical and numerical study of effects in ionospheric plasma associated with earthquakes and volcano eruptions, Intern. Workshop Electromagn. Phenomena Related to Earthquake Prediction (Tokyo, Univ. of Electro-Communication 1993) p. 90
144. A.A. Belashova and V.Yu. Belashov: Large-scale wave disturbances generated by the eclipse in the ionosphere and EMC problems, Proc. 1989 Int. Symp. on EMC. Vol. 1 (Nagoya 1989) p. 226
145. Y. Kuramoto and T. Tsuzuki: Persistent propagation of concentration waves in dissipative media far from thermal equilibrium, Prog. Theor. Phys. **55**, 356 (1976)
146. T. Yamada and Y. Kuramoto: A reduced model showing chemical turbulence, Prog. Theor. Phys. **56**, 681 (1976)
147. G.I. Sivashinsky: Instabilities, pattern formation, and turbulence in flames, Annu. Rev. Fluid Mech. **15**, 179 (1983)
148. V.I. Karpman and V.Yu. Belashov: Evolution of three-dimensional nonlinear pulses in weakly dispersive media, Phys. Lett. A **154**, 140 (1991)
149. E. Ott and R.N. Sudan: Damping of solitary waves, Phys. Fluids **13**, 1432 (1970)
150. J.P. Boris: Phys. Fluids **16**, 855 (1973)

151. V.Yu Belashov: Dynamics of KP equation solitons in media with low-frequency wave field stochastic fluctuations, Phys. Lett. A **197**, 282 (1995)
152. M. Wadati: Stochastic Korteweg–de Vries equation, J. Phys. Soc. Jap. **52**, 2642 (1983)
153. D.K. Arrowsmith and C.M. Place: Dynamical systems : differential equations, maps, and chaotic behaviour, (Chapman and Hall, London 1992)
154. N.N. Bautin and E.A. Leontovich: Metody i priemy kachestvennogo issledovaniya dinamicheskikh sistem na ploskosti (Methods and rules for the qualitative study of dynamical systems on the plane) (Moscow, Nauka 1990) [in Russian]
155. A.A. Andronov, A.A. Vitt, and S.E. Khaikin: Theory of oscillators (Dover, New York 1987)
156. N.N. Bautin: Povedenie dinamicheskikh sistem vblizi granits oblasti ustojchivosti (The behavior of dynamical systems near the boundaries of the domain of stability) (Moscow, Nauka 1984) [in Russian]
157. S.V. Vladimirov: On electric forces in a time-dependent medium, Phys. Lett. A **219**, 233 (1996)
158. V.E. Zakharov and A.B. Shabat: Exact theory of two-dimensional self-focusing and one-dimensional self-modulation of waves in nonlinear media, Sov. Phys. JETP **34**, 62 (1972) [ZhETF **61**, 118 (1971)]
159. P.D. Lax: Integrals of nonlinear equations of evolution and solitary waves, Comm. Pure Appl. Math. **21**, 467 (1968)
160. V.E. Zakharov: Collapse of Langmuir waves, Sov. Phys. JETP **35**, 908 (1972) [ZhETF **62**, 1745 (1972)]
161. L.I. Rudakov and V.N. Tsytovich: Strong Langmuir turbulence, Phys. Rep. **40C**, 1, (1978)
162. S.G. Thornhill and D. ter Haar: Langmuir turbulence and modulational instability, Phys. Rep. **43C**, 43 (1978)
163. D. ter Haar and V.N. Tsytovich: Modulation instabilities in astrophysics, Phys. Rep. **73**, 175 (1981)
164. M.V. Goldman: Strong turbulence of plasma waves, Rev. Mod. Phys. **56**, 709 (1984)
165. P.A. Robinson: Nonlinear wave collapse and strong turbulence, Rev. Mod. Phys. **69**, 507 (1997)
166. S.A. Boldyrev, S.V. Vladimirov, and V.N. Tsytovich: Coupled Langmuir and ion-acoustic solitons, Sov. J. Plasma Phys. **18**, 727 (1992) [Fiz. Plazmy **18**, 1409 (1992)]
167. V.G. Makhankov: On stationary solutions of the Schrödinger equation with a self-consistent potential satisfying Boussinesq's equation (Langmuir solitons), Phys. Lett. A, **50**, 42 (1974)
168. K. Nishikawa, H. Hojo, K. Mima, and H. Ikezi: Coupled nonlinear electron-plasma and ion-acoustic waves, Phys. Rev. Lett. **33**, 148 (1974)
169. S.A. Boldyrev, V.N. Tsytovich, and S.V. Vladimirov: On dissipative acceleration of near-sonic solitons, Comm. Plasma Phys. Contr. Fusion **15**, 1 (1992)
170. O.A. Pokhotelov, V.A. Pilipenko, E.N. Fedorov E.N., L. Stenflo, and P.K. Shukla: Induced electromagnetic turbulence in the ionosphere and the magnetosphere, Phys. Scr. **50**, 600 (1994)
171. M.B. Gokhberg, V.A. Pilipenko, O.A. Pokhotelov, and S. Parthasarati: Acoustic disturbance from an underground nuclear explosion as a source of electrostatic turbulence in the magnetosphere, Sov. Phys. Doklady **313**, 31 (1992) [DAN SSSR **313**, 568 (1992)]

172. M.J. Ablowitz, D.J. Kaup, A.C. Newell, and H. Segur: The inverse scattering transform-Fourier analysis for nonlinear problems, Stud. Appl. Math. **53**, 249 (1974)
173. E. Mjølhus: A note on the modulational instability of long Alfvén waves parallel to the magnetic field, J. Plasma Phys. **19**, 437 (1978)
174. S.R. Spangler, J.P. Sheerin, and Q.L. Payne: A numerical study of nonlinear Alfvén waves and solitons, Phys. Fluids **28**, 104 (1985)
175. V.Yu. Belashov: The problem of stability for three-dimensional Alfvén waves propagating in magnetized plasma, Doklady Phys. **44**, 327 (1999) [DAN Russ. **366**, 465 (1999)]
176. S.P. Dawson and C.F. Fontán: Extension of the Ablowitz–Ladik method to the derivative nonlinear Schrödinger equation, J. Comput. Phys. **76**, 192 (1988)
177. T.R. Taha and M.J. Ablowitz: Analytical and numerical aspects of certain nonlinear evolution equation, J. Comput. Phys. **55**, 192 (1984)
178. D. Potter: Computational Physics (Wiley, London 1973).
179. G.A. Korn and T.M. Korn: Mathematical handbook for scientists and engineers (McGraw-Hill, New York 1968)
180. N.S. Bakhvalov: Méthodes numériques. Analyse, algèbre, équations différentielles ordinaires (Moscow, Mir 1976)
181. F.J. Harris: On the use of windows for harmonic analysis with the discrete Fourier transform, Proc. IEEE **66**, 51 (1978)
182. M.D. Spektor: Stability of cnoidal waves in media with positive and negative dispersion, Sov. Phys. JETP **67**, 104 (1988) [ZhETF **94**, 186 (1988)]
183. E. Infeld, G. Rowlands, and M. Hen: Three dimensional stability of Korteweg–de Vries waves and solitons, Acta Phys. Polon. **54A**, 131 (1978)
184. E. Infeld: Three dimensional stability of Korteweg–de Vries waves and solitons. III. Lagrangian methods, KdV with positive dispersion, Acta Phys. Polon. **60A**, 623 (1981)
185. A.B. Mikhailovskii, S.V. Makurin, and A.I. Smolyakov: Stability of nonlinear periodic waves in weakly dispersive media, Sov. Phys. JETP **62**, 928 (1985) [ZhETF **89**, 1603 (1985)]
186. S.P. Burtsev: Instability of a periodic chain of two-dimensional solitons, Sov. Phys. JETP **61**, 959 (1985) [ZhETF **88**, 1609 (1985)]
187. V.Yu. Belashov: Spetsialnye funkcii i algoritmy ikh vychisleniya (Special functions and algorithms of their computation) (Intern. Pedagog. Univ., Magadan, 1997) [in Russian]
188. A.A. Zaitsev: Formation of stationary nonlinear waves by superposition of solitons, Sov. Phys. Doklady **28**, 720 (1983) [DAN SSSR **272**, 583 (1983)]
189. Yu.A. Danilov and V.I. Petviashvili: Solitony v plazme In: Itogi nauki i tekhniki. Fizika plazmy (Solitons in a plasma). Vol. 4 (VINITI, Moskva 1983) p. 5 [in Russian]
190. V.I. Petviashvili: Multidimensional and dissipative solitons, Physica D **3**, 329 (1981)
191. G.I. Marchuk: On the theory of the spliting-up method, In: Numerical solution of partial differential equations. II. SYNSPADE-1970 (Academic, New York 1971) p. 16
192. G.I. Marchuk: Methods of numerical mathematics, (Springer, Berlin Heidelberg New York 1982)
193. S.L. Musher and A.V. Shafarenko: The method of numerical modeling of wave collapses in media with weak dispersion, Private communication (1986)
194. V.Yu. Belashov: Stability of two- and three-dimensional solitons in weakly dispersive media, Sov. Physi. Doklady **36**, 626 (1991) [DAN SSSR **320**, 85 (1991)]

195. V.Yu. Belashov: Numerical study of dynamics of 3D ion-acoustic and FMS nonlinear waves in plasma using spectral approach, Proc. 5th Int. School/Symp. for Space Simulation (ISSS-5) (RASC, Kyoto Univ., Kyoto 1997) p. 118.
196. V.Yu. Belashov: Nonlinear effects for FMS waves propagating in magnetized plasma, Plasma Phys. Control. Fusion **36**, 1661 (1994)
197. A.G. Litvak, T.A. Petrova, A.M. Sergeev, and A.D. Yunakovskii: A form of wave self-action in a plasma, Sov. J. Plasma Phys. **9**, 327 (1983) [Fiz. Plazmy **9**, 495 (1983)]
198. V.Yu. Belashov and V.I. Karpman: 2D and 3D disturbances dynamics in the weakly dispersive media with dissipation, Proc. XX Int. Conf. Phen. in Ionized Gases. Contr. papers, Vol. 6 (Pisa 1991) p. 1239
199. V.Yu. Belashov: Dissipation's effect on structure and evolution of nonlinear waves and solitons in plasma, Proc. XI Intern. Symp. on EMC. Vol. 2 (Wroclaw 1992) p. 591
200. V.G. Grudnickii and Yu.A. Prohorchuk: A way of constructing difference schemes with arbitrary order of approximation for partial differential equations, Sov. Phys. Doklady **22**, 845 (1977) [DAN SSSR **234**, 1249 (1977)]
201. V.Yu. Belashov: Dynamics of KP equation solitons in media with low-frequency stochastic fluctuations, Proc.XXIV URSI General Assembly (Kyoto 1993) p. 656
202. V.A. Gorodtsov: Stochastic Kadomtsev–Petviashvili equation, JETP **117**, 1270 (2000) [ZhETF **117**, 1526 (2000)]
203. V.Yu. Belashov: Soliton Evolution in Media with Variable Dispersion, Ann. Geophys. Suppl. **14**, 91 (1996)
204. V.Yu. Belashov: Evolution of the FMS waves in plasma with variable dispersion, Proc. XXV URSI Gen. Assembly (Lille 1996) p. 475
205. V.Yu. Belashov: Dynamics of the 3D FMS waves beam in plasma with stochastic fluctuations of magnetic field, Proc. 1996 Int. Conf. on Plasma Physics. Contributed Papers, Vol.1 (Nagoya 1996) p. 950
206. V.Yu. Belashov: Dynamics of nonlinear waves and solitons in plasma with wave field stochastic fluctuations, Proc. Worksh. Theory and Observations of Nonlin. Processes in Near-Earth Environment. Abstracts (Warsaw 1995) p. 4.1
207. Y. Nejon: A two-dimensional ion acoustic solitary wave in a weakly relativistic plasma, J. Plasma Phys. **38**, 439 (1987)
208. M. Temerin, K. Cerny, W. Lotko, and F.S. Mozed: Observations of double layers and solitary waves in the auroral plasma, Phys. Rev. Lett. **48**, 1175 (1982)
209. A.A. Belashova, V.Yu. Belashov, and I.I. Poddelskiy: Composite studies of the dynamics of wave disturbances of the ionosphere in the far-eastern USSR, Geomagn. and Aeron. **30**, 543 (1990) [Geomagn. i Aeron. **30**, 647 (1990)]
210. V.Yu. Belashov: Some results of common analysis of ionospheric and seismic data obtained in the periods of seismic activity on the Russian North East Station Network, Proc. Intern. Workshop on Seismo Electromagnetics (IWSE-97, NASDA, Tokyo 1997) p. 234

Index

Springer Series in
SOLID-STATE SCIENCES

Series Editors:
M. Cardona P. Fulde K. von Klitzing R. Merlin H.-J. Queisser H. Störmer

90 **Earlier and Recent Aspects of Superconductivity**
Editor: J.G. Bednorz and K.A. Müller

91 **Electronic Properties and Conjugated Polymers III**
Editors: H. Kuzmany, M. Mehring, and S. Roth

92 **Physics and Engineering Applications of Magnetism**
Editors: Y. Ishikawa and N. Miura

93 **Quasicrystals**
Editor: T. Fujiwara and T. Ogawa

94 **Electronic Conduction in Oxides**
2nd Edition By N. Tsuda, K. Nasu, A. Fujimori, and K. Siratori

95 **Electronic Materials**
A New Era in Materials Science
Editors: J.R. Chelikowski and A. Franciosi

96 **Electron Liquids**
2nd Edition By A. Isihara

97 **Localization and Confinement of Electrons in Semiconductors**
Editors: F. Kuchar, H. Heinrich, and G. Bauer

98 **Magnetism and the Electronic Structure of Crystals**
By V.A. Gubanov, A.I. Liechtenstein, and A.V. Postnikov

99 **Electronic Properties of High-T_c Superconductors and Related Compounds**
Editors: H. Kuzmany, M. Mehring and J. Fink

100 **Electron Correlations in Molecules and Solids**
3rd Edition By P. Fulde

101 **High Magnetic Fields in Semiconductor Physics III**
Quantum Hall Effect, Transport and Optics By G. Landwehr

101 **High Magnetic Fields in Semiconductor Physics III**
Quantum Hall Effect, Transport and Optics By G. Landwehr

102 **Conjugated Conducting Polymers**
Editor: H. Kiess

103 **Molecular Dynamics Simulations**
Editor: F. Yonezawa

104 **Products of Random Matrices**
in Statistical Physics By A. Crisanti, G. Paladin, and A. Vulpiani

105 **Self-Trapped Excitons**
2nd Edition By K.S. Song and R.T. Williams

106 **Physics of High-Temperature Superconductors**
Editors: S. Maekawa and M. Sato

107 **Electronic Properties of Polymers**
Orientation and Dimensionality of Conjugated Systems Editors: H. Kuzmany, M. Mehring, and S. Roth

108 **Site Symmetry in Crystals**
Theory and Applications
2nd Edition By R.A. Evarestov and V.P. Smirnov

109 **Transport Phenomena in Mesoscopic Systems**
Editors: H. Fukuyama and T. Ando

110 **Superlattices and Other Heterostructures**
Symmetry and Optical Phenomena 2nd Edition
By E.L. Ivchenko and G.E. Pikus

111 **Low-Dimensional Electronic Systems**
New Concepts
Editors: G. Bauer, F. Kuchar, and H. Heinrich

112 **Phonon Scattering in Condensed Matter VII**
Editors: M. Meissner and R.O. Pohl

Springer Series in

SOLID-STATE SCIENCES

Series Editors:
M. Cardona P. Fulde K. von Klitzing R. Merlin H.-J. Queisser H. Störmer

113 **Electronic Properties of High-T_c Superconductors**
Editors: H. Kuzmany, M. Mehring, and J. Fink

114 **Interatomic Potential and Structural Stability**
Editors: K. Terakura and H. Akai

115 **Ultrafast Spectroscopy of Semiconductors and Semiconductor Nanostructures**
By J. Shah

116 **Electron Spectrum of Gapless Semiconductors**
By J.M. Tsidilkovski

117 **Electronic Properties of Fullerenes**
Editors: H. Kuzmany, J. Fink, M. Mehring, and S. Roth

118 **Correlation Effects in Low-Dimensional Electron Systems**
Editors: A. Okiji and N. Kawakami

119 **Spectroscopy of Mott Insulators and Correlated Metals**
Editors: A. Fujimori and Y. Tokura

120 **Optical Properties of III–V Semiconductors**
The Influence of Multi-Valley Band Structures By H. Kalt

121 **Elementary Processes in Excitations and Reactions on Solid Surfaces**
Editors: A. Okiji, H. Kasai, and K. Makoshi

122 **Theory of Magnetism**
By K. Yosida

123 **Quantum Kinetics in Transport and Optics of Semiconductors**
By H. Haug and A.-P. Jauho

124 **Relaxations of Excited States and Photo-Induced Structural Phase Transitions**
Editor: K. Nasu

125 **Physics and Chemistry of Transition-Metal Oxides**
Editors: H. Fukuyama and N. Nagaosa

126 **Physical Properties of Quasicrystals**
Editor: Z.M. Stadnik

127 **Positron Annihilation in Semiconductors**
Defect Studies. By R. Krause-Rehberg and H.S. Leipner

128 **Magneto-Optics**
Editors: S. Sugano and N. Kojima

129 **Computational Materials Science**
From Ab Initio to Monte Carlo Methods. By K. Ohno, K. Esfarjani, and Y. Kawazoe

130 **Contact, Adhesion and Rupture of Elastic Solids**
By D. Maugis

131 **Field Theories for Low-Dimensional Condensed Matter Systems**
Spin Systems and Strongly Correlated Electrons. By G. Morandi, P. Sodano, A. Tagliacozzo, and V. Tognetti

132 **Vortices in Unconventional Superconductors and Superfluids**
Editors: R.P. Huebener, N. Schopohl, and G.E. Volovik

133 **The Quantum Hall Effect**
By D. Yoshioka

134 **Magnetism in the Solid State**
By P. Mohn

135 **Electrodynamics of Magnetoactive Media**
By I. Vagner, B.I. Lembrikov, and P. Wyder